Lecture Notes in Mathematics

Edited by A. Dold and B. Eckmann

865

Roelof W. Bruggeman

Fourier Coefficients of Automorphic Forms

Springer-Verlag
Berlin Heidelberg New York 1981

Author

Roelof W. Bruggeman
Mathematisch Instituut, Rijksuniversiteit Utrecht
Budapestlaan 6, Utrecht, The Netherlands

AMS Subject Classifications (1980): 10D15, 10D40, 22E40, 42A16, 43A65, 44A15

ISBN 3-540-10839-4 Springer-Verlag Berlin Heidelberg New York
ISBN 0-387-10839-4 Springer-Verlag New York Heidelberg Berlin

Printing and binding: Beltz Offsetdruck, Hemsbach/Bergstr.
2141/3140-543210

Table of contents

1. Introduction.

1.1. A sum formula for the modular case

To give an idea of the subject of these notes I discuss in this subsection a simple case.

Holomorphic modular cusp forms are well known; of many possible references I mention [23], [39], [42]. These cusp forms are holomorphic functions on the upper half plane $\mathcal{h} = \{z \in \mathbb{C} | \operatorname{Im} z > 0\}$, satisfying the transformation property

$$(1.1.1) \qquad f\left(\frac{\alpha z+\beta}{\gamma z+\delta}\right) = (\gamma z+\delta)^b f(z), \qquad \alpha,\beta,\gamma,\delta \in \mathbb{Z},\ \alpha\delta-\beta\gamma = 1,$$

for some positive even number b, the weight, and having a Fourier series expansion

$$(1.1.2) \qquad f(z) = \sum_{n=1}^{\infty} a_n e^{2\pi i n z}.$$

The numbers a_n are called the Fourier coefficients of f. For a fixed weight b the space of such cusp forms has finite dimension; it may be provided with the Petersson scalar product. For fixed n the map $f \to a_n$ is linear on such a space, hence it may be described as $f \mapsto (f,p_n)_p$ for some cusp form p_n. By $(.,.)_p$ I denote the Petersson scalar product. Up to a multiple this p_n is given by a Poincaré series. Petersson has given a formula for the Fourier coefficients of these Poincaré series, see [28] or [23], p.298. The n-th Fourier coefficient of the m-th Poincaré series of weight b equals

$$(1.1.3) \qquad \delta_{m,n} + 2\pi(-1)^{\frac{1}{2}b}(n/m)^{\frac{1}{2}b-\frac{1}{2}} \sum_{c=1}^{\infty} S(n,m;c)c^{-1} J_{b-1}(4\pi c^{-1}\sqrt{mn}).$$

J_{b-1} is a Bessel function, and $S(n,m;c)$ is a Kloosterman sum:

$$(1.1.4) \qquad S(n,m;c) = \sum_{\substack{x \bmod c \\ (x,c)=1}} e^{2\pi i c^{-1}(nx+m\tilde{x})}$$

with $x\tilde{x} \equiv 1(c)$.

Let $f_{b,1},\dots,f_{b,r_b}$ be an orthonormal basis for the space of cusp forms of weight b (with respect to the Petersson scalar product). Denote the n-th Fourier coefficient of $f_{b,j}$ by $a_{b,j}(n)$. Then formula (1.1.3) implies

$$(1.1.5) \qquad (b-2)!(4\pi m)^{1-b} \sum_{j=1}^{r_b} \overline{a_{b,j}(m)} a_{b,j}(n) =$$
$$\delta_{m,n} + 2\pi(-1)^{\frac{1}{2}b}(n/m)^{\frac{1}{2}b-\frac{1}{2}} \sum_{c=1}^{\infty} c^{-1}S(n,m;c)J_{b-1}(4\pi c^{-1}\sqrt{mn}).$$

This is easily derived from the fact that the m-th Poincaré series gives $(b-2)!(4\pi m)^{1-}$ times the cusp form p_m mentioned above.

Formula (1.1.5) gives an explicit relation which the Fourier coefficients of an orthonormal basis of a space of cusp forms have to satisfy.

Let us turn now to weight zero; then condition (1.1.1) means just invariance under the modular group $\Gamma = Sl_2(\mathbb{Z})$. For b = 0 any non-constant holomorphic function on $\mathcal{h}$

satisfying (1.1.1) has a Fourier series expansion like the one in (1.1.2), but with at least one non-zero term $a_n e^{2\pi i\, nz}$ with $n < 0$. So it grows exponentially for $z \to i\infty$. For $b = 0$ the Petersson scalar product is obtained by integrating over $\Gamma\backslash h$ with respect to the invariant measure $y^{-2}dxdy$. We cannot hope for convergence when considering exponentially growing funtions.

A sensible generalization is due to Maass. As I do not want to go into representations at this point, or into Dirichlet series, I refrain from explaining why it is sensible. As references I mention [24] Ch.IV and V, [14], [42], [36], [37]. The condition of holomorphy is dropped, and replaced by an eigenvector equation:

(1.1.6) $$-y^2\left(\frac{\partial^2}{\partial x^2} + \frac{\partial^2}{\partial y^2}\right)f = \lambda f \qquad \text{for some } \lambda \in \mathbb{C}.$$

Here $x+iy = z \in h$. The differential operator in (1.1.6) is the Laplace operator for the hyperbolic metric on h. Γ-invariant functions satisfying (1.1.6) and some growth condition one calls real analytic modular forms of weight zero. The analyticity is due to ellipticity of the Laplace operator. To state which of these modular forms are cusp forms we need a Fourier series expansion. One cannot expect exponential functions in the terms of this expansion. For $n \neq 0$ the terms are of the form $B_n(y)e^{2\pi inx}$, where B_n satisfies a differential equation which is easily transformed into the Whittaker differential equation. A real analytic modular form is a cusp form if it may be expanded in the following way

(1.1.7) $$f(x+iy) = \sum_{n\neq 0} a_n W_{0,s}(4\pi|n|y)e^{2\pi inx};$$

the s in the Whittaker function $W_{0,s}$ is determined by $\lambda = \frac{1}{4}-s^2$. So a cusp form has no zero-th term in its Fourier series expansion and the other terms are quickly decreasing for $y \to \infty$ (this last fact characterizes the multiples of $W_{0,s}$ among the solutions of the Whittaker differential equation).

These cusp forms are square integrable for the Petersson scalar product; they generate a closed subspace ${}^0L^2(\Gamma\backslash h, y^{-2}dxdy)$ in $L^2(\Gamma\backslash h, y^{-2}dxdy)$. One has proved that ${}^0L^2(\Gamma\backslash h, y^{-2}dxdy)$ has a countably infinite orthonormal basis $f_1, f_2, \ldots$ consisting of cusp forms with corresponding eigenvalues $\lambda_1 \leq \lambda_2 \leq \lambda_3 \leq \ldots$. These eigenvalues are all positive (even $> \frac{1}{4}$ in the modular case), only finite multiplicities may occur. So there exist a lot of modular cusp forms of weight 0, but no one has been explicitly constructed up till now; for computer results see [42].

Denote the n-th Fourier coefficient a_n in (1.1.7) for f_j by $a_j(n)$. One may ask whether one may find a formula for $\sum_{j,\ \lambda_j=\lambda} \overline{a_j(m)}a_j(n)$ like the one in (1.1.5). Certainly one may express "taking the n-th Fourier coefficient" by taking the scalar product with a fixed cusp form. One might write down a Poincaré series for this cusp form, but this series diverges. It is possible, however, to relate the cusp form with a Poincaré series, but only after analytical continuation; see [26]. This is not enough to obtain something like (1.1.5). The only hope of generalizing (1.1.5) for real analytic modular cusp forms is to handle all f_j together. This is done by the sum formula which is the subject of these notes. For the case of this subsection

one may find it in [15], thm. 2, p.40, [18], thm. 1 and in [1], thm. (3.29). I shall state it as given by Kuznetsov in [15], but slightly modified.
We need a test function h on $(0,\infty)$; let us restrict ourselves to $h \in C_c^\infty(0,\infty)$. Not only h occurs in the sum formula, but also its Bessel transform

(1.1.8) $$\widetilde{h}(s) = \frac{1}{\sin \pi s} \int_0^\infty h(y)(J_{-2s}(y) - J_{2s}(y))\frac{dy}{y} .$$

Further we need the Riemann zeta function ζ and

(1.1.9) $$\sigma_{2s}(n) = \sum_{d|n} d^{2s}$$

(1.1.10) $$\lambda_j = \tfrac{1}{4} - s_j^2 .$$

Let $n,m \geqslant 1$, then everything in the following equality converges absolutely:

(1.1.11) $$\sum_{j=1}^{\infty} \widetilde{h}(s_j)(\cos \pi s_j)^{-1} \overline{a_j(m)}\, a_j(n)$$
$$+ \sum_{b\equiv 0(2),b\geqslant 12} \widetilde{h}(\tfrac{1}{2}(b-1))2^{-2b}\pi^{-1-b}(mn)^{-\frac{1}{2}b}(b-1)! \sum_{j=1}^{r_b} \overline{a_{bj}(m)}a_{bj}(n)$$
$$+ \frac{1}{4\pi i} \int_{\mathrm{Re}\, s=0} \widetilde{h}(s)n^{-\frac{1}{2}-s}m^{-\frac{1}{2}+s}|\zeta(1+2s)|^{-2}\sigma_{2s}(n)\sigma_{-2s}(m)ds =$$
$$= (2\pi^2\sqrt{mn})^{-1}\delta_{m,n} \cdot 0$$
$$+ (2\pi\sqrt{mn})^{-1} \sum_{c=1}^{\infty} c^{-1}S(n,m;c)h(4\pi\sqrt{mn}\, c^{-1}).$$

This is the sum formula generalizing (1.1.5). The zero term in the right hand side has not been omitted to stress the resemblance to (1.1.5). For $h \notin C_c^\infty(0,\infty)$ something nontrivial may appear. The first term in the left hand side concerns the cusp forms of weight zero; in the second term the holomorphic cusp forms turn up. The last term in the left hand side is due to the orthogonal complement of ${}^0L^2(\Gamma\backslash \mathfrak{h},y^{-2}dxdy)$ in $L^2(\Gamma\backslash \mathfrak{h},y^{-2}dxdy)$, which may be described with help of Eisenstein series. (1.1.11) may be derived easily from Kuznetsov's sum formula [15], thm. 2, p.40. The sum formula in [1] is similar , but I did not make the Bessel transform explicit (and gave the Eisenstein term the wrong factor $(2\pi i)^{-1}$).

The sum formula connects Fourier coefficients of cusp forms with Kloosterman sums. One may use it to obtain information concerning either Fourier coefficients or Kloosterman sums.
In the modular case the Fourier coefficients of cusp forms are closely related to eigenvalues of Hecke operators. I do not go into these operators, but remark that eigenvalues of Hecke operators have been studied with the sum formula as a tool. In [1], §4 an asymptotic formula is given for the distribution of the eigenvalues of a fixed Hecke operator on real analytic cusp forms. Proskurin, [32], gave estimates of eigenvalues of Hecke operators on these cusp forms not only for the full modular group, but also for some congruence subgroups. I refer to [32] and [1] for the results and the methods, and only remark that to me the Hecke operators seem essential. For discrete groups without a ring of Hecke operators the methods break down. In [15] Kuznetsov attacked the Ramanujan-Petersson conjecture for real

analytic cusp forms of weight zero by combining (1.1.11) with another sum formula; up till now the attack has not succeeded.
Kuznetsov, [15], §7, and Proskurin [33], [34] §7 have also used the sum formula to estimate averages of Kloosterman sums.
Other uses of the sum formula may be found in [18], [16], [17], [19].

In these notes I am interested in the sum formula itself, not in its applications. The notes try to answer two questions:
Viewing (1.1.11) as generalization of (1.1.5) I understand that there is no hope to obtain a formula like (1.1.5) for $\sum_j \overline{a_j(m)} a_j(n)$ with j running through a finite set, for the Poincaré series do not converge in the right region. So I have to consider all f_j together. It is sensible to use the spectral decomposition of the Laplace operator as a tool, so also the other part of $L^2(\Gamma\backslash\mathfrak{h}, y^{-2}dxdy)$, given by Eisenstein series, turns up.
Question 1: Why do the holomorphic cusp forms enter the picture?
Formula (1.1.11) has the structure of a Poisson formula. In the right hand side a sum of function values at certain points is given, it equals an expression containing a Bessel transform of the function.
Question 2: Where does this Bessel transformation come from?
The latter question also asks for an explanation of the J_{b-1} in Petersson's formula (1.1.3). In subsection 1.3 I shall say more concerning the aim of these notes.

1.2 Sum formula

In this subsection I discuss the more general sum formula which I shall prove in these notes. Generalization of (1.1.11) is possible in several directions:
(a) n and/or m may be negative.
(b) $Sl_2(\mathbb{Z})$ may be replaced by other groups.
(c) One may look at non-even weights.

Generalization (a) may already be found in [15], thm. 7, p.49 and [1], thm. (3.29). Remark that the cases n = 0 or m = 0 give no information concerning cusp forms, they concern Eisenstein series only.
Generalization (b). In [31] Proskurin considers general Fuchsian groups of the first kind, with at least one cusp; he puts some condition on the growth of the Kloosterman sums, but this condition may be removed. Proskurin uses the method of [15]; the proof in [1] relies more heavily on properties of the modular group.
Generalization (c). In [34] Proskurin considers arbitrary real weights in the case n and m positive. In these notes I shall prove the sum formula under generalizations (a), (b) and (c) by a method which differs from those in [15], [1], [31] and [34]. Generalizations that might be possible, but about which I have not thought

seriously are:

(d) do everything for Sl_2 over adeles of a number field.

(e) consider not only Fourier series at cusps, but also at, for instance, elliptic fixed points, or along closed geodetics in $\Gamma\backslash h$.

I shall now state the general sum formula to be proved in these notes. To avoid introducing a lot of concepts and notations which I shall discuss later on in another way, I refer freely to Roelcke's work on real analytic automorphic forms, [36], [37]. Going over directly to subsection 1.3 will cause no harm.

Let Γ be a discrete subgroup of $Sl_2(\mathbb{R})$, acting discontinuously on the upper plane h, such that $\Gamma\backslash h$ is not compact, but has finite volume. Fix $\tau \in [0,1]$ and suppose that v is a multiplier system for the weight τ, see [36], p.296. The automorphic forms I consider have values in $\mathbb{C}$. In §2 of [36] Fourier series expansions of automorphic forms are discussed. A Fourier coefficient depends on two data: its place in the Fourier series expansion and the cusp at which the form is expanded. In the sum formula we shall be concerned with two Fourier coefficients. So I fix two cusps $A^{-1}\infty$ and $B^{-1}\infty$, with A and B as in [36], §2, and two non-zero real numbers n and m with $n \sim \tau_A$ and $m \sim \tau_B$, see [36], (2.13) and (2.6). If the cusps $A^{-1}\infty$ and $B^{-1}\infty$ are equivalent, I suppose that A = B.

Let S be the set of those $\lambda \in \mathbb{C}$ for which there are square integrable automorphic forms in $F_\tau(\Gamma, \mathbb{C}, v)$, see [36], p.297, and let $S_g = S\backslash\{\frac{1}{2}\tau-\frac{1}{4}\tau^2\}$. For each $\lambda \in S_g$ I fix an orthonormal basis $\{f_{\lambda,j}\}$ for the space of these square integrable forms. Now let the term specified by n in the Fourier series expansion of $f_{\lambda,j}$ at the cusp $A^{-1}\infty$ be

(1.2.1) $$c_{\lambda,j}(n)W_{\frac{1}{2}\tau \operatorname{sgn} n,\sqrt{\frac{1}{4}-\lambda}}(4\pi|n|y)e^{2\pi inx},$$

and let $d_{\lambda,j}(m)$ correspond to $B^{-1}\infty$ and m in a similar way.

Let $\{A_\alpha^{-1}.\infty\}$ represent the classes of singular cusps, see [37], p. 294. Let

(1.2.2) $$E_\alpha(z,s) = E(z,s,1;A_\alpha,\Gamma,\tau,\mathbb{C},v)$$

be the Eisenstein series defined in [37], (10.1).

Let $c_\alpha(s,n)$ and $d_\alpha(s,m)$ be the Fourier coefficients of $E_\alpha(z,s)$ specified by $A^{-1}.\infty, n$ and $B^{-1}.\infty, m$. I also need classical holomorphic automorphic forms, see definition 1.2 of [36]. Let $\eta = \pm 1$, $b \equiv \eta\tau(2)$, $b > 0$. Let $\{f_j^\eta\}$ be an orthonormal basis of the space of square integrable holomorphic automorphic forms for the group Γ, weight b and multiplier system v if $\eta = 1$, $\bar{v}$ if $\eta = -1$. (The square integrability refers to the Petersson scalar product. If $\tau = 0$ or 1 and v is real we may take $f_j^1 = f_j^{-1}$.)

At the cusp $A^{-1}.\infty$ we have a Fourier series expansion

(1.2.3) $$f_j^\eta(A^{-1}.z)(c_A z+d_A)^b = \sum_{l\geq 0,\, l\equiv\eta n(1)} a_l e^{2\pi i l n}.$$

If $\eta n > 0$ put $c_j^\eta(b,n) = a_{\eta n}$. Define $d_j^\eta(b,m)$ similarly. In the sum formula these Fourier coefficients are related to generalized Kloosterman sums.

Let C be the set of positive numbers c for which there is an element $\gamma \in \Gamma$ such that

(1.2.4) $\quad A\gamma B^{-1} = \begin{pmatrix} a & * \\ c & d \end{pmatrix}$.

For a given $c \in C$ there are only a finite number of d as in (1.2.4) such that $0 \leqslant d < c$. For each of those d we pick a $\gamma \in \Gamma$ satisfying (1.2.4) and define

(1.2.5) $\quad S(c) = \sum_{\gamma} e^{2\pi i c^{-1}(na+md)} v(\gamma)^{-1}$.

This is a generalized Kloosterman sum; it does not depend on the choice of the γ.

To formulate the sum formula I separate the cases $nm > 0$ and $nm < 0$. First take <u>$\operatorname{sgn} n = \operatorname{sgn} m = \varepsilon$</u>. In (1.1.11) the function h is the principal test function; here I fix $\phi = \widetilde{h}$, which determines h.

The test functions ϕ are defined on $\{s \in \mathbb{C} \mid |\operatorname{Re} s| \leqslant \sigma\} \cup \{\tfrac{1}{2}(b-1) \mid b \equiv \varepsilon\tau(2),\ b > 1\}$ for some $\sigma > \tfrac{1}{2}$ and satisfy

(1.2.6) $\quad \phi$ is an even holomorphic function on $|\operatorname{Re} s| \leqslant \sigma$

(1.2.7) $\quad |\phi(s)| \ll (1 + |\operatorname{Im} s|)^{-a}$ for some $a > 2$

(1.2.8) $\quad \sum_{b\geqslant 1,\, b\equiv\varepsilon\tau(2)} (b-1)\,|\phi(\tfrac{1}{2}(b-1))| < \infty$

We define a Bessel transform, for $y > 0$:

(1.2.9) $$f(y) = \frac{1}{2\pi i} \int_{\operatorname{Re} s=0} \phi(s) J_{2s}(y)\, 2\pi s e^{\frac{1}{2}\pi i\varepsilon\tau} (\sin \pi(\tfrac{1}{2}+s-\tfrac{1}{2}\varepsilon\tau))^{-1} ds$$
$$+ \sum_{b>1,\, b\equiv\varepsilon\tau(2)} (b-1) e^{\frac{1}{2}\pi i b} \phi(\tfrac{1}{2}(b-1)) J_{b-1}(y).$$

This function f is the h in (1.1.11). Further we fix

(1.2.10) $$\phi^0 = \frac{1}{2\pi i} \int_{\operatorname{Re} s=0} \phi(s)(-\pi s) \sin 2\pi s (\cos 2\pi s + \cos \pi\tau)^{-1} ds$$
$$+ \sum_{b>1,\, b\equiv\varepsilon\tau(2)} \tfrac{1}{2}(b-1)\phi(\tfrac{1}{2}(b-1))$$

(1.2.11) $$\delta = \begin{cases} 1 & \text{if } A = B,\ n = m \\ 0 & \text{otherwise} \end{cases}$$

Then all sums and integrals in the following equality converge absolutely; this equality is called the <u>sum formula</u>.

(1.2.12) $$\frac{1}{4\pi i} \int_{\operatorname{Re} s=0} \phi(s)\, |\Gamma(\tfrac{1}{2}+s+\tfrac{1}{2}\varepsilon\tau)|^2 \sum_{\alpha} \overline{c_\alpha(\tfrac{1}{2}+s,n)}\, d_\alpha(\tfrac{1}{2}+s,m)\, ds$$
$$+ \sum_{\lambda\in S_g} \phi(\sqrt{\tfrac{1}{4}-\lambda})\,\Gamma(\tfrac{1}{2}+\tfrac{1}{2}\varepsilon\tau+\sqrt{\tfrac{1}{4}-\lambda})\,\Gamma(\tfrac{1}{2}+\tfrac{1}{2}\varepsilon\tau-\sqrt{\tfrac{1}{4}-\lambda}) . \sum_j \overline{c_{\lambda,j}(n)}\, d_{\lambda,j}(m)$$
$$+ \sum_{\substack{b>0 \\ b\equiv\varepsilon\tau(2)}} (4\pi)^{-b} |nm|^{-\frac{1}{2}b} \Gamma(b) . \begin{cases} \sum_j \overline{c_j^1(b,n)}\, d_j^1(b,m) & \text{if } \varepsilon = 1 \\ \sum_j \overline{c_j^{-1}(b,n)}\, d_j^{-1}(b,m) & \text{if } \varepsilon = -1 \end{cases}$$
$$= \tfrac{1}{2} e^{-\frac{1}{2}\pi i\tau(1+\varepsilon)} |nm|^{-\frac{1}{2}} \sum_{c\in C} c^{-1} S(c) f(4\pi |nm|^{\frac{1}{2}} c^{-1}) + \frac{1}{2\pi}\, \delta\phi^0 |n|^{-1}.$$

In the case <u>$\operatorname{sgn} n = -\operatorname{sgn} m = \varepsilon$</u> the test function ϕ is even and holomorphic on a strip $|\operatorname{Re} s| \leqslant \sigma$ with $\sigma > \tfrac{1}{2}$ and satisfies

(1.2.13) $\quad |\phi(s)| \ll (1 + |\operatorname{Im} s|)^{-a} e^{-\pi|\operatorname{Im} s|}$

for some $a > 2$. The Bessel transform is in this case

(1.2.14) $$f(y) = \frac{1}{2\pi i} \int_{\operatorname{Re} s=0} \phi(s) K_{2s}(y) (-2) e^{-\frac{1}{2}\pi i\varepsilon\tau} s \sin 2\pi s\, ds.$$

The sum formula is

$$(1.2.15)\qquad \frac{1}{4\pi i}\int_{\mathrm{Re}\, s=0} \phi(s) \sum_{\alpha} \overline{c_\alpha(\tfrac{1}{2}+s,n)}\, d_\alpha(\tfrac{1}{2}+s,m)\,ds$$
$$+ \sum_{\lambda\in S} \phi(\sqrt{\tfrac{1}{4}-\lambda}) \sum_j \overline{c_{\lambda,j}(n)}\, d_{\lambda,j}(m)$$
$$= \frac{1}{2\pi}\, e^{\frac{1}{2}\pi i\tau(\varepsilon-1)} |mn|^{-\frac{1}{2}} \sum_{c\in C} c^{-1} S(c) f(4\pi|nm|^{\frac{1}{2}}c^{-1}).$$

1.3 The aim of these notes.

If one considers the proof of the sum formula in [15] or [1], one sees an analogy with the Selberg trace formula, see e.g. [14], appendix. In both cases we have an operator in $L^2(\Gamma\backslash \mathfrak{h})$, given by a kernel. In the trace formula the trace of this operator is computed in two ways: according to the spectral decomposition of $L^2(\Gamma\backslash \mathfrak{h})$ with respect to the Laplace operator, and geometrically. The resulting equality is the trace formula. One may proceed similarly for a double-sided Fourier coefficient of the kernel. Taking a suitable limit of the resulting formula gives the sum formula.

In the formula obtained in this way, e.g. thm. 2 on p.40 of [15], or (3.29) of [1], one has on one side Fourier coefficients of automorphic forms of weight zero and on the other side Kloosterman sums and a δ-term. This is quite understandable from the way the formula has been obtained. But it is a short step from Kuznetsov's formula in [15] to (1.1.11) in subsection 1.1; in fact, Kuznetsov alludes to it in 6.8 of [15]. In §5 of [1] I modified the sum formula a bit, and ended up with Fourier coefficients of holomorphic cusp forms. These holomorphic forms have nothing to do with weight zero. So it might be better not to stick to one weight, when dealing with the sum formula. After all, there are differential operators which map automorphic forms of a certain weight onto forms of another weight. These are the operators K_k and Λ_k described in [36], §3; they go back to Maass; I shall explain them in subsection 1.4. The important fact is that K_k and Λ_k leave the Fourier coefficients essentially unchanged.

The natural objects to study are not individual automorphic forms, but systems of automorphic forms connected by the differential operators K_k and Λ_k. In another language this means studying representations of the Lie algebra $\underline{sl}_2$ in spaces of automorphic forms. One may compare this with [13], where L-functions are no longer associated to automorphic forms, but to representations.

My aim in these notes is to derive the sum formula working from a representational viewpoint. The principal objects are representations; no weight should be privileged. I try to connect the things I do with automorphic forms on the upper half plane in the language of Roelcke, [36], [37]. In describing this connection I stick to the weights used in subsection 1.2.

I try to understand integral transforms as giving the spectral decomposition of representation spaces. In this way the Bessel transforms in the sum formula can be connected with the Plancherel formula of a certain convolution algebra. This possibility has been suggested to me by Prof. T.A. Springer. I work in a fairly general case: arbitrary Fuchsian groups of the first kind, with cusps, and arbitrary real weights. The sum formula is essentially non-arithmetic; hence I have not tried to make the proof adelic.

1.4 The modular case again

To make clear the principles of the proof of the sum formula to be carried out in these notes I give in this subsection a sketch of the proof of (1.1.11). Here the central objects are automorphic models, to be defined in 1.4.1, no longer automorphic forms. Anybody new to this game may profit from checking with explicit calculations the statements I make on the connection between both concepts.

I first fix some notations, valid in subsection 1.4 only, and mention some results on representations of the Lie algebra $\underline{sl}_2(\mathbb{R})$ (see e.g. Ch. VI of [21]).

$G = Sl_2(\mathbb{R})/\{1,-1\}$, $\Gamma = Sl_2(\mathbb{Z})/\{1,-1\}$,

$n(x) = \pm\begin{pmatrix}1 & x\\ 0 & 1\end{pmatrix}$, $a(y) = \pm\begin{pmatrix}\sqrt{y} & 0\\ 0 & 1/\sqrt{y}\end{pmatrix}$, $k(\theta) = \pm\begin{pmatrix}\cos\theta & \sin\theta\\ -\sin\theta & \cos\theta\end{pmatrix}$

$N = \{n(x)\}$, $A = \{a(y)\}$, $K = \{k(\theta)\}$.

The Lie algebra $\underline{g}_{\mathbb{R}} = \underline{sl}_2$ of G is complexified to $\underline{g}$; this complex Lie algebra is spanned by

$\underline{W} = \begin{pmatrix}0 & 1\\ -1 & 0\end{pmatrix}$, $\underline{E}^+ = \begin{pmatrix}1 & i\\ i & -1\end{pmatrix}$ and $\underline{E}^- = \begin{pmatrix}1 & -i\\ -i & -1\end{pmatrix}$.

For $s \in \mathbb{C}$ a well known representation space of $\underline{g}$ is $H(s)$, given by

(1.4.1) $\quad H(s) = \bigoplus_{n\in\mathbb{Z}} \mathbb{C}\cdot\phi_{2n}(s)$

$$\underline{W}\phi_{2n}(s) = 2in\phi_{2n}(s),\quad \underline{E}^{\pm}\phi_{2n}(s) = (2s+1\pm 2n)\phi_{2n\pm 2}(s),$$

see [21], VI §5.

For Re $s = 0$ there is a unitary structure on $H(s)$ for which the elements of $\underline{g}_{\mathbb{R}}$ are skew symmetric; $\{\phi_{2n}(s)\}$ is an orthonormal basis for this structure.

If $b \geq 2$, b even, then $H(\tfrac{1}{2}(b-1))$ has irreducible subspaces; I only need one, D_b^+, given by

(1.4.2) $\quad D_b^+ = \bigoplus_{n\geq\frac{1}{2}b} \mathbb{C}\,\phi_{2n}(\tfrac{1}{2}(b-1))$.

It has a unitary structure for which

(1.4.3) $\quad \langle \phi_{2n}(\tfrac{1}{2}(b-1)),\phi_{2m}(\tfrac{1}{2}(b-1))\rangle = \delta_{m,n}(n-\tfrac{1}{2}b)!/(n+\tfrac{1}{2}b-1)!$

If $2s+1 \notin 2\mathbb{Z}$, then $H(s)$ is irreducible.

1.4.1 Automorphic models

By a model of a representation space U of $\underline{g}$ I understand an operator T from U onto a subspace of $C^\infty(G)$ which intertwines the $\underline{g}$-action on U with the $\underline{g}$-action on $C^\infty(G)$ given by right differentation. So for $u \in U$, $\underline{X} \in \underline{g}_{\mathbb{R}}$, $g \in G$

(1.4.4) $(T\,\underline{X}\,u)(g) = \frac{d}{dt}(Tu)(ge^{t\underline{X}})|_{t=0}$.

A model of U I call <u>automorphic</u> if the functions in TU satisfy

(1.4.5) $f(\gamma g) = f(g)$ for $\gamma \in \Gamma$

(1.4.6) $|f(na(y)k)| \ll y^A$ for $y \to \infty$

for some A, uniformly in $n \in N$, $k \in K$.

If the Casimir operator acts as a scalar in U, the functions in an automorphic model satisfy a differential equation. This is the case for all U we consider. If T is an automorphic model of H(s) and $\phi_{2n}(s) \in H(s)$, then

(1.4.7) $f(x+iy) = T\phi_{2n}(s)(n(x)a(y))$

defines a real analytic automorphic form of weight 2n. [First check this for n = 0. Use that the Casimir operator is given by $-y^2(\frac{\partial^2}{\partial x^2} + \frac{\partial^2}{\partial y^2}) + y\frac{\partial^2}{\partial x\partial\theta}$ in the coordinates $g = n(x)a(y)k(\theta)$.] One may take (1.4.7) as definition of real analytic automorphic forms of even weight. It coincides with Roelcke's definition in [36]. The elements $\underline{E}^+$ and $\underline{E}^-$ of $\underline{g}$ give rise to maps on automorphic forms which shift the weight over ±2. They turn out to be multiples of K_{2n} and Λ_{2n} from [36], §3. So indeed, an automorphic model of H(s) describes a system of real analytic automorphic forms, connected by differential operators.

The same holds for an automorphic model of D_b^+; only weights $\geqslant$ b occur in this case. In weight b it is interesting to consider

(1.4.8) $g(x+iy) = y^{-\frac{1}{2}b}T\phi_b(\frac{1}{2}(b-1))(n(x)a(y))$.

This function g is a holomorphic automorphic form of weight b.

These maps from automorphic models to forms are bijective [the surjectivity is the hardest fact to check; this is discussed in section 4].

1.4.2. <u>Whittaker models</u>.

The functions in a <u>Whittaker model</u> satisfy

(1.4.9) $f(n(x)g) = e^{\frac{1}{2}ix}f(g)$.

If one also asks that the functions satisfy a growth condition, one gets the usual Whittaker model of H(s):

(1.4.10) $W(s)\phi_{2n}(s)(n(x)a(y)k(\theta)) = e^{\frac{1}{2}ix}\Gamma(\frac{1}{2}-s-n)W_{n,s}(y)e^{2in\theta}$.

Another Whittaker model of H(s) is

(1.4.11) $M(s)\phi_{2n}(s)(n(x)a(y)k(\theta)) = e^{\frac{1}{2}ix}M_{n,s}(y)e^{2in\theta}$.

$M_{\cdot,\cdot}$ and $W_{\cdot,\cdot}$ are the usual Whittaker functions. A Whittaker model D^b of D_b^+ is given by

(1.4.12) $D^b = M(\frac{1}{2}(b-1))|D_b^+$;

(1.4.13) $\underset{s\to\frac{1}{2}(b-1)}{\mathrm{res}} W(s)|D_b^+ = \frac{1}{(b-1)!}D^b$.

1.4.3 <u>Fourier coefficients</u>.

The importance of Whittaker models one sees when considering Fourier coefficients of automorphic models.

Let $m \geq 1$, m integral. If f satisfies (1.4.5) then define

(1.4.14) $\quad F_m f(g) = \int_0^1 e^{-2\pi imx} f(n(x)a(4\pi m)^{-1}g)dx.$

F_m is the operator "taking the m-th Fourier coefficient".

Clearly $F_m f$ satisfies (1.4.9).

Now if T is an automorphic model of H(s) there exists a constant $c_s(m,T)$ such that

(1.4.15) $\quad F_m T = c_s(m,T)W(s).$

Similarly for automorphic models of D_b^+:

(1.4.16) $\quad F_m T = c_b^+(m,T)D^b.$

These constants I call the <u>Fourier coefficients</u> of the automorphic model. They are easily expressed in the Fouriercoefficients of the automorphic forms mentioned in (1.4.7) and (1.4.8).

It is important to note that the Fourier coefficients are attributes of the representation, not of the individual automorphic forms.

1.4.4 <u>The space S.</u>

This space consists of those functions in $C^\infty(G)$ which satisfy

(1.4.17) $\quad f(n(x)g) = e^{\frac{1}{2}ix} f(y)$

(1.4.18) $\quad |f| \in C_c(N\backslash G)$

(1.4.19) $\quad$ the right translates of f under K span a finite dimensional space.

<u>g</u> acts on S by right differentiation. On S we have got a unitary structure given by

(1.4.20) $\quad \langle f,g \rangle_{N\backslash G} = \int_{N\backslash G} f(x)\overline{g(x)}d\dot{x} = \frac{1}{2}\int_0^\infty \int_0^\pi f(a(y)k(\theta))\overline{g(a(y)k(\theta))}d\theta y^{-2}dy.$

The elements of S will be auxiliary test functions in the proof of the sum formula.

There are intertwining operators $\omega(s)$ and π_b:

(1.4.21) $\quad \omega(s) : S \to H(s)$

$\omega(s)f = \sum_n \langle f, W(-\bar{s})\phi_{2n}(-\bar{s}) \rangle_{N\backslash G} \phi_{2n}(s)$

(1.4.22) $\quad \pi_b : S \to D_b^+$

$\pi_b f = \sum_{n \geq \frac{1}{2}b} \frac{\sqrt{2}(n+\frac{1}{2}b-1)!\sqrt{b-1}}{(b-1)!(n-\frac{1}{2}b)!} \langle f, D^b\phi_{2n}(\frac{1}{2}(b-1)) \rangle \phi_{2n}(\frac{1}{2}(b-1)).$

By an inversion formula for Whittaker forms one gets for $f,g \in S$:

(1.4.23) $\quad \langle f,g \rangle_{N\backslash G} = \frac{1}{2\pi i}\int_0^{i\infty} \langle \omega(s)f, \omega(s)g \rangle |\Gamma(2s)|^{-2}ds + \sum_{\substack{b \geq 2 \\ b \text{ even}}} \langle \pi_b f, \pi_b g \rangle.$

By $(0,i\infty)$ I mean $\{ix \mid x > 0\}$.

In the spectral decomposition one finds the principal series and the "holomorphic part" of the discrete series, see [21], Ch. VI, §6.

1.4.5 <u>Convolution operators.</u>

Q is the set of those functions in $C^\infty(G)$ which satisfy

(1.4.24) $\quad q(n(x)gn(x_1)) = e^{\frac{1}{2}i(x+x_1)} q(g)$

(1.4.25) $\quad |q| \in C_c(N\backslash G/N)$

An element $q \in Q$ is determined by a function $a_q \in C_c^\infty(0,\infty)$:

(1.4.26) $\quad q(a(y)w) = y^{\frac{1}{2}}a_q(y^{\frac{1}{2}})$

with $w = k(\frac{1}{2}\pi)$. (The choice of $y^{\frac{1}{2}}a_q(y^{\frac{1}{2}})$ instead of $a_q(y)$ makes that we arrive at Bessel transforms.)

The convolution operator C_q is defined by

(1.4.27) $\quad C_q f(g) = q * f(g) = \int_{G/N} q(x)f(x^{-1}g)d\dot{x};$

it makes sense for $f \in S$. It is unimportant whether $C_q f \in S$.

These convolution operators are the principal test functions in the sum formula.

Let $d(t)$ be the distribution on $C_c^\infty(0,\infty)$, with respect to $\frac{dx}{x}$,

(1.4.28) $\quad \phi \mapsto \pi t^{\frac{1}{2}}\phi(t^{\frac{1}{2}})$.

If we build q as in (1.4.26) with a_q replaced by $d(t)$, and C_q as in (1.4.27), we get an operator $\delta(t)$, acting on S. We have

(1.4.29)
$$\begin{aligned}\delta(t)f(g) &= \int_{G/N} q(x)f(x^{-1}g)d\dot{x}\\ &= \frac{1}{2\pi}\int_0^\infty\int_{-\infty}^{+\infty} q(n(x)a(y)w)f(w^{-1}a(y)^{-1}n(x)^{-1}g)dxy^{-2}dy\\ &= \frac{1}{2\pi}\int_{-\infty}^{+\infty} e^{-\frac{1}{2}ix}\int_0^\infty y^{\frac{1}{2}}d(t)(y^{\frac{1}{2}})f(w^{-1}a(y)^{-1}n(x)g)y^{-2}dydx\\ &= \frac{1}{\pi}\int_{-\infty}^{\infty} e^{-\frac{1}{2}ix}\int_0^\infty d(t)(y)f(w^{-1}a(y^{-2})n(x)g)y^{-2}dydx\\ &= \int_{-\infty}^{+\infty} e^{-\frac{1}{2}ix} f(w^{-1}a(t^{-1})n(x)g)dx\ .\end{aligned}$$

For $q \in Q$ or $q = d(t)$ I define

(1.4.30) $\quad \hat{q}(s) = \frac{-1}{\sin \pi s}\int_0^\infty (J_{2s}(y)-J_{-2s}(y))a_q(y)y^{-1}dy,$

so in particular

(1.4.31) $\quad \delta(t)\hat{\ }(s) = \frac{-\pi t^{\frac{1}{2}}}{\sin \pi s}(J_{2s}(t^{\frac{1}{2}})-J_{-2s}(t^{\frac{1}{2}}))$.

Now the inversion theorem for Bessel transforms proved by Kuznetsov, [15], theorem 5, p.13. implies that for $q \in Q$:

(1.4.32)
$$q(a(t)w) = \frac{1}{2\pi i}\int_0^{i\infty} \hat{q}(s)\overline{\delta(t)\hat{\ }(s)}\,\frac{-2s\sin 2\pi s\, ds}{\cos 2\pi s + 1} + \frac{1}{2\pi}\sum_{\substack{b\geq 2\\ b \text{ even}}} (b-1)\hat{q}(\tfrac{1}{2}(b-1))\overline{\delta(t)\hat{\ }(\tfrac{1}{2}(b-1))}$$

The action of C_q on S (with $q \in Q$ or $q = d(t)$) may be described as follows:

(1.4.33)
$$\langle C_q f, g\rangle_{N\backslash G} = \frac{1}{2\pi i}\int_0^{i\infty} \hat{q}(s)\langle\omega(s)f,\omega(s)g\rangle|\Gamma(2s)|^{-2}ds + \sum_{\substack{b\geq 2\\ b \text{ even}}} \hat{q}(\tfrac{1}{2}(b-1))\langle \pi_b f,\pi_b g\rangle.$$

Here it becomes clear why we need the Bessel transform in (1.4.30); it describes the spectral decomposition of C_q in S.

The function $s \mapsto \langle\omega(s)f,\omega(s)g\rangle$ has a meromorphic continuation to $\mathbb{C}$, which I indicate by $s \mapsto \{\omega f,\omega g\}(s)$.

An important modification of (1.4.33) is

(1.4.34)
$$\langle C_q f,\delta(t)g\rangle = \frac{4t^{\frac{1}{2}}}{2\pi i}\int_{\operatorname{Re} s=\sigma} s\cos \pi s\, J_{2s}(t^{\frac{1}{2}})\hat{q}(s)\{\omega f,\omega g\}(s)ds + 2\pi t^{\frac{1}{2}}\sum_{\substack{b>2\sigma+1\\ b \text{ even}}} (-1)^{\frac{1}{2}b}J_{b-1}(t^{\frac{1}{2}})\hat{q}(\tfrac{1}{2}(b-1))\langle \pi_b f,\pi_b g\rangle\ .$$

This expression is $\mathcal{O}(t^{\frac{1}{2}+\sigma})$ for $t \downarrow 0$.

1.4.6 Theta operators.

The operator F_m has an adjoint, the theta operator Θ_m. For $f \in S$, or for f satisfying some less heavy conditions:

(1.4.35) $$\Theta_m f(g) = \sum_{\gamma \in (N \cap \Gamma)\backslash \Gamma} f(a(4\pi m)\gamma g).$$

This is the same construction as used to obtain the holomorphic Poincaré series. In the next formula we see that Θ_m indeed describes "taking the m-th Fourier coefficient".

For $g \in C^\infty(\Gamma\backslash G)$ and $f \in S$:

(1.4.36) $$< \Theta_m f, g >_{\Gamma\backslash G} = 4\pi m < f, F_m g >_{N\backslash G} .$$

For positive integers n and m there is the following important formula

(1.4.37) $$F_m\Theta_n = \delta_{nm}\,\mathrm{Id} + \frac{1}{4\pi m}\sum_{c=1}^{\infty} S(m,n;c)\delta(16\pi^2 nmc^{-2}),$$

with the Kloosterman sum:

(1.4.38) $$S(n,m;c) = \sum_{\substack{x=1 \\ (x,c)=1}}^{c} e^{2\pi i c^{-1}(nx+m\tilde{x})}$$

$(\tilde{x}x \equiv 1(c))$.

For $f,g \in S$, $q \in Q$ one obtains, using (1.4.36):

(1.4.39) $$< \Theta_m C_q f, \Theta_n g > = 4\pi m\delta_{mn} < C_q f, g > + \sum_{c=1}^{\infty} S(m,n;c) < C_q f, \delta(16\pi^2 nmc^{-2})g >$$

To see that the sum over c may be taken outside the scalar product one has to use (1.4.34) with $\sigma > \frac{1}{2}$.

1.4.7 Spectral decomposition.

In (1.4.39) $< \Theta_m C_q f, \Theta_n g >$ has been computed in one way; now I compute it in another way. The spectral decomposition of $L^2(\Gamma\backslash G)$ may be described with the help of automorphic models. To express $< \Theta_m C_q f, \Theta_n g >$ one needs the following ones:

1) The Eisenstein models $\underline{E}(s)$ of $H(s)$, forming a meromorphic family. For $\mathrm{Re}\, s > \frac{1}{2}$

(1.4.40) $$\underline{E}(s)\phi_0(s)(n(x)a(y)) = \sum_{\gamma \in (\Gamma \cap N)\backslash \Gamma} (\mathrm{Im}\,\gamma.(x+iy))^{\frac{1}{2}+s} .$$

2) A sequence $\{T_j\}$ of square integrable automorphic models of $H(s_j)$ with $\mathrm{Re}\, s_j = 0$.

(1.4.41) $$f_j(x+iy) = T_j\phi_0(s_j)(n(x)a(y))$$

defines a square integrable automorphic form of weight zero. The f_j form the orthonormal system in ${}^0L^2(\Gamma\backslash h)$ mentioned in 1.1. The T_j give rise to orthonormal systems in other even weights as well.

3) For each even $b \geq 2$ a finite set $\{T_{b,j}\}$ of square integrable automorphic models of D_b^+.

(1.4.42) $$g_{b,j}(x+iy) = y^{-\frac{1}{2}b}T_{b,j}\phi_b(\tfrac{1}{2}(b-1))(n(x)a(y))$$

is a holomorphic cusp form of weight b; the $g_{b,j}$ form an orthogonal basis of the cusp forms with respect to the Petersson scalar product; their norms are $((b-1)!)^{-\frac{1}{2}}$; so they are multiples of the $f_{b,j}$ in 1.1. In fact, in the modular case we only need

to consider $b \geq 12$.
Automorphic models of D_b^- exist, but are not needed here. In the modular case there are no square integrable automorphic models of representations from the complementary series.

One may prove the following equality:

(1.4.43) $$< \Theta_m C_q f, \Theta_n g > (16\pi^2 mn)^{-1} =$$
$$\frac{1}{2\pi i}\int_0^{i\infty} \overline{c_s(m,\underline{E}(s))}\, c_s(n,\underline{E}(s))\hat{q}(s)\{\omega f,\omega g\}(s)ds$$
$$+\sum_j \overline{c_{s_j}(m,T_j)}\, c_{s_j}(n,T_j)\hat{q}(s_j)\{\omega f,\omega g\}(s_j)$$
$$+ \tfrac{1}{2}\sum_{\substack{b\geq 2\\ b \text{ even}}} (b-1)!(b-2)!\sum_j \overline{c_b^+(m,T_{b,j})}\, c_b^+(n,T_{b,j}).\hat{q}(\tfrac{1}{2}(b-1)) < \pi_b f, \pi_b g >.$$

To see how we arrive at (1.4.43) remark that the projection of $\Theta_n g$ onto the subspace of $L^2(\Gamma\backslash G)$ generated by $T_j H(s_j)$ will be

(1.4.44) $$\sum_{k\in\mathbb{Z}} < \Theta_n g, T_j\phi_{2k}(s_j) >_{\Gamma\backslash G} T_j\phi_{2k}(s_j) .$$

From (1.4.36), (1.4.15) and (1.4.21) follows

(1.4.45) $$< \Theta_n g, T_j\phi_{2k}(s_j) >_{\Gamma\backslash G} =$$
$$4\pi n\, \overline{c_{s_j}(n,T_j)} < g, W(s_j)\phi_{2k}(s_j) >_{N\backslash G} =$$
$$4\pi n\, \overline{c_{s_j}(n,T_j)} < \omega(s_j)g, \phi_{2k}(s_j) >.$$

After proving that this goes through for $g \in C_q S$, it is not difficult to obtain the term in (1.4.43) corresponding to T_j.

1.4.8 Sum formula

(1.4.39) and (1.4.43) give an equality, containing as test functions $q\in Q$ and $f,g \in S$. Let us try to get rid of f and g.
Rewrite (1.4.39) using (1.4.33):

(1.4.46) $$< \Theta_m C_q f, \Theta_n g > =$$
$$4\pi m\delta_{mn}\Big[\frac{1}{2\pi i}\int_0^{i\infty} \hat{q}(s)\{\omega f,\omega g\}(s)|\Gamma(2s)|^{-2}ds + \sum_{b\geq 2}\hat{q}(\tfrac{1}{2}(b-1)) < \pi_b f, \pi_b g >\Big]$$
$$+ \sum_{c=1}^{\infty} S(n,m;c)\Big[\frac{1}{2\pi i}\int_0^{i\infty} \hat{q}(s)\overline{\delta(16\pi^2 nmc^{-2})^{\wedge}(s)}\{\omega f,\omega g\}(s)|\Gamma(2s)|^{-2}ds$$
$$+ \sum_{b\geq 2}\hat{q}(\tfrac{1}{2}(b-1))\overline{\delta(16\pi^2 nmc^{-2})^{\wedge}(\tfrac{1}{2}(b-1))} < \pi_b f, \pi_b g >\Big]$$

Now suppose that we could find $f,g \in S$ such that

(1.4.47) $$\{\omega f,\omega g\}(s) = \Gamma(2s)\Gamma(-2s)\frac{-2s\sin 2\pi s}{\cos 2\pi s + 1} = \frac{\pi}{2(\cos \pi s)^2}$$
$$< \pi_b f, \pi_b g > = \frac{1}{2\pi}(b-1).$$

Then we would obtain from (1.4.46), (1.4.32) and (1.4.43):

(1.4.48) $$\sum_{c=1}^{\infty} c^{-1}S(n,m;c)4\pi\sqrt{mn}\; a_q(4\pi\sqrt{mn}\;c^{-1}) + 4\pi m\delta_{m,n}\Big[\frac{1}{2\pi i}\int_0^{i\infty}\hat{q}(s)\frac{-2s\sin 2\pi s}{\cos 2\pi s + 1}ds +$$

$$+ \sum_{\substack{b\geq 2\\ b\text{ even}}} \frac{1}{2\pi}(b-1)\hat{q}(\tfrac{1}{2}(b-1))]$$

$$= \frac{16\pi^2 nm}{2\pi i}\int_0^{i\infty} \overline{c_s(m,\underline{E}(s))}c_s(n,\underline{E}(s))\hat{q}(s)\frac{\pi}{\cos 2\pi s + 1}\,ds$$

$$+ 16\pi^2 nm \sum_j \overline{c_{s_j}(m,T_j)}c_{s_j}(n,T_j)\frac{\pi}{\cos 2\pi s_j + 1}\hat{q}(s_j)$$

$$+ 4\pi nm \sum_{\substack{b\geq 2\\ b\text{ even}}} (b-1)!^2 \sum_j \overline{c_b^+(m,T_{b,j})}c_b^+(n,T_{b,j})\hat{q}(\tfrac{1}{2}(b-1)).$$

This is the sum formula relating Fourier coefficients of automorphic models and Kloosterman sums.

But f,g $\in$ S satisfying (1.4.47) do not exist. To obtain (1.4.48) I have to be more careful. It is necessary to admit that f,g and q may have no compact support. The easiest way to impose the conditions is not to work with the functions themselves, but with their Whittaker- or Bessel-transforms.

I consider $\{H(s)\mid s\in\mathbb{C}\}$ as a bundle H of g-spaces. $s\mapsto\omega(s)f$ may be viewed as a meromorphic section of H, decreasing quicker than $(1+|\mathrm{Im}\ s|)^{-a}e^{-\pi|\mathrm{Im}\ s|}$ for each $a\in\mathbb{R}$ on strips $|\mathrm{Re}\ s|\leq\sigma$, outside a neighbourhood of the poles. This section satisfies a relation between $\omega(s)f$ and $\omega(-s)f$ and further

(1.4.49) $$\operatorname*{res}_{s\to\frac{1}{2}(b-1)} \omega(s)f = \frac{1}{\sqrt{2}}(b-1)^{-\frac{1}{2}}\pi_b f.$$

Now I consider arbitrary sections of H satisfying these conditions. To such sections ξ and η may be associated functions $\Phi\xi$, $\Phi\eta$ such that

(1.4.50) $$<\Phi\xi,\Phi\eta>_{N\backslash G} = \frac{1}{2\pi i}\int_0^{i\infty} <\xi(s),\eta(s)>|\Gamma(2s)|^{-2}ds + \sum_{b\geq 2} 2(b-1)<\operatorname*{res}_{\frac{b-1}{2}}\xi,\operatorname*{res}_{\frac{b-1}{2}}\eta>.$$

I prove that (1.4.39) and (1.4.43) hold for $\Phi\xi$ and $\Phi\eta$. Even after this extension I cannot satisfy (1.4.47) exactly, but I can easily approximate it. For instance, I can find ξ_v and η_v such that

(1.4.51) $$\{\xi_v,\eta_v\}(s) = \frac{\pi}{2(\cos\pi s)^2}B_v(s)$$

$$2(b-1)<\operatorname*{res}_{s\to\frac{b-1}{2}}\xi_v(s),\operatorname*{res}_{s\to\frac{1}{2}(b-1)}\eta_v(s)> = \frac{1}{2\pi}(b-1)B_v(\tfrac{1}{2}(b-1))$$

$$B_v(s) = \begin{cases} e^{2vs^2} & \text{if } |\mathrm{Re}\ s|\leq v^{-\frac{1}{2}}\\ 0 & \text{otherwise}\end{cases}$$

Now I get from (1.4.34) and (1.4.39):

(1.4.52) $$<\Theta_m C_q\Phi\xi_v,\Theta_n\eta_v> =$$

$$4\pi\delta_{mn}m[\frac{1}{2\pi i}\int_0^{i\infty}\hat{q}(s)\frac{-2s\sin 2\pi s}{\cos 2\pi s+1}B_v(s)ds + \frac{1}{2\pi}\sum_{b\geq 2}\hat{q}(\tfrac{1}{2}(b-1))(b-1)B_v(\tfrac{1}{2}(b-1))]$$

$$+\sum_{c=1}^{\infty}c^{-1}S(n,m;c)[\frac{8\pi\sqrt{nm}}{2\pi i}\int_{\mathrm{Re}\ s=\sigma}\frac{\pi s}{\cos\pi s}J_{2s}(4\pi\sqrt{nm}c^{-1})\hat{q}(s)B_v(s)ds$$

$$+ 4\pi\sqrt{mn}\sum_{b>2\sigma+1}(-1)^{\frac{1}{2}b}(b-1)J_{b-1}(4\pi\sqrt{nm}c^{-1})\hat{q}(\tfrac{1}{2}(b-1))B_v(\tfrac{1}{2}(b-1))].$$

If $\sigma>\frac{1}{2}$ then there is no big problem to show that the limit of (1.4.52) for $v\downarrow 0$

is given by the left hand side of (1.4.48). In the second term the limit may be taken inside first, after that the integral is moved back to $(0,i\infty)$.
Now we know that the limit for $v \downarrow 0$ of

$$(1.4.53)\qquad 16\pi^2 nm \frac{1}{2\pi i}\Big[\int_0^{i\infty} \overline{c_s(m,\underline{E}(s))}\, c_s(n,\underline{E}(s))\hat{q}(s)\frac{\pi}{\cos 2\pi s+1} B_v(s)\,ds$$
$$+ \sum_j \overline{c_{s_j}(m,T_j)}\, c_{s_j}(n,T_j)\hat{q}(s_j) \frac{\pi}{\cos 2\pi s_j+1} B_v(s_j) +$$
$$+ \frac{1}{4\pi} \sum_{b\geq 2} |(b-1)!|^2 \sum_j \overline{c_b^+(m,T_{b,j})}\, c_b^+(n,T_{b,j})\hat{q}(\tfrac{1}{2}(b-1)) B_v(\tfrac{1}{2}(b-1))\Big]$$

exists and is equal to the expression in the left hand side of (1.4.48).
I extend the class of test functions q by imposing a condition on $\hat{q}(s)$.
Let ϕ be holomorphic on $|\mathrm{Re}\, s| \leq \sigma$, let

$$(1.4.54)\qquad |\phi(s)| \ll (1 + |\mathrm{Im}\, s|)^{-a}$$

for some $a > 2$, and let ϕ also be defined on $\frac{1}{2} + \mathbb{Z}$, satisfying

$$(1.4.55)\qquad \sum_{b\geq 2} (b-1)|\phi(\tfrac{1}{2}(b-1))| < \infty.$$

It can be shown that such a ϕ is the Bessel transform $\hat{q}_\phi$ of a function q_ϕ satisfying (1.4.24). Everything done up till now is valid for such a q_ϕ. Furthermore, such a ϕ may be majorized by a ϕ_0, satisfying the same conditions and being positive in all points occurring in (1.4.53).
We take $n = m$ and $\phi = \phi_0$. Everything in (1.4.53) is positive. Application of Fatou's lemma shows that the sums and integral in (1.4.53) converge absolutely if we take $B_v(s) = 1$. So we have a majorant to apply Lebesgue's theorem, and we obtain (1.4.48) in the case $n = m$, $\phi = \phi_0$. In the general case we observe

$$(1.4.56)\qquad |\bar{a}b| \leq |a|^2 + |b|^2,$$

hence we may use the terms of the case $n = m$, $\phi = \phi_0$ as a majorant.

1.4.9 Remarks

At this point I try to answer the questions at the end of subsection 1.1. Working with functions on $\Gamma\backslash G$ we may expect that all square integrable automorphic models enter the sum formula, provided they have nonzero Fourier coefficients of the right order. In the approach just sketched the principal test functions are the convolution operators C_q. In the spectral decomposition of these operators not only representations from the principal series occur, but also from the discrete series. One may try to suppress the latter by working in weight zero, but they keep turning up in the formulas. At the beginning of subsection 1.3 I explained that weight zero proofs of the sum formula are based on computations of two-sided Fourier coefficients of an operator in $L^2(\Gamma\backslash \mathfrak{h})$. This operator is the projection into weight zero of a convolution operator C_q. If one studies the spectral decomposition of these operators, the Bessel differential equation is a natural consequence of the form of the Casimir operator in certain coordinates.

1.5 The general case.

Giving full proofs for all statements in subsection 1.4 takes quite some space. In the general case the situation is still more complicated.

In 1.4 I have considered positive m and n only. For negative m or n one needs also Whittaker models consisting of functions transforming according to the character $n(x) \mapsto e^{-\frac{1}{2}ix}$. In $Sl_2(\mathbb{R})$ these models are not readily reduced to those considered in 1.4.2. Therefore I work on $Sl_2(\mathbb{R}) \cup \begin{pmatrix} -1 & 0 \\ 0 & 1 \end{pmatrix} Sl_2(\mathbb{R})$.

Automorphic forms on the upper half plane h of integral weight may be lifted to functions on $\Gamma\backslash Sl_2(\mathbb{R})$, see (1.4.7). For non-integral weights one needs covering groups of $Sl_2(\mathbb{R})$. Therefore I work on the universal covering group G of $Sl_2(\mathbb{R}) \cup \begin{pmatrix} -1 & 0 \\ 0 & 1 \end{pmatrix} Sl_2(\mathbb{R})$; see section 2.

In section 3 I discuss representations. As G is not connected, it is not sufficient to consider the action of the Lie algebra only; also the action of j, covering $\begin{pmatrix} -1 & 0 \\ 0 & 1 \end{pmatrix}$ has to be taken into account.

Automorphic models and the relation with automorphic forms on h are discussed in section 4, square integrable automorphic models in section 10.

Whittaker models are given in section 5; the Fourier coefficients of automorphic models are defined in section 7. As Γ may have several inequivalent cusps I introduce a notation specifying a cusp and a term in the Fourier series expansion at that cusp, see definition 7.2.1.

The spectral decomposition of the space S discussed in 1.4.4. is based on an inversion theorem for Whittaker transforms, proved in section 13. In this section I also study the set H_σ^τ of sections, generalizing the Whittaker transforms of elements of ${}^{\frac{1}{2}}S_\tau$ (the space S of 1.4.4).

In section 6 I start the study of the convolution operators mentioned in 1.4.5. I do this in a bit more general situation, to get at the same time formulas to be used at other places. The Bessel transformations as given in (1.4.31), (1.4.32) I treat in section 14.

Section 8 contains the material on theta operators, see 1.4.6.

The spectral decomposition of $L^2(\Gamma\backslash G,\chi)$ gave me a lot of work. One needs the meromorphic continuation of Eisenstein series. This is well known in the case of weight zero, see e.g. [14], [21], Ch. XIV. Selberg has obtained this continuation in the generality needed here, but as far as I know a complete proof for this general case is not in the literature. I have undertaken to write it down in the language of automorphic models. I follow Selberg's method, as found in [22], [11]. The Eisenstein models are defined in section 9; section 11 contains the analytic continuation. In section 12 I discuss the spectral decomposition of $L^2(\Gamma\backslash G,\chi)$.

Section 15 contains the proof of the equations corresponding to (1.4.39) and (1.4.43). I do not use Fourier coefficients of automorphic models in the formulation, but introduce "Poincaré elements" in spaces of square integrable automorphic models.

This avoids the necessity of choosing an orthonormal basis.
Section 16 contains the proof of the sum formula along the lines indicated in 1.4.8. I reformulate it and I compare it with the sum formulas in the literature.

The amount of work necessary to obtain the sum formula in this way has turned out to be considerable. Partly it is a matter of language. Maass, Roelcke and Elstrodt, [24], [36], [37], [2], [3], [4], have done much work on real analytic automorphic forms, but this had to be reformulated from a representational viewpoint. The generality I have chosen for adds complications at several places. The proof of the analytical continuation of Eisenstein series is time consuming anyhow.
I hope that the representational viewpoint and the generality will make up for the size and the intricacy.

1.6. Acknowledgements

I have learned much from the work of Kuznetsov and Proskurin. Although I propose another road to the sum formula than theirs, I would not have written these notes without knowledge of their work.
I have also profited from D. Zagier's proof of the sum formula, which he showed to me.
Most of the work on these notes I have done while supported by the research-pool of Utrecht University.
At the Mathematical Institute at Utrecht I got help and interest from many people, for which I am grateful. Especially I thank Prof. F. van der Blij and Prof. T.A. Springer for their interest and stimulation.
I am very grateful to Ans van Hoof, Sineke Koorn, Petra van der Kuilen, Renske Kuipers, Joke Stalpers and Marjan Visser who have done the laborious job of typing these notes.

1.7. Remarks.

1.7.1. Notation

$\mathbb{R}$, $\mathbb{C}$, $\mathbb{Z}$, $\mathbb{N}$, C^∞, C_c^∞ have their standard meaning.
$(0,i\infty) = \{ix \in \mathbb{C} \mid x > 0\}$.
$\Gamma(.)$, $\zeta(.)$: gamma- and zetafunction
$Sl_2(\mathbb{R})$: real 2×2 matrices with determinant 1.
$\langle f,g \rangle$ means $\int_X f(x)\overline{g(x)}dx$, the space X and the measure dx should be clear from the context.
$f\{\phi\}$ is the notation for the distribution f acting on the function ϕ.

Crossreferences. All formulas are numbered by (s,ss,n), with s sectionnumber, ss number of subsection. The same numbering s,ss,n, but without brackets, is used for definitions, propositions, etc.

1.7.2 Computations

Without explicitely mentioning them I often use formulas like:

(1.7.1) $\Gamma(s)\Gamma(1-s) = \pi(\sin \pi s)^{-1}$

(1.7.2) $\Gamma(\frac{1}{2}+s+\frac{1}{2}\varepsilon r)\Gamma(\frac{1}{2}-s+\frac{1}{2}\eta r)(\cos 2\pi s+\cos \pi r) =$
$2\pi^2\Gamma(\frac{1}{2}+s-\frac{1}{2}\varepsilon r)^{-1}\Gamma(\frac{1}{2}-s-\frac{1}{2}\eta r)^{-1}$ with $\varepsilon,\eta = \pm 1$.

Stirling's formula

(1.7.3) $\Gamma(z) \sim \sqrt{2\pi}\, z^{z-\frac{1}{2}}e^{-z}$ for $|z| \to \infty$, $|\arg z| < \pi$,

see [25], p.12, is often needed in estimates; especially in the form

(1.7.4) $|\Gamma(z)| \sim \sqrt{2\pi}\, e^{-\frac{1}{2}\pi|\mathrm{Im}\, z|}(1+|\mathrm{Im}\, z|)^{\mathrm{Re}\, z-\frac{1}{2}}e^{-\mathrm{Re}\, z}$ for $|\mathrm{Im}\, z| \to \infty$

for $x_0 \leqslant \mathrm{Re}\, z \leqslant x_1$.

1.7.3 Prerequisites

The reader I have had in mind while writing these notes has seen automorphic forms before, both holomorphic and real analytic ones (e.g. [23], [39], [36], [37], [14], [21] Ch.XIV, [42]), has read from Lang's book [21] Ch.II §1, VI §1-6, XII §1,2, and is willing to accept occasionally results or look them up.

Although I think it is good to study automorphic forms from a representational point of view, I do not want to cut loose from automorphic forms as functions on the upper half plane. So I shall repeatedly connect automorphic models with automorphic forms as studied by Roelcke. I have chosen [36], [37] as comprehensive point of reference concerning automorphic forms; it is certainly not necessary to study it completely before starting these notes.

The use of bundles in sections 3, 7 and others makes some formulations easier; no results on bundles are used.

2. The universal covering group G.

2.1 Introduction

In §4 of [36] Roelcke shows how automorphic forms may be considered as functions on the universal covering group of $Sl_2(\mathbb{R})$ instead of functions on the upper half plane. Instead of $Sl_2(\mathbb{R})$ I use the universal covering group of $Sl_2(\mathbb{R}) \cup \begin{pmatrix}-1 & 0\\ 0 & 1\end{pmatrix} Sl_2(\mathbb{R})$. This group I introduce in 2.2; I fix notations for subgroups and invariant measures in 2.3 and consider the Lie algebra in 2.5. The connection between functions on this group and functions on the upper half plane is discussed in 2.3, 2.6 and 2.7.

Everything in this section is known. Most of it may be found in [36].

2.2 The group G

Let G be the group generated by

(2.2.1) $\{j, n(x), a(y), k(\theta) \mid x,\theta \in \mathbb{R},\ y > 0\}$

subjected to the relations

(2.2.2) $a(y_1)a(y_2) = a(y_1y_2)$

(2.2.3) $n(x_1+x_2) = n(x_1)n(x_2)$

(2.2.4) $a(y)n(x) = n(yx)a(y)$

(2.2.5) $j^2 = 1$

(2.2.6) $jp(z) = p(-\bar{z})j \qquad$ for $z \in h$

(2.2.7) $k(\theta_1+\theta_2) = k(\theta_1)k(\theta_2)$

(2.2.8) $jk(\theta) = k(-\theta)j$

(2.2.9) $k(\theta)p(z) = p\left(\frac{z\cos\theta + \sin\theta}{-z\sin\theta + \cos\theta}\right)k\left(\theta - \arg_{(-\pi,\pi)} e^{i\theta}(-z\sin\theta + \cos\theta)\right)$

for $z \in h$,

with $p(x+iy) = n(x)a(y)$ and $h = \{z \in \mathbb{C} \mid \operatorname{Im} z > 0\}$, the upper half plane.

The map $(z,\theta) \mapsto p(z)k(\theta)$ gives a bijection $h \times \mathbb{R} \to G$. The natural structure of analytic manifold on $h \times \mathbb{R}$ is transfered to G.

I denote by $g \longmapsto \bar{g}$ the natural projection $G \to Sl_2(\mathbb{R}) \cup \begin{pmatrix}-1 & 0\\ 0 & 1\end{pmatrix} Sl_2(\mathbb{R})$. We have

(2.2.10) $\bar{j} = \begin{pmatrix}-1 & 0\\ 0 & 1\end{pmatrix}$, $\overline{n(x)} = \begin{pmatrix}1 & x\\ 0 & 1\end{pmatrix}$,

$$\overline{a(y)} = \begin{pmatrix}\sqrt{y} & 0\\ 0 & 1/\sqrt{y}\end{pmatrix}, \quad \overline{k(\theta)} = \begin{pmatrix}\cos\theta & \sin\theta\\ -\sin\theta & \cos\theta\end{pmatrix}.$$

For $\operatorname{Im} z > 0$

(2.2.11) $\arg_{(-\pi,\pi)} e^{i\theta}(-z\sin\theta + \cos\theta) = \arg_{(-\pi,\pi]}(-z\sin\theta + \cos\theta) - \arg_{(-\pi,\pi]} e^{-i\theta}$.

A section $Sl_2(\mathbb{R}) \to G$ is given by

(2.2.12) $\begin{pmatrix}a & b\\ c & d\end{pmatrix} \mapsto \widetilde{\begin{pmatrix}a & b\\ c & d\end{pmatrix}} = p\left(\frac{ai+b}{ci+d}\right)k\left(\arg_{[-\pi,\pi)}(-ci+d)\right)$.

For this section

(2.2.13) $\tilde{g}_1\tilde{g} = \widetilde{g_1 g}\, k(-2\pi \underline{w}(g_1,g))$

where $\underline{w}$ is the 2-cocycle given by Petersson in [29], §2; see also [36], §1.

The group G consists of two simply connected components. One of them is the universal covering group of $Sl_2(\mathbb{R})$, the other one covers $\begin{pmatrix}-1 & 0\\ 0 & 1\end{pmatrix}Sl_2(\mathbb{R})$.

2.3 Functions on $\mathfrak{h}$.

Let $r \in \mathbb{R}$. A function F on G is said to have weight r if

(2.3.1) $F(gk(\theta)) = F(g)e^{ir\theta}$.

A function F of weight r is fully determined by the following two functions F^+ and F^- on $\mathfrak{h}$:

(2.3.2) $F^+(z) = F(p(z))y^{-\frac{1}{2}r}$

$F^-(z) = \overline{F(p(z)j)}y^{-\frac{1}{2}r}$

Let $x = \begin{pmatrix}a & b\\ c & d\end{pmatrix} \in Sl_2(\mathbb{R})$ and let $F_1(g) = F(\tilde{x}g)$; then F_1 has weight r as well. We have

(2.3.3) $F_1^{\pm}(z) = F^{\pm}(\frac{az+b}{cz+d})(cz+d)^{-r}$,

where $(cz+d)^{-r}$ is computed using

(2.3.4) $\arg(cz+d) \in (-\pi,\pi]$.

This transformation is used in the definition of holomorphic automorphic forms, see e.g. [23], p.268. To get the forms studied by Roelcke remark that F of weight r is also fully determined by the following two functions on $\mathfrak{h}$:

(2.3.5) ${}^{R}F^{+}(z) = F(p(z))$, ${}^{R}F^{-}(z) = \overline{F(p(z)j)}$

The relation with $F^{\pm}$ is given by:

(2.3.6) ${}^{R}F^{\pm}(z) = F^{\pm}(z)y^{\frac{1}{2}r}$.

With F_1 as above:

(2.3.7) ${}^{R}F_1^{\pm} = {}^{R}F^{\pm}|[\begin{pmatrix}a & b\\ c & d\end{pmatrix},r]$,

the transformation Roelcke uses, see [36], (1.12).

2.4 Subgroups and integral formulas.

Some subgroups of G are

(2.4.1) $N = \{n(x) | x \in \mathbb{R}\}$

$A = \{a(y) | y > 0\}$

$K_0 = \{k(\theta) | \theta \in \mathbb{R}\}$

$K = K_0 \cup jK_0$

$G_0 = NAK_0$

$Z_0 = \{k(\pi l) | l \in \mathbb{Z}\}$, the center of G_0

$H_0 = Z_0A$

$Z = Z_0 \cup jZ_0$, Z is non-abelian

$H = H_0 \cup jH_0$

N has an invariant measure dn corresponding to the Lebesgue measure on $\mathbb{R}$ under $x \mapsto n(x)$. Likewise A has an invariant measure $da = y^{-1}dy$, with $a = a(y)$. K_0 has an invariant measure $dk = (2\pi)^{-1}d\theta$, with $k = k(\theta)$.

Let

(2.4.2) $\alpha(a(y)) = y$,

then

(2.4.3) $d(ana^{-1}) = \alpha(a)dn$

An invariant measure on G_0 is given by

(2.4.4) $dg = \alpha(a)^{-1}dkda\, dn$

with $g = nak$.

Let for $\varepsilon = 1$ or -1

(2.4.5) $j_\varepsilon = \begin{cases} 1 & \text{if } \varepsilon = 1 \\ j & \text{if } \varepsilon = -1. \end{cases}$

An invariant measure on K, also denoted by dk is given by

(2.4.6) $\int_K \phi(k)dk = \sum_{\varepsilon=\pm1} \int_{K_0} \phi(j_\varepsilon k)dk$

The formula in (2.4.4) gives also an invariant measure on G. An invariant measure on H is given by

(2.4.7) $\int_H \phi(h)dh = \sum_{z\in Z} \int_A \phi(az)da.$

Often I shall wish to integrate functions which are left-invariant under Z_0. The quotient measure on $Z_0\backslash K$ I also denote by dk and the measure on $Z_0\backslash G$ by dg.

If ϕ is a function on G, left-invariant under Z_0N then

(2.4.8) $\int_{Z_0N\backslash G} \phi(g)d\dot{g} = \int_A \int_{Z_0\backslash K} \phi(ak)dk\alpha(a)^{-1}da.$

The Bruhat decomposition $G = NH \amalg NHwN$, with

(2.4.9) $w = k(\tfrac{1}{2}\pi)$,

leads to another integral formula:

(2.4.10) $\int_G \phi(g)dg = \frac{1}{2\pi} \int_N \int_H \int_N \phi(n_1 h\, w\, n_2)dn_1\alpha(h)^{-1}dh\, dn_2;$

here α has been extended to H by defining it equal to 1 on Z.

On G/N we get

(2.4.11) $\int_{G/N} \phi(g)d\dot{g} = \frac{1}{2\pi} \int_H \int_N \phi(nhw)\alpha(h)^{-1}dh\, dn.$

2.5 Lie algebra.

The Lie algebra of G is the same as the Lie algebra of $Sl_2(\mathbb{R})$. I denote by $\underline{g}_{\mathbb{R}}$ the real Lie algebra with basis

(2.5.1) $\underline{W} = \begin{pmatrix} 0 & 1 \\ -1 & 0 \end{pmatrix}, \quad \underline{H} = \begin{pmatrix} 1 & 0 \\ 0 & -1 \end{pmatrix}, \quad \underline{X} = \begin{pmatrix} 0 & 1 \\ 0 & 0 \end{pmatrix},$

and $\underline{g} = \underline{g}_{\mathbb{R}} \otimes_{\mathbb{R}} \mathbb{C}$. Further notations:

(2.5.2) $\underline{V} = 2\underline{X} - \underline{W}, \quad \underline{E}^{\pm} = \underline{H} \pm i\underline{V}.$

The Lie algebra acts on functions on G by right differentation. The following formulas describe this action with respect to the coordinates:

(2.5.3) $(x,y,\varepsilon,\theta) = n(x)a(y)j_\varepsilon k(\theta).$

(2.5.4) $\underline{W} = \frac{\partial}{\partial\theta}$

$\underline{E}^{\pm} = e^{\pm 2i\theta}(\pm 2i\varepsilon y \frac{\partial}{\partial x} + 2y\frac{\partial}{\partial y} \mp i\frac{\partial}{\partial\theta}).$

The Casimir operator $\underline{\omega} = -\tfrac{1}{4}\underline{E}^{+}\underline{E}^{-} + \tfrac{1}{4}\underline{W}^2 - \tfrac{1}{2}i\underline{W}$ acts as

(2.5.5) $\underline{\omega} = -y^2(\frac{\partial^2}{\partial x^2} + \frac{\partial^2}{\partial y^2}) + \varepsilon y \frac{\partial^2}{\partial x\partial\theta}.$

In the coordinates $(x,\varepsilon,l,y,u) = n(x)j_\varepsilon k(\pi l)a(y)wn(u)$

(2.5.6) $\quad \underline{\omega} = -y^2\frac{\partial^2}{\partial y^2} + \varepsilon y \frac{\partial^2}{\partial u \partial x}$.

2.6 Differential operators on the upper half plane.

For a given function F of weight r the action of $\underline{W}$ is just multiplication by ir. The action of $\underline{E}^{\pm}$ gives differential operators on h if one uses (2.3.2) or (2.3.5). From (2.5.4) follows that $\underline{E}^{\pm}$ maps F into the functions of weight r $\pm$ 2. In the case of (2.3.2):

(2.6.1) $\quad (\underline{E}^{+}F)^{\pm}(z) = 2ry^{-1}F^{\pm}(z) + 2y(\frac{\partial}{\partial y} + i\frac{\partial}{\partial x})F^{\pm}(z)$

$\quad (\underline{E}^{-}F)^{\pm}(z) = 2y^2(\frac{\partial}{\partial y} - i\frac{\partial}{\partial x})F^{\pm}(z).$

For the other case

(2.6.2) $\quad {}^{R}(\underline{E}^{+}F)^{\pm}(z) = +r\,{}^{R}F^{\pm}(z) + 2y(\frac{\partial}{\partial y} + i\frac{\partial}{\partial x})\,{}^{R}F^{\pm}(z)$

$\quad = 2K_r\,{}^{R}F^{\pm}(z)$

$\quad {}^{R}(\underline{E}^{-}F)^{\pm}(z) = -r\,{}^{R}F^{\pm}(z) + 2y(\frac{\partial}{\partial y} - i\frac{\partial}{\partial x})\,{}^{R}F^{\pm}(z)$

$\quad = -2\Lambda_r\,{}^{R}F^{\pm}(z);$

here K_r and Λ_r are the operators given by Roelcke in [36], 3.1, 3.2.

2.7 Examples.

The theta function.

Let

(2.7.1) $\quad \theta(z) = \sum_{n\in\mathbb{Z}} e^{\pi i z n^2}$

be the well-known theta function on h. A function F_θ on G of weight $\frac{1}{2}$ is defined by

(2.7.2) $\quad F_\theta^{+} = \theta,\ F_\theta^{-} = 0.$

From the transformation properties of θ follows

(2.7.3) $\quad F_\theta(n(2)g) = F_\theta(g)$

$\quad F_\theta(wg) \quad = F_\theta(g)e^{\frac{1}{4}\pi i}.$

Let Γ_θ be the group generated by n(2) and w. Let χ_θ be the character of Γ_θ defined by

(2.7.4) $\quad \chi_\theta(n(2)) = 1; \quad \chi_\theta(w) = e^{\frac{1}{4}\pi i}.$

This is indeed a character, for the only relations between $\overline{n(2)}$ and $\overline{w}$ in $Sl_2(\mathbb{R})$ are given by

(2.7.5) $\quad \overline{w}^2$ central in $\overline{\Gamma}_\theta$

$\quad \overline{w}^4 = 1,$ (see [23], p.365)

and one can check from this that in G the only relation between n(2) and w is the centrality of w^2. For each $\gamma \in \Gamma_\theta$, $g \in G$:

(2.7.6) $\quad F_\theta(\gamma g) = \chi_\theta(\gamma)F_\theta(g).$

Remark that $\underline{E}^{-}F_\theta = 0$, $\underline{\omega}F_\theta = \frac{3}{16}F_\theta$.

Modular forms of weight zero.

$\Gamma(1)$ is the group generated by n(1) and w. It is the full covering of the modular group in $Sl_2(\mathbb{R})$. Let f_1 and f_{-1} be two $\overline{\Gamma(1)}$-invariant functions on h, satisfying

(2.7.7) $\quad -y^2(\frac{\partial^2}{\partial x^2} + \frac{\partial^2}{\partial y^2})f_{\pm 1} = \lambda f_{\pm 1}$,

and some growth condition for $y \to \infty$. The function F on G defined by

(2.7.8) $\quad F^{\pm}(z) = f_{\pm 1}(z)$

satisfies ·

(2.7.9) $\quad F(\gamma g) = F(g) \quad$ for all $\gamma \in \Gamma(1)$

(2.7.10) $\quad \underline{\omega}F = \lambda F$

For weight zero ${}^{R}F^{\pm} = F^{\pm}$.

This kind of functions $f_{\pm 1}$ are real analytic modular forms as discussed in 1.1.

3. Representations.

3.1 Introduction.

The operators Λ_r and K_r of Maass and Roelcke, mentioned in 2.6 connect automorphic forms. If one knows the Fourier coefficients of one automorphic form, one can compute those of all automorphic forms derived from the first one by successive application of Λ_r and K_r. In 2.6 one sees that Λ_r and K_r are in fact elements of the Lie algebra $\underline{g}$. As I want to study systems of automorphic forms instead of individual forms, I have to study representations of $\underline{g}$. Actually I shall use representations of $\underline{g}$ and K together.

In 3.2 I define the kind of representation I consider. Next I exhibit the most important irreducible ones, 3.3. Some of them posses a unitary structure, 3.4. The group G has two components. In the spectral decomposition of several big representation spaces in later sections there will be a kind of doubleness. Some notations to handle this situation I fix in 3.5. Finally I mention two more examples of representations. Some of the results of this section are given by Pukánszky, [35], for G_0. Everything is a straightforward extension of the results for $Sl_2(\mathbb{R})$; see e.g. Lang, [21], Ch. VI.

3.2 $\underline{g}$-K-spaces.

The representations I consider are representations of K and $\underline{g}$. As Z_0 is an abelian subgroup of K, its irreducible representations are given by characters

(3.2.1) $\quad k(\pi l) \mapsto e^{\pi i \upsilon l}$

with $\upsilon \in (-1,1]$. As $j \in K$ and $jk(\pi l) = k(-\pi l)j$ the occurrence of υ in a representation space of K implies the occurrence of $-\upsilon$, unless $\upsilon = 0$ or 1. If $\tau \in [0,1]$, then I call a representation of Z of type τ if only $\upsilon = \tau$ occurs in the case $\tau = 0$ or 1, and if only $\upsilon = \tau$ and $\upsilon = -\tau$ occur in the case $0 < \tau < 1$. The letter τ I reserve for indicating the type.

3.2.1 Definition. *A complex vector space V is called a $\underline{g}$-K-space, if $\underline{g}$ and K act on it such that*

(3.2.2) $\quad$ V *consists of K-finite vectors;*

(3.2.3) $\quad$ Z *acts according to type* τ;

(3.2.4) $\quad$ *the actions of $\underline{g}$ and K are compatible, i.e.* $\mathrm{ad}(k)\underline{X}.v = k.\underline{X}.k^{-1}.v$ *for all* $k \in K$, $\underline{X} \in \underline{g}$, $v \in V$.

K-finiteness of $v \in V$ means that the subspace spanned by K.v is finite dimensional. A $\underline{g}$-K-space is not necessarily admissible (see Jacquet-Langlands, [13], p. 154). It is admissible if and only if all irreducible representations of K occur with finite multiplicity.

I do not use a special symbol to denote the action of $\underline{g}$ and K; I write k.v, and not $\rho(k)v$. For each $\underline{g}$-K-space V

(3.2.5) $\quad V_r = \{v \in V | k(\theta).v = e^{ir\theta}v \text{ for all } k(\theta) \in K_0\}.$

Then

(3.2.6) $\quad V = \bigoplus_{r \equiv \pm\tau(2)} V_r.$

3.3 Non-unitary principal series.

All irreducible $\underline{g}$-K-spaces I need are contained in the non-unitary principal series. I give them as abstract spaces. One may obtain them as in [21], Ch. VI, §5. Let $\tau \in [0,1]$ and a complex number s be given; let $\delta = 1$ or -1 if $\tau = 0$ or 1.

3.3.1 Definition. I define if $\tau = 0$ or 1:

(3.3.1) $\quad H_\tau^\delta(s) = \bigoplus_{r \equiv \tau(2)} \mathbb{C}\, \phi_r^\delta(s),$

if $0 < \tau < 1$:

(3.3.2) $\quad H_\tau(s) = \bigoplus_{r \equiv \pm\tau(2)} \mathbb{C}\, \phi_r(s).$

$\{\phi_r^\delta(s)\}$ or $\{\phi_r(s)\}$ is called the standard basis of $H_\tau^\delta(s)$ or $H_\tau(s)$. The $\underline{g}$-K-action is defined by

(3.3.3) $\quad j.\phi_r^\delta(s) = \delta\phi_{-r}^\delta(s), \; j.\phi_r(s) = \phi_{-r}(s)$

(3.3.4) $\quad k(\theta).\phi_r^{(\delta)}(s) = e^{ir\theta}\phi_r^{(\delta)}(s)$

(3.3.5) $\quad \underline{W} \,.\, \phi_r^{(\delta)}(s) = ir\, \phi_r^{(\delta)}(s)$

(3.3.6) $\quad \underline{E}^\pm \,.\, \phi_r^{(\delta)}(s) = (1+2s\pm r)\phi_{r\pm 2}^{(\delta)}(s).$

These $\underline{g}$-K-spaces are admissible. In general they are irreducible. By the methods of [35] or [21], Ch. VI §5 one sees that the exceptions to the irreducibility occur if $s = \frac{1}{2}(b-1)$ with $b \equiv \pm\tau(2)$, $b \neq 1$. The reducibility is due to the existence of $\phi_r^{(\delta)}(s)$ with $\underline{E}^+\phi_r^{(\delta)}(s) = 0$ or $\underline{E}^-\phi_r^{(\delta)}(s) = 0$. In the case $\tau = b = 1$ the action of j makes the whole space irreducible. Lang calls it the 'mock discrete series', [21], p.119, case 2. Let $b \equiv \pm\tau(2)$, $b \neq 1$. In $H_\tau^{(\delta)}(\frac{1}{2}(b-1))$ one finds the irreducible subspace $D_b^{(\delta)}$, given by

(3.3.7) $\quad D_b^{(\delta)} = \bigoplus_{\substack{\alpha=\pm 1,\, \alpha r \geq b \\ \alpha r \equiv b(2)}} \mathbb{C}\, \phi_r^{(\delta)}(\tfrac{1}{2}(b-1))$

if $b > 1$ or if $0 < \tau < 1$, and in the case $b < 1$, $\tau = 0,1$ given by

(3.3.8) $\quad D_b^\delta = \bigoplus_{\substack{r \equiv b(2) \\ |r| \leq -b}} \mathbb{C}\, \phi_r^\delta(\tfrac{1}{2}(b-1)$

If $\tau = 0$ or 1 and $b > 1$ then D_b^1 and D_b^{-1} are isomorphic by

(3.3.9) $\quad \phi_r^1(\tfrac{1}{2}(b-1)) \leftrightarrow \alpha\phi_r^{-1}(\tfrac{1}{2}(b-1))$

with $\alpha = \pm 1$, $\alpha r \geq b$, $\alpha r \equiv b(2)$. The same formula gives an isomorphism $H_1^1(0) \cong H_1^{-1}(0)$. If $\tau = 0,1$ and $b \geq 1$ I indicate by D_b the abstract space isomorphic to $D_b^1 \cong D_b^{-1}$ (or to $H_1^1(0) \cong H_1^{-1}(0)$ if $b = 1$). By $\phi_r[b]$ I denote the element corresponding to

$\phi_r^1(\frac{1}{2}(b-1))$ and $\alpha\phi_r^{-1}(\frac{1}{2}(b-1))$ as in (3.3.9). In the case $0 < \tau < 1$ I denote $\phi_r[b] = \phi_r(\frac{1}{2}(b-1))$. If $\tau = 0,1$ and $b < 1$, then the spaces D_b^δ have finite dimension.

3.3.2. Isomorphisms. A complex number s I call general if $2s+1 \not\equiv \pm\tau(2)$. For general s there are isomorphisms. Let

(3.3.10) $$\gamma_r(s) = \frac{\Gamma(\frac{1}{2}-s-\frac{1}{2}r)}{\Gamma(\frac{1}{2}+s-\frac{1}{2}r)} ;$$

extend γ_r holomorphically as far as possible.

It $\tau = 0,1$, $\delta = \pm 1$, then

(3.3.11) $$\iota_\tau^\delta(s) : H_\tau^\delta(s) \to H_\tau^{\delta(-1)^\tau}(-s)_\tau$$
$$\iota_\tau^\delta(s)\phi_r^\delta(s) = \gamma_r(s)\phi_r^{\delta(-1)^\tau}(-s).$$

For general s this is an isomorphism of $\underline{g}$-K-spaces. Let $b \equiv \tau(2)$. $b > 1$. Then the residue of $\iota_\tau^\delta(s)$ in $\frac{1}{2}(b-1)$ and its extension in $\frac{1}{2}(1-b)$ induce isomorphisms:

(3.3.12) $$\iota_b^* : H_\tau^\delta(\tfrac{1}{2}(b-1)) \bmod D_b^\delta \to D_{2-b}^{\delta_1}$$
(3.3.13) $$\iota_{2-b}^* : H_\tau^\delta(\tfrac{1}{2}(b-1)) \bmod D_{2-b}^\delta \to D_b^{\delta_1}$$
$\Bigg\}$ with $\delta_1 = \delta(-1)^\tau$

For $0 < \tau < 1$, $\delta = \pm 1$, I define the map $\iota_\tau^\delta(s)$:

(3.3.14) $$\iota_\tau^\delta(s) : H_\tau(s) \to H_\tau(-s)$$
$$\iota_\tau^\delta(s)\phi_r(s) = \gamma_{\alpha r}(s)\phi_r(-s)$$

with $\alpha = \pm 1$, $\alpha r \equiv \delta\tau(2)$.

Remark that

(3.3.15) $$\iota_\tau^\delta(s) = c_\delta(s)\iota_\tau^1(s)$$

with

(3.3.16) $$c_\delta(s) = \frac{\sin \pi(\frac{1}{2}+s+\frac{1}{2}\tau)}{\sin \pi(\frac{1}{2}+s+\frac{1}{2}\delta\tau)} .$$

For general s these maps are isomorphisms. If $b \equiv \delta\tau(2)$, then $\iota_\tau^\delta(\frac{1}{2}(b-1))$ is well defined and induces an isomorphism

(3.3.17) $$\iota_b^* : H_\tau(\tfrac{1}{2}(b-1)) \bmod D_b \to D_{2-b}.$$

This covers both cases that had to be separated for $\tau = 0,1$.

(3.3.18) $$\operatorname*{res}_{s\to\frac{1}{2}(b-1)} \iota_\tau^{-\delta}(s) = \pi^{-1} \sin \pi b\ \iota_\tau^\delta(\tfrac{1}{2}(b-1)).$$

For all τ we have for general s

(3.3.19) $$\iota_\tau^\delta(s) \,.\, \iota_\tau^{\delta_1}(-s) = \text{identity}$$

with $\delta_1 = \delta$, unless $\tau = 1$; in that case $\delta_1 = -\delta$.

Remark that $\iota_1^\delta(s)$ extends to $\iota_1^\delta(0)$, giving minus the isomorphism $H_1^\delta(0) \cong H_1^{-\delta}(0)$ mentioned in (3.3.9).

3.3.3 Duality.

There is a form relating $H_\tau^{(\delta)}(s)$ and $H_\tau^{(\delta)}(-\bar{s})$; it is linear in the first factor and antilinear in the second one:

(3.3.20) $$\{\Sigma_r a_r\phi_r^{(\delta)}(s), \Sigma_r b_r\phi_r^{(\delta)}(-\bar{s})\}_s = \Sigma_r a_r\bar{b}_r .$$

Properties ($\xi \in H_\tau^{(\delta)}(s)$, $\eta \in H_\tau^{(\delta)}(-\bar{s})$):

(3.3.21) $$\{k.\xi,k.\eta\}_s = \{\xi,\eta\}_s \quad \text{for all } k \in K$$

(3.3.22) $$\{\underline{X}.\xi,\eta\}_s + \{\xi,\underline{X}\eta\}_s = 0 \text{ for all } \underline{X} \in \underline{g}_{\mathbb{R}}$$

(3.3.23) $$\overline{\{\xi,\eta\}_s} = \{\eta,\xi\}_{-\bar{s}}$$

(3.3.24) $\{\iota_\tau^\delta(s)\xi,\iota_\tau^\delta(-\bar{s})\eta\}_{-s} = \{\xi,\eta\}_s$

3.3.4 The spaces E_1^δ

I shall not always be able to avoid the following unpleasant g-K-spaces, containing $H_1^\delta(0)$:

(3.3.25)
$$E_1^\delta = H_1^\delta(0) \oplus \bigoplus_{r\equiv 1(2)} \mathbb{C}\,\hat{\phi}_r^\delta(0)$$
$$\underline{W}\,\hat{\phi}_r^\delta(0) = ir\,\hat{\phi}_r^\delta(0), \quad k(\theta)\hat{\phi}_r^\delta(0) = e^{ir\theta}\hat{\phi}_r^\delta(0)$$
$$j\,\hat{\phi}_r^\delta(0) = \delta\hat{\phi}_{-r}^\delta(0)$$
$$\underline{E}^{\pm}\hat{\phi}_r^\delta(0) = (1 \pm r)\hat{\phi}_{r\pm2}^\delta(0) + 2\phi_{r\pm2}^\delta(0)$$

Clearly E_1^δ is a g-K-space containing $H_1^\delta(0)$ as a non-trivial irreducible subspace. The Casimir operator acts in E_1^δ as multiplication by $\frac{1}{4}$. The endomorphism ring of E_1^δ has dimension two. A non-trivial element is the projection

(3.3.26)
$$p_E^\delta : E_1^\delta \to H_1^\delta(0) : \begin{cases} \hat{\phi}_r^\delta(0) \mapsto \phi_r^\delta(0) \\ \phi_r^\delta(0) \mapsto 0 \end{cases}$$

An isomorphism $\iota_E^\delta : E_1^\delta \to E_1^{-\delta}$ is given by

(3.3.27)
$$\iota_E^\delta\,\hat{\phi}_r^\delta(0) = \operatorname{sgn} r\,(\hat{\phi}_r^{-\delta}(0) + 2\Gamma'(\tfrac{1}{2}(1+|r|))\Gamma(\tfrac{1}{2}(1+|r|))^{-1}\phi_r^{\delta}(0))$$
$$\iota_E^\delta\phi_r^\delta(0) = -\operatorname{sgn} r\ \phi_r^\delta(0) \qquad (= \iota_1^\delta(0)\phi_r^\delta(0)).$$

One may view E_1^δ as the fiber over 0 of a bundle over $\mathbb{C}$, with basis $\{\hat{\phi}_r^\delta,\phi_r^\delta \mid r \equiv 1(2)\}$. For $s \neq 0$ the fiber is $H_1^\delta(s) \oplus H_1^{+\delta}(-s)$ and

(3.3.28)
$$\phi_r^\delta(s) = \phi_r^\delta(s) \oplus 0$$
$$\hat{\phi}_r^\delta(s) = \frac{1}{2s}\,\phi_r^\delta(s) \oplus \frac{-1}{2s}\phi_r^\delta(-s).$$

The isomorphism ι_E^δ corresponds to the matrix $\begin{pmatrix} 0 & \iota_1^{+\delta}(-s) \\ \iota_1^\delta(s) & 0 \end{pmatrix}$ for $s \neq 0$.

I do not fully understand the significance of this g-K-space, but I could not get around it when dealing with the Maass-Selberg relation in the case $\tau = 1$, $s = 0$; see the remark in 7.6.

3.4 Unitary g-K-spaces

A g-K-space V I call unitary if it is provided with a positive definite scalar product satisfying

(3.4.1) $< k.v,k.w > = < v,w >$ for all $k \in K$

(3.4.2) $< \underline{X}.v,w > + < v,\underline{X}^*.w > = 0$ for all $\underline{X} \in \underline{g}$.

$\underline{X}^*$ is the conjugate of $\underline{X}$ with respect to the real subspace $\underline{g}_{\mathbb{R}}$ of $\underline{g}$.

I give a list of irreducible unitary g-K-spaces:

3.4.1 The unitary principal series consists of the $H_\tau^{(\delta)}(s)$ with Re $s = 0$. The scalar product is given by

(3.4.3) $< \xi,\eta >_s = \{\xi,\eta\}_s$.

(Remark that $-\bar{s} = s$.) The maps $\iota_\tau^\delta(s)$ are unitary.

3.4.2 The complementary series consists of the $H_0^\delta(s)$ with $\delta = \pm 1$, $s \in \mathbb{R}$, $0 < |s| < \frac{1}{2}$ if $\tau = 0$, and of the $H_\tau(s)$ with $s \in \mathbb{R}$, $0 < |s| < \frac{1}{2}(1-\tau)$ if $0 < \tau < 1$. It is absent for $\tau = 1$. The scalar product is given by

$$(3.4.4)\qquad [\xi,\eta]_s = \begin{cases} \{\xi,\iota_\tau^\delta(s)\eta\}_s & \text{if } \tau = 0 \\ \{\xi,\iota_\tau^1(s)\eta\}_s & \text{if } 0 < \tau < 1. \end{cases}$$

The maps $\iota_0^\delta(s)$ resp. $\iota_\tau^1(s)$ are unitary.

3.4.3 The discrete series consists of the D_b with $b > 1$, $b \equiv \pm\tau(2)$. There are also some D_b with unitary structure for $b < 1$: D_0^1, D_0^{-1} and D_b with $0 < b < 1$, $b = \tau$. This last set of spaces has no name, as far as I know.

For $b \neq 0$ the scalar product is given by

$$(3.4.5)\qquad < \phi_r[b],\phi_q[b] >_b = \frac{(\frac{1}{2}(\alpha r-b))!}{\Gamma(\frac{1}{2}(\alpha r+b))}\,\delta_{q,r}$$

with $\alpha r \equiv b(2)$, $\alpha = \pm 1$. The one-dimensional space D_0^δ has $\{\phi_0^\delta(-\frac{1}{2})\}$ as orthonormal basis.

These scalar products can also be related to the form $\{.,.\}_{\frac{1}{2}(1-b)}$. This form is defined on $H_\tau^{(\delta)}(\frac{1}{2}(1-b)) \times H_\tau^{(\delta)}(\frac{1}{2}(b-1))$. It may be restricted to $H_\tau^{(\delta)}(\frac{1}{2}(1-b)) \times D_b^{(\delta)}$. This restriction induces a form on $(H_\tau^{(\delta)}(\frac{1}{2}(1-b)) \bmod D_{2-b}^{(\delta)}) \times D_b^{(\delta)}$. Using the isomorphism ι^*_{2-b} we get a form $B(.,.)$ on $D_b^{(\delta_1)} \times D_b^{(\delta)}$ satisfying

$$(3.4.6)\qquad B(\iota^*_{2-b}\phi,\psi) = \{\phi,\psi\}_{\frac{1}{2}(b-1)}$$

for $\phi \in H_\tau^{(\delta)}(\frac{1}{2}(1-b))$, $\psi \in D_b^\delta$. Here $\delta_1 = \delta$, unless $\tau = 1$, then $\delta_1 = -\delta$. In the case $\tau = 0,1$, $\delta = \pm 1$, $b \equiv \tau(2)$, $b < 1$ we find

$$(3.4.7)\qquad B(\phi_r^{\delta_1}(\tfrac{1}{2}(1-b)),\ \phi_q^\delta(\tfrac{1}{2}(b-1))) = (-1)^{1+\frac{1}{2}(r-b)}(-\tfrac{b+r}{2})!\,(\tfrac{r-b}{2})!\,\delta_{q,r}.$$

If $\tau = 0$ this is only definite for $b = 0$. In this case

$$(3.4.8)\qquad B(\phi,\psi) = -< \phi,\psi >_0.$$

For $\tau = 1$ the spaces $D_b^{\delta_1}$ and D_b^δ are non-isomorphic.

In the case $\tau = 0,1$, $\delta = \pm 1$, $b \equiv \tau(2)$, $b > 1$ we find

$$(3.4.9)\qquad B(\phi_r^{\delta_1}(\tfrac{1}{2}(b-1)),\phi_q^\delta(\tfrac{1}{2}(b-1))) = ((-1)^\tau,-\mathrm{sgn}\ r)\frac{(\frac{1}{2}(|r|-b))!}{(\frac{1}{2}(b+|r|)-1)!}\,\delta_{q,r}.$$

So for $\tau = 0$

$$(3.4.10)\qquad B(\phi,\psi) = \pm< \phi,\psi >_b \text{ on } D_b^\delta \cong D_b$$

In the case $\tau = 1$ denote for the moment by $\phi \mapsto \phi^*$ the isomorphism in (3.3.9); then

$$(3.4.11)\qquad B(\phi^*,\psi) = -< \phi,\psi >_b \text{ on } D_b^\delta \cong D_b.$$

Finally, in the case $0 < \tau < 1$, $b \equiv \pm\tau$; $\alpha r \equiv b(2)$, we find

$$(3.4.12)\qquad B(\phi_r(\tfrac{1}{2}(b-1)),\ \phi_q(\tfrac{1}{2}(b-1))) = \Gamma(\tfrac{1}{2}(\alpha r+b))^{-1}(\tfrac{1}{2}(\alpha r-b))!\,\delta_{r,q}.$$

If $b < 0$ the factor $\Gamma(\frac{1}{2}(\alpha r+b))$ changes sign when r varies. For $b > 0$ we get on D_b

$$(3.4.13)\qquad B(\phi,\psi) = < \phi,\psi >_b$$

3.4.4 Comparison with Pukánszky's list of irreducible unitary representations of G_0; see [35], p.102.

Re $s = 0$, $\tau = 0$ or 1, $\delta=\pm 1$ $\quad H_\tau^\delta(s)$ corresponds to $C^{(\frac{1}{2}\tau)}_{\frac{1}{4}-s^2}$

Re $s = 0$, $0 < \tau < 1$ $\quad H_\tau(s)$ corresponds to $C^{(\frac{1}{2}\tau)}_{\frac{1}{4}-s^2} \oplus C^{(1-\frac{1}{2}\tau)}_{\frac{1}{4}-s^2}$

$b > 0,\ b \neq 1$	D_b	corresponds to	$D^+_{\frac{1}{2}b} \oplus D^-_{\frac{1}{2}b}$
$s \in \mathbb{R},\ 0 < \lvert s\rvert < \frac{1}{2},\ \delta = \pm 1$	$H^{\delta}_{0}(s)$	" "	$E^{(0)}_{\frac{1}{4}-s^2}$
$s \in \mathbb{R},\ 0 < \lvert s\rvert < \frac{1}{2}(1-\tau),\ 0 < \tau < 1$	$H_{\tau}(s)$	" "	$E^{(\frac{1}{2}\tau)}_{\frac{1}{4}-s^2} \oplus E^{(1-\frac{1}{2}\tau)}_{\frac{1}{4}-s^2}$

3.4.5 Proposition. Let π be an irreducible unitary representation of G in a Hilbert space V. Then the space V_K of K-finite vectors in V is isomorphic to one of the g-K-spaces mentioned in 3.4.1 - 3.4.3.

Proof. The corresponding result for G_0 has been proved by Pukánszky, [35], p.102. To make the step from G_0 to G one may follow the reasoning of Gelbart in lemma 4.3-4.5 of [7], p.78, 79.

3.5 Bundles of g-K-spaces

When one looks at the definition of $H^{(\delta)}_{\tau}(s)$ in 3.3.1 the dependence on s looks rather holomorphic. Also the isomorphisms $\iota^{\delta}_{\tau}(s)$ look meromorphic in s. To make this precise I use the terminology of bundles.

3.5.1 Definition. H^{τ} is a bundle of g-K-spaces over $\mathbb{C}$. As bundle of vector spaces it is trivial; it has the following algebraic basis:

(3.5.1) $\{\phi^{\eta}_{r} \mid r \equiv \pm\tau(2),\ \eta = \pm 1\}$.

The g-K-structure is given by

(3.5.2) $(j\phi^{\eta}_{r})(s) = \begin{cases} \eta\phi^{\eta}_{-r}(s) & \text{if } \tau = 0 \text{ or } 1 \\ \phi^{\eta}_{-r}(s) & \text{if } 0 < \tau < 1 \end{cases}$

$(k(\theta)\phi^{\eta}_{r})(s) = e^{ir\theta}\phi^{\eta}_{r}(s)$

$(\underline{W}\phi^{\eta}_{r})(s) = ir\phi^{\eta}_{r}(s)$

$(\underline{E}^{\pm}\phi^{\eta}_{r})(s) = (1 + 2s \pm r)\phi^{\eta}_{r\pm 2}(s)$

So for every $s \in \mathbb{C}$

(3.5.3) $H^{\tau}(s) = \begin{cases} H^{1}_{\tau}(s) \oplus H^{-1}_{\tau}(s) & \text{if } \tau = 0 \text{ or } 1 \\ H_{\tau}(s) \oplus H_{\tau}(s) & \text{if } 0 < \tau < 1 \end{cases}$

The basis in (3.5.1) I call the standard basis of H^{τ}. Sections of H^{τ} over $U \subset \mathbb{C}$ can be written as finite linear combinations of elements of this basis; a section is called holomorphic or meromorphic on U if all coefficient functions in that linear combination are holomorphic or meromorphic.

In (3.5.3) one sees that $H^{(\delta)}_{\tau}(s)$ can be found inside a fiber of H^{τ}. The spaces $D^{(\delta)}_{b}$ I also consider as embedded in fibers of H^{τ}. Let

(3.5.4) $Y_{\tau} = \{b \in \mathbb{R} \mid b \equiv \pm\tau(2),$ if $\tau = 0,1$ then $b > 1\}$

(3.5.5) $Y^{f}_{\tau} = \{b \in \mathbb{R} \mid b \equiv \tau(2),\ b < 1\}$ for $\tau = 0,1$

If $b \in Y_{\tau}$ or $b = \tau = 1$, then D_b is considered as subspace of $H^{\tau}(\frac{1}{2}(b-1))$ by identifying $\phi_r[b]$ with

(3.5.6) $\begin{cases} \frac{1}{\sqrt{2}}(\phi_r^1(\frac{1}{2}(b-1)) + \alpha\phi_r^{-1}(\frac{1}{2}(b-1))) & \text{if } \tau = 0,1 \\ \phi_r^{\eta}(\frac{1}{2}(b-1)) & \text{if } 0 < \tau < 1,\ b = \eta\tau(2), \end{cases}$

for $r \equiv \alpha b(2)$, $\alpha r \geqslant b$. This will turn out to be the most convenient embedding of D_b. Another embedding has a basis consisting of

(3.5.7) $\tilde{\Phi}_r[b] = \begin{cases} \frac{1}{\sqrt{2}}(\phi_r^1(\frac{1}{2}(b-1)) - \alpha\phi_r^{-1}(\frac{1}{2}(b-1))) & \text{if } \tau = 0,1 \\ \phi_r^{-\eta}(\frac{1}{2}(b-1)) & \text{if } 0 < \tau < 1,\ b \equiv \eta\tau(2), \end{cases}$

for $r \equiv \alpha b(2)$, $\alpha r \geqslant b$. The resulting subspace of $H^\tau(\frac{1}{2}(b-1))$ I call $\tilde{D}_b$. I define $< .,. >_b$ on D_b and $\tilde{D}_b$ by prescribing the elements in (3.5.6) and (3.5.7) to satisfy (3.4.5).

If $\tau = 0,1$, $b \in Y_\tau^f$, $\delta = \pm 1$, then D_b^δ is embedded in $H^\tau(\frac{1}{2}(b-1))$ by its inclusion in $H_\tau^\delta(\frac{1}{2}(b-1))$.

If ξ is a section of H^τ, its decomposition with respect to (3.5.3) I denote by

(3.5.8) $\xi = \xi^1 + \xi^{-1}$.

3.5.2 Definition. $\iota : H^\tau \to H^\tau$ <u>is a meromorphic map</u>; <u>on the fibers it is given by</u> $\iota(s) : H^\tau(s) \to H^\tau(-s)$,

(3.5.9) $\iota(s) = \begin{pmatrix} \iota_0^1(s) & 0 \\ 0 & \iota_0^{-1}(s) \end{pmatrix}$ <u>if</u> $\tau = 0$

$\iota(s) = \begin{pmatrix} 0 & \iota_1^{-1}(s) \\ \iota_1^1(s) & 0 \end{pmatrix}$ <u>if</u> $\tau = 1$

$\iota(s) = \begin{pmatrix} \iota_\tau^1(s) & 0 \\ 0 & \iota_\tau^{-1}(s) \end{pmatrix}$ <u>if</u> $0 < \tau < 1$.

The matrices refer to the decomposition in (3.5.3). The choice of $\iota(s)$ in the case $0 < \tau < 1$ will turn out to be sensible when we study Whittakers models.

Remark that if ξ is a section of H^τ, then

(3.5.10) $(\iota\xi)(s) = \iota(-s)\xi(-s)$.

ι is meromorphic, i.e. it maps meromorphic sections onto meromorphic ones.

If $\tau = 0,1$, then ι is holomorphic outside $\{\frac{1}{2}(b-1) | b \in Y_\tau\}$. For $b \in Y_\tau$:

(3.5.11) $(\operatorname{res}_{s \to \frac{1}{2}(b-1)} \iota(s))H^\tau(\frac{1}{2}(b-1)) = D_{2-b}^1 + D_{2-b}^{-1}$;

$\operatorname{res}_{s \to \frac{1}{2}(b-1)} \iota(s)|D_b + \tilde{D}_b = 0$

$\iota(\frac{1}{2}(1-b))H^\tau(\frac{1}{2}(1-b)) = D_b + \tilde{D}_b$

$\iota(\frac{1}{2}(1-b))|D_b^1 + D_b^{-1} = 0$.

If $0 < \tau < 1$ then the poles of ι are situated in $\{\frac{1}{2}(b-1) | b \in Y_\tau\}$.

For $b \in Y_\tau$:

(3.5.12) $\operatorname{res}_{s \to \frac{1}{2}(b-1)} \iota(s)D_b = 0$, even

$\lim_{s \to \frac{1}{2}(b-1)} \iota(s)|D_b = 0$

$$(\operatorname*{res}_{s\to\frac12(b-1)} \iota(s))H^{\tau}(\tfrac12(b-1)) = D_{2-b}$$
$$\operatorname*{res}_{s\to\frac12(b-1)} \iota(s)\ \tilde{D}_b = 0.$$

3.5.3 Definition. If ξ is a section of H^{τ} over $U \subset \mathbb{C}$ and η is one over $V \subset \mathbb{C}$, then the function $\{\xi,\eta\} : U \cap (-\overline{V}) \to \mathbb{C}$ is defined by

(3.5.13) $$\{\xi,\eta\}(s) = \sum_{\delta=\pm1} \{\xi^{\delta}(s),\eta^{\delta}(-\overline{s})\}_s$$

If ξ and η are holomorphic (meromorphic), then so is $\{\xi,\eta\}$.

If ξ is decomposed as

(3.5.14) $$\xi(s) = \sum_{\delta,r} \xi^{\delta}_r(s)\phi^{\delta}_r(s),$$

and η similarly, then

(3.5.15) $$\{\xi,\eta\}(s) = \sum_{\delta,r} \xi^{\delta}_r(s)\overline{\eta^{\delta}_r}(-s).$$

Some properties:

(3.5.16) $$\overline{\{\xi,\eta\}}(s) = \{\eta,\xi\}(-s)$$

(3.5.17) $$\{\iota\xi,\iota\eta\}(s) = \{\xi,\eta\}(-s)$$

If $0 < \tau < 1$ each element ψ of the standard basis of H^{τ}, see (3.5.1), satisfies

(3.5.18) $$(\iota\psi)(s) = \gamma_{\psi}(-s)\psi(-s)$$

for some meromorphic function γ_{ψ}.

In the case $\tau = 0$ or 1 one may take the basis

(3.5.19) $$\{\tfrac{1}{\sqrt{2}}(\phi^1_r + \eta\phi_r^{-1}) \mid r \equiv \tau,\ \eta = \pm1\}.$$

This basis has the property (3.5.18).

If one uses this basis to decompose ξ as in (3.5.14), then the formula corresponding to (3.5.15) keeps valid.

3.6 Other examples of g-K-spaces

Let $a = 0$ or $\frac12$. The character ψ_a of N is defined by

(3.6.1) $$\psi_a(n(x)) = e^{iax}$$

3.6.1. Definition. ${}^aS_{\tau}$ is the space of functions in $C^{\infty}(G)$ satisfying:

(3.6.2) $|f| \in C^{\infty}_c(Z_0N\backslash G)$

(3.6.3) $f(ng) = \psi_a(n)f(g)$ for all $n \in N$, $g \in G$

(3.6.4) the right translates of f under Z span a space in which Z acts irreducibly according to τ (see 3.2)

(3.6.5) f is K-finite under right translation

${}^aS_{\tau}$ is a g-K-space under right action; it is not admissible. In the case $\tau = 0,1$ a unitary structure is given by

(3.6.6) $$\langle f,g\rangle = \int_{Z_0N\backslash G} f(x)\overline{g(x)}\,d\dot{x}$$

In the case $0 < \tau < 1$ each $f \in {}^aS_{\tau}$ may be written

(3.6.7) $\quad f = \sum_{\varepsilon=\pm 1} P_\varepsilon f$

with

(3.6.8) $\quad P_\varepsilon f(k(\pi l)g) = e^{\pi i \tau \varepsilon l} P_\varepsilon f(g) \qquad$ for $g \in G$, $l \in \mathbb{Z}$.

In this case a unitary structure is given by

(3.6.9) $\quad \langle f,g \rangle = \sum_{\varepsilon=\pm 1} \int_{Z_0 N\backslash G} P_\varepsilon f(x) \overline{P_\varepsilon g(x)} d\dot{x}$

Let $f \in {}^a S_\tau$ be of weight r. The functions ${}^R f^+$ and ${}^R f^-$ on h defined in (2.3.5) have to be of the form

(3.6.10) $\quad {}^R f^\pm(x+iy) = e^{\pm iax} c^\pm(y)$

with $c^\pm \in C_c^\infty(0,\infty)$.

If $g \in {}^a S_\tau$ is also of weight r and determines $d^\pm \in C_c^\infty(0,\infty)$, then

(3.6.11) $\quad \langle f,g \rangle = \frac{1}{2}\int_0^\infty (c^+(y)\overline{d^+(y)} + \overline{c^-(y)\, d^-(y)})y^{-2} dy.$

4. Automorphic models

4.1 Introduction

The central object of this paper are systems of automorphic forms, the elements of which are connected by differential operators, see 2.6. In subsection 4.5 I make this more precise. I define an automorphic model to be a representation of a $\underline{g}$-K-space satisfying some conditions.
To do this I need to fix a discrete subgroup Γ of G_0 and a character χ of Γ; these are discussed in subsections 4.2 and 4.3. The relation between automorphic models and automorphic forms on the upper half plane I consider in subsections 4.4 and 4.5. The idea that certain representations are connected with automorphic forms one finds in [9], Ch. I. It is also the underlying idea in [13].

4.2 The discrete subgroup Γ

I take a group Γ satisfying:

4.2.1 Assumption. Γ is a discrete subgroup of G_0 containing Z_0;
$\Gamma\backslash G_0$ has finite volume with respect to the quotient measure;
Γ has a finite, non-zero, number of cusps.
The term cusp calls for some explanation. A conjugate ${}^gN = gNg^{-1}$ of the unipotent group N is called a cuspidal subgroup of G_0 if ${}^gN/({}^gN \cap \Gamma)$ is compact; this amounts to ${}^gN \cap \Gamma$ containing more than one element. Γ acts by conjugation on the cuspidal subgroups. Each Γ-orbit under this action is called a cusp (see e.g. [21], XII §1). If c is a cusp then I fix an element $g_c \in G_0$ such that $g_c N g_c^{-1}$ represents c. There is some freedom in the choice of g_c; any element of $\Gamma g_c H_0 N$ will do. An additional condition will be introduced in (7.2.1) and (7.2.2).

4.2.2 The action of Γ/Z_0 on h
By $\gamma \mapsto \overline{\gamma}$ the group Γ is mapped onto the discrete subgroup $\overline{\Gamma}$ of $Sl_2(\mathbb{R})$. As $Sl_2(\mathbb{R})/\{1,-1\}$ is the group of orientation preserving conformal homeomorphisms of the upper halfplane h, we may view Γ/Z_0 as a group of homeomorphisms of h. This group is discontinuous, see Lehner, [23], p.99. The fundamental domain of Γ/Z_0 has finite volume, so Γ/Z_0 is finitely generated (Siegel; [41], theorem 5). Hence Γ is finitely generated as well.
On the other hand, if a discontinuous group of orientation preserving conformal homeomorphism of h, with a non-compact fundamental domain with finite volume, is lifted to G_0, one obtains a group Γ satisfying assumption 4.2.1.

4.2.3 Examples

In 2.7 I mentioned already $\Gamma(1)$, generated by $n(1)$ and w, and its subgroup Γ_θ, generated by $n(2)$ and w.
All congruence subgroups of the modular group $Sl_2(\mathbf{Z})$ may be lifted to give subgroups of $\Gamma(1)$, satisfying 4.2.1. Lifting of the Hecke groups gives $\Gamma[\lambda]$, generated by $n(\lambda)$ and w; $\lambda = 2\cos \pi q^{-1}$, $q = 3,4,5,\ldots$, see [12]. Remark that $\Gamma(1) = \Gamma[1]$.
The groups $\Gamma[\lambda]$ have one cusp, which I call ∞; I choose $g_\infty = a(\sqrt{\lambda})$.
The theta group Γ_θ has two cusps; I choose $g_\infty = a(2)$, $g_1 = wn(-1)w^{-1}$.
The names ∞ and 1 of the cusps in these examples refer to the classical meaning of the word cusp. A cuspidal subgroup N_1 acts on $\mathfrak{h}$ and also on its boundary $\mathbf{R} \cup \{\infty\}$. It has exactly one fixed point, situated in $\mathbf{R} \cup \{\infty\}$; classically a cusp is a $\overline{\Gamma}$-class of such points. If one restricts oneself to considering a fundamental domain for $\overline{\Gamma}$ one may choose points in $\mathbf{R} \cup \{\infty\}$ representing the cusps. This is done in the examples.

4.3 The character χ.

In 2.7 I mentioned modular forms of weight zero; the corresponding functions on G are left-invariant under $\Gamma(1)$. As $Z_0 \subset \Gamma(1)$ they are right-invariant under Z_0, which is all right, for the weight is zero. In general left invariance under Γ is only possible for even weights.
For a given τ the characters of Z_0 occurring under right translation are $k(\pi l) \mapsto e^{\pi i \upsilon l}$ with $\upsilon = \pm\tau$ ($\upsilon = \tau$ if $\tau = 0$ or 1). The same characters will occur under left translation. So one looks not for left-Γ-invariance, but for left-Γ-equivariance according to a character χ of Γ.

4.3.1 Assumption. Let χ be a character of Γ such that

$$|\chi(\gamma)| = 1 \qquad \text{for all } \gamma \in \Gamma$$
$$\chi(k(\pi l)) = e^{\pi i l \tau} \qquad \text{for all } k(\pi l) \in Z_0$$

So in the case $0 < \tau < 1$ I have singled out one of the two possible characters of Z_0. The consequences of this choice I discuss in 4.4.

4.3.2 Examples

The character χ_θ of Γ_θ given in (2.7.4) satisfies 4.3.1 for $\tau = \frac{1}{2}$.
For the Hecke group $\Gamma[2\cos \pi q^{-1}]$, see 4.2.3, the character χ has to satisfy

(4.3.1) $\qquad \chi(w)^2 = e^{\pi i \tau}$

(4.3.2) $\qquad (\chi(n(2\cos \pi q^{-1}))\chi(w))^q = e^{\pi i \tau(q-1)}$

To see this, remark that $(wZ_0)^2 = Z_0$ and $(n(2\cos \pi q^{-1})wZ_0)^q = Z_0$ generate the relations in $\Gamma[2\cos \pi q^{-1}]/Z_0$. It is easy to see that the first relation lifts to (4.3.1). To get (4.3.2) take $\rho = e^{\pi i q^{-1}}$ and work out that

(4.3.3) $\qquad p(\rho)^{-1} n(2\cos \pi q^{-1}) w p(\rho) = k(\pi(1-q^{-1}))$.

Let $q = 3$, so $\Gamma[2 \cos \pi 3^{-1}] = \Gamma(1)$; the character connected with the Dedekind η-function, see [23] p. 343, is given by

$$(4.3.4)\qquad \chi(w) = e^{\frac{1}{4}\pi i}$$
$$\chi(n(1)) = e^{\frac{1}{12}\pi i}$$

This satisfies (4.3.1), (4.3.2) for $\tau = \frac{1}{2}$.

4.4 Γ-equivariance

Often I shall consider functions f on G which are left-Γ-equivariant according to χ, i.e.

$$(4.4.1)\qquad f(\gamma g) = \chi(\gamma) f(g) \text{ for all } \gamma \in \Gamma,\ g \in G.$$

If such a function f has weight r, $r \equiv \varepsilon\tau(2)$, $\varepsilon = \pm 1$, then the corresponding functions f^+ and f^- on h have familiar properties.

In the case $\tau = 0,1$ one derives easily from (2.3.3):

$$(4.4.2)\qquad f^+\left(\frac{az+b}{cz+d}\right) = (cz+d)^r \chi\left(\widetilde{\begin{pmatrix} a & b \\ c & d \end{pmatrix}}\right) f^+(z)$$
$$f^-\left(\frac{az+b}{cz+d}\right) = (cz+d)^r \overline{\chi\left(\widetilde{\begin{pmatrix} a & b \\ c & d \end{pmatrix}}\right)} f^-(z)$$

for all $\begin{pmatrix} a & b \\ c & d \end{pmatrix} \in \overline{\Gamma}$. So (4.4.1) is equivalent to f^+ and f^- satisfying the transformation property of classical automorphic forms of weight r for the group $\overline{\Gamma}$ and multiplier system v^+ or v^- given by

$$(4.4.3)\qquad v^+(x) = \chi(\tilde{x});\qquad v^-(x) = \overline{\chi(\tilde{x})}$$

for $x \in \overline{\Gamma}$; see e.g. [23], p.268.

In the case $0 < \tau < 1$ there is a problem: for each $k(\pi 1) \in Z_0$ we have for $g \in G_0 j_\eta$, $\eta = \pm 1$:

$$(4.4.4)\qquad e^{\pi i \tau 1} f(g) = f(k(\pi 1) g) = f(g k(\eta \pi 1))$$
$$= f(g) e^{\pi i r \eta 1} = f(g)\, e^{\pi i \tau \varepsilon \eta 1}$$

So f lives only on $G_0 j_\varepsilon$.

In this case (4.4.1) is equivalent to

$$(4.4.5)\qquad f^- = 0$$
$$f^+\left(\frac{az+b}{cz+d}\right) = \chi\left(\widetilde{\begin{pmatrix} a & b \\ c & d \end{pmatrix}}\right)(cz+d)^r f(z)$$

for all $\begin{pmatrix} a & b \\ c & d \end{pmatrix} \in \overline{\Gamma}$, if $r \equiv \tau(2)$, and to

$$(4.4.6)\qquad f^+ = 0$$
$$f^-\left(\frac{az+b}{cz+d}\right) = \overline{\chi\left(\widetilde{\begin{pmatrix} a & b \\ c & d \end{pmatrix}}\right)}(cz+d)^r f(z)$$

for all $\begin{pmatrix} a & b \\ c & d \end{pmatrix} \in \overline{\Gamma}$, if $r \equiv -\tau(2)$.

So for $0 < \tau < 1$ the assumption that χ restricted to Z_0 behaves like τ makes that we loose something. But if χ_1 is a character of Γ satisfying

$$(4.4.7)\qquad \chi_1(k(\pi 1)) = e^{-\pi i \tau 1},$$

and f_1 satisfies (4.4.1) with χ replaced by χ_1, then f defined by $f(g) = f_1(jg)$ satisfies

$$(4.4.8)\qquad f(\gamma g) = \chi(\gamma) f(g) \qquad \text{for all } \gamma \in j\Gamma j,\ g \in G$$

with $\chi(\gamma) = \chi_1(j\gamma j)$. As $j\Gamma j$ and χ satisfy 4.2.1 and 4.3.1, nothing is really lost.

The multiplier systems derived above from χ are general. If a multiplier system v for the weight r, with absolute value one, is given, then one may define a character χ_v of Γ by

(4.4.9) $\chi_v(\tilde{x}k(\pi l)) = v(x)e^{\pi i r l}$

for $x \in \overline{\Gamma}$ and $k(\pi l) \in Z_0$. This follows from the property of multiplier systems

(4.4.10) $v(xy) = e^{2\pi i \underline{w}(x,y)} v(x)v(y)$;

use (2.2.13).

If $\tau = 0$ or 1, then χ_v satisfies 4.3.1. If $0 < \tau < 1$ then it satisfies 4.3.1 if $r \equiv \tau(2)$ and $\overline{\chi_v}$ does if $r \equiv -\tau(2)$.

The corresponding results for ${}^{R}f^{+}$ and ${}^{R}f^{-}$ are similar. See (2.3.7) and [36], (1.16).

4.5 Automorphic models

I have stated before that I want to consider systems of automorphic forms connected by differential operators. So I might now define an automorphic model as a representation of an irreducible g-K-space in a space of functions satisfying (4.4.1). For technical reasons, mainly to be able to handle Eisenstein series conveniently, I state a bit wider definition.

4.5.1 Definition. Let U be a g-K-space contained in $H^{\tau}(s)$ for some $s \in \mathbb{C}$ or let $U = E_1^{\delta}$. An automorphic model of U is an operator $T : U \to C^{\infty}(G)$ intertwining the g-K-action in U with the right g-K-action in $C^{\infty}(G)$, such that all functions $f \in TU$ satisfy

(4.5.1) $f(\gamma g) = \chi(\gamma)f(g)$ for all $\gamma \in \Gamma$, $g \in G$

(4.5.2) $|f(g_c n a(y)k)| \ll y^d$ for $y \to \infty$

for all cusps c, uniformly in $n \in N$, $k \in K$, for some $d \in \mathbb{R}$.

Remark that the g-K-action in $C^{\infty}(G)$ is defined pointwise: for $k \in K$:

(4.5.3) $(kf)(g) = f(gk)$;

for $\underline{X}, \underline{Y} \in \underline{g}_{\mathbb{R}}$:

(4.5.4) $(\underline{X}f)(g) = \frac{d}{dt}f(ge^{t\underline{X}})|_{t=0}$

(4.5.5) $(\underline{X} + i\underline{Y})f = \underline{X}f + i\underline{Y}f$

The growth condition (4.5.2) is usually imposed on automorphic forms. It admits the Eisenstein series, but excludes the real analytic Poincaré series at parabolic points studied by Neunhöffer, [26].

4.5.2 Automorphic forms.

The representation space TU of an automorphic model T indeed contains what are usually called automorphic forms. The space TU is spanned by weight functions, i.e. functions satisfying

(4.5.6) $f(gk(\theta)) = e^{ir\theta}f(g)$ for all $k(\theta) \in K_0$, $g \in G$.

for some weight $r \equiv \pm\tau(2)$. These weight functions are automorphic forms. As $U \subset H^{\tau}(s)$ or E_1^{δ}, the Casimir operator $\underline{\omega}$ acts in U as multiplication by $\frac{1}{4} - s^2$ or $\frac{1}{4}$. So all elements of TU are eigenfunctions of $\underline{\omega}$. In (2.5.5) one sees that for functions of fixed weight the operator $\underline{\omega}$ is elliptic. So automorphic forms are analytic functions. For ${}^{R}f^{+}$ and ${}^{R}f^{-}$ one obtains

(4.5.7) $$(-y^2\frac{\partial^2}{\partial x^2} - y^2\frac{\partial^2}{\partial y^2} + iry\frac{\partial}{\partial x})\,{}^{R}f^{\pm} = \begin{cases} (\frac{1}{4}-s^2)\,{}^{R}f^{+} \\ (\frac{1}{4}-\bar{s}^2)\,{}^{R}f^{-} \end{cases}$$

Together with (4.5.2) and the discussion in 4.4 this implies that ${}^{R}f^{\pm}$ are automorphic forms in the sense of Roelcke, [36], p.297. In his notation:

(4.5.8) $${}^{R}f^{+} \in F_{r,\frac{1}{4}-s^2}(\bar{\Gamma},\mathbb{C},v^{+})$$
$${}^{R}f^{-} \in F_{r,\frac{1}{4}-\bar{s}^2}(\bar{\Gamma},\mathbb{C},v^{-})$$

If $0 < \tau < 1$ one of both functions is zero.

For convenience of the reader I state Roelcke's definition of automorphic forms:

4.5.3 Definition. Let $r \in \mathbb{R}$, $\lambda \in \mathbb{C}$, v a multiplier system. Then $F_{r,\lambda}(\bar{\Gamma},\mathbb{C},v)$ consists of the functions f on h satisfying

(4.5.9) $$(-y^2\frac{\partial^2}{\partial x^2} - y^2\frac{\partial^2}{\partial y^2} + iry\frac{\partial}{\partial x})f = \lambda f$$

(4.5.10) $$e^{-ir\,\arg(cz+d)}f(\frac{az+b}{cz+d}) = v(\begin{smallmatrix} a & b \\ c & d\end{smallmatrix})f(z) \quad \text{for } (\begin{smallmatrix} a & b \\ c & d\end{smallmatrix}) \in \bar{\Gamma}$$

(4.5.11) $$f(\bar{g}_c \cdot z) = O(y^{\kappa}) \text{ for } y \to \infty,$$
for each cusp c, for some $\kappa \in \mathbb{R}$.

In the case that $U = D_b$ with $b \in Y_{\tau}$, we may consider $g = T\phi_b(\frac{1}{2}(b-1))$. As $\underline{E}^{-}g = 0$ we have by (2.6.1):

(4.5.12) $$(\frac{\partial}{\partial y} - i\frac{\partial}{\partial x})g^{\pm} = 0,$$

so g^{+} and g^{-} are holomorphic on h. Together with (4.4.2) and (4.5.2) we see that g^{+} and g^{-} are classical holomorphic automorphic forms of weight b. In the notation of Lehner, [23] p. 268:

(4.5.13) $$g^{\pm} \in \{\bar{\Gamma},-b,v^{\pm}\}.$$

Again, if $0 < \tau < 1$ one of both is zero.

4.5.4 Definition. For U as in 4.5.1 the set of all automorphic models of U is denoted by $A(U)$.

Clearly $A(U)$ is a linear space over $\mathbb{C}$. If $\psi \in U$ has weight r, then

(4.5.14) $$T \mapsto ({}^{R}(T\psi)^{+}, {}^{R}(T\psi)^{-})$$

gives an $\mathbb{R}$-linear map

(4.5.15) $$A(U) \to F_{r,\frac{1}{4}-s^2}(\bar{\Gamma},\mathbb{C},v^{+}) \oplus F_{r,\frac{1}{4}-\bar{s}^2}(\bar{\Gamma},\mathbb{C},v^{-}).$$

If U is irreducible, then this map is injective; this is true even in the more general case that ψ generates U as $\underline{g}$-K-space. For the irreducible $\underline{g}$-K-spaces U I fix a weight vector $\psi \in U$, to get a correspondence between $A(U)$ and spaces of automorphic forms. I make the following choices:

$$
\begin{aligned}
&(4.5.16)\qquad U = H_\tau^\delta(s),\ \tau = 0,1,\ \delta = \pm 1,\ 2s+1 \not\equiv \tau(2):\ \psi = \phi_\tau^\delta(s),\ r = \tau\\
&\qquad\qquad\ U = H_\tau(s),\ 0 < \tau < 1,\ 2s+1 \not\equiv \pm\tau(2) \qquad :\ \psi = \phi_\tau(s),\ r = \tau\\
&\qquad\qquad\ U = D_b,\ b \in Y_\tau \text{ or } b = \tau = 1 \qquad :\ \psi = \phi_b[b],\ r = b\\
&\qquad\qquad\ U = D_b^\delta,\ \tau = 0,1,\ \delta = \pm 1,\ b \in Y_\tau^f \qquad :\ \psi = \phi_b^\delta(\tfrac12(b-1)),\ r = b
\end{aligned}
$$

These choices could be replaced by other ones.

If $0 < \tau < 1$, then only one of the summands in the codomain in (4.5.15) is important. In the case $\tau = 0$ or 1, $U = H_\tau^\delta(s)$, this is also the case, as I now show. Let $\tau = 0$, $T \in A(H_\tau^\delta(s))$. Then $T\psi(gj) = \delta T\psi(g)$, so

$$(4.5.17)\qquad {}^R(T\psi)^-(z) = \delta\, {}^R(T\psi)^+(z).$$

Let $\tau = 1$, $T \in A(H_\tau^\delta(s))$. Then $j\underline{E}^-\psi = 2\delta s\psi$, so

$$(4.5.18)\qquad 2\delta s\, \overline{{}^R(T\psi)^-(z)} = \left(-r+2y\frac{\partial}{\partial y} - 2iy\frac{\partial}{\partial x}\right){}^R(T\psi)^+(z).$$

If $\tau = 0,1$, $b \equiv \tau(2)$, $b < 1$, $\delta = \pm 1$ and $U = D_b^\delta$, then $j(\underline{E}^+)^{-b}\phi_b^\delta(\tfrac12(b-1)) = \delta \prod_{j=0}^{-b-1}(1+2s+b+2j)\phi_b^\delta(\tfrac12(b-1))$ gives rise to a similar relation between ${}^R(T\psi)^+$ and ${}^R(T\psi)^-$. As these D_b^δ do not occur in the sum formula, I don't go into their automorphic models, except for the case $b = 0$. In this case $T\psi$ has to be constant on G_0 and on $G_0 j$. This is only possible if $\chi = 1$.

4.5.5. Proposition. If $\chi \neq 1$, then $\dim A(D_0^\delta) = 0$; if $\chi = 1$, then $\dim A(D_0^\delta) = 1$. In the last case $A(D_0^\delta)$ is generated by T with

$$(4.5.19)\qquad T\phi_0^\delta(-\tfrac12)(g_0 j_\varepsilon) = (\delta,\varepsilon) \qquad \text{for } g_0 \in G_0,\ \varepsilon = \pm 1.$$

Here $(1,1) = (1,-1) = (-1,1) = 1$, $(-1,-1) = -1$.

For the other cases we have the following relations between automorphic models and automorphic forms:

4.5.6. Proposition. Let U be $H_\tau^\delta(s)$ or $H_\tau(s)$, with s general. Let ψ be as in (4.5.16). Then $T \mapsto {}^R(T\psi)^+$ defines an isomorphism $A(U) \cong F_{\tau,\frac14 - s^2}(\bar\Gamma,\mathbb{C},v^+)$.

4.5.7. Proposition. Let $U = D_b$, $b \in Y_\tau$, $\tau = 0,1$ or $b = \tau = 1$, ψ as in (4.5.16). Then

$$T \mapsto ((T\psi)^+,\ (T\psi)^-)$$

defines a bijective $\mathbb{R}$-linear map $A(D_b) \to \{\bar\Gamma,-b,v^+\} \oplus \{\bar\Gamma,-b,v^-\}$.

4.5.8. Proposition. Let $U = D_b$, $b \in Y_\tau$, $0 < \tau < 1$, $b \equiv \pm\tau(2)$. Let ψ be as in (4.5.16). Then

$$T \mapsto (T\psi)^\pm$$

defines a bijective $\mathbb{R}$-linear map $A(D_b) \to \{\bar\Gamma,-b,v^\pm\}$.

Proofs. From the discussion above it is clear that only the surjectivity still needs to be proved. If an automorphic form or a pair of automorphic forms is given, then it is easy to reconstruct a function f on G satisfying (4.5.1), (4.5.2) and

(4.5.20) $\underline{\omega}f = (\frac{1}{4}-s^2)f$

with $s = \frac{1}{2}(b-1)$ in the cases of 4.5.7 and 4.5.8. This function has the weight r defined in (4.5.16). Let V be the $\underline{g}$-K-subspace of $C^\infty(G)$ generated by f.

I have to prove: 1) V is $\underline{g}$-K-isomorphic to U; 2) all elements of V satisfy (4.5.1) and 3) all elements of V satisfy (4.5.2).

As the $\underline{g}$-K-action in $C^\infty(G)$ is on the right it is clear that 2) holds.

To prove 1) remark that V is spanned by elements of the form

(4.5.21) $j_\varepsilon(\underline{E}^{\pm})^l f$,

with $\varepsilon = 1$ or -1, l integral, $l \geq 0$. The weight of such an element is $\varepsilon(r \pm 2l)$.

If $0 < \tau < 1$, then all these weights are different; if $\tau = 0,1$ there may be relations.

Let us first consider the case $0 < \tau < 1$. If $2s+1 \not\equiv \pm\tau(2)$, then $V \cong H_\tau(s)$. Indeed, an isomorphism T may be constructed step by step. Put $T\phi_\tau(s) = f$; then $T\phi_{-\tau}(s) = Tj\phi_\tau(s) = jf$, $T\phi_{\tau+2}(s) = (1+2s+\tau)^{-1}T\underline{E}^+\phi_\tau(s) = (1+2s+\tau)^{-1}\underline{E}^+f$, etc.

Condition (4.5.20) guarantees that we get a $\underline{g}$-K-isomorphism. In the case $U = D_b$ we follow the same procedure; only we have to use the additional information that from the construction of f follows that $j_\varepsilon(\underline{E}^-)^l f = 0$ for all $l > 0$.

In the case $\tau = 0,1$, $2s+1 \not\equiv \tau(2)$, there are two elements $j_\varepsilon(\underline{E}^{\pm})^l f$ with the same weight. But f has been constructed in such a way that $j(\underline{E}^-)^\tau f$ is exactly the right multiple of f. An isomorphism $U \to V$ again may be defined stepwise. If $U = D_b$, then $j_\varepsilon(\underline{E}^-)^l f = 0$ for all $l > 0$. So there are no relations to worry about.

To prove 3) remark first that property (4.5.2) stays unchanged under right translation with elements of K. The question is whether right differentiation might change this condition.

Let $f_1 \in V$ satisfy (4.5.2). For each $h \in C_c^\infty(G_0)$ also $f_1 * h$ satisfies (4.5.2). Indeed, let S be the support of h. Then $KS^{-1} \subset N_c A_c K$, where N_c and A_c are compact subsets of N and A. So

(4.5.22) $f_1 * h(g_c na(y)k) = \int_S f_1(g_c na(y)kx^{-1})h(x)dx$

satisfies (4.5.2).

Now we apply a theorem of Harish Chandra; [10], thm.1 on p.18. As f_1 is a K-finite eigenfunction of $\underline{\omega}$ there are functions $h_\varepsilon \in C_c^\infty(G_0)$ such that $f_1 * h_\varepsilon = f_1$ on $G_0 j_\varepsilon$.

Now for each $\underline{X} \in \underline{g}$ we have

(4.5.23) $\underline{X}f_1 = f_1 * \underline{X}h_\varepsilon$ on $G_0 j_\varepsilon$.

But $\underline{X}h_\varepsilon \in C_c^\infty(G)$; hence $\underline{X}f_1$ satisfies (4.5.2). This completes the proof of 4.5.6-4.5.8.

4.5.9 Remark. If $\tau = 0$ or 1, $2s+1 \not\equiv \tau(2)$, then $A(H_\tau^1(s))$ and $A(H_\tau^{-1}(s))$ are both isomorphic to the same space of automorphic forms, hence isomorphic to each other.

4.5.10. Remark. If $b \in \mathbb{Z}$, $b > 1$ then $A(D_b)$ is isomorphic to the direct sum of two spaces of holomorphic automorphic forms. If χ has real values, then these two spaces are the same.

4.5.11 <u>Remark</u>. If $T \in A(E_1^{\delta})$, then

(4.5.24) $\quad {}^{R}(T\hat{\Phi}_1)^{\pm} \in F_{1,\frac{1}{4}}(\overline{\Gamma}, \mathbb{C}, v^{\pm})$.

As $\underline{E}^{-}\hat{\Phi}_1 \neq 0$ these automorphic forms do not correspond to holomorphic ones. Compare Roelcke's examples 1 and 2 on p.297-298 of [36].

5. Standard and Whittaker models.

5.1. Introduction.

Suppose that the character χ of Γ is trivial on $g_c N g_c^{-1} \cap \Gamma$, for some cusp c. If T is an automorphic model of a g-K-space U, then for each $\psi \in U$

$$(5.1.1) \qquad x \mapsto T\psi(g_c n(x) g)$$

is a periodic function. As $T\psi$ is an analytic function, we may expand (5.1.1) in a Fourier series. If χ is non-trivial on $g_c N g_c^{-1} \cap \Gamma$ the situation is essentially the same.

This Fourier series expansion is well known in the case of holomorphic automorphic forms; for a discussion of the real-analytic case see [36], (2.9) - (2.17). As the g-K-action is on the right, the Fourier expansions for different $\psi \in U$ are related. To describe this relation I use realizations of U in $C^\infty(G)$ consisting of functions transforming on the left according to a character of N.

The trivial character of N gives realizations of U in spaces of functions on $N\backslash G$. These are the models of the induced representation from NH to G. I call it the standard model. All other characters of N can be transformed into each other by conjugation by elements of H. They lead to the Whittaker models. I choose the character $n(x) \mapsto e^{\frac{1}{2}ix}$ to get the nicest formulas.

The general differential equation for these models is given in 5.2. The standard models are introduced in 5.3, the Whittaker models in 5.4. Relations between these models are discussed in 5.5. I do this using the language of the bundle H^τ. The standard models, connected with induced representations are well known. The Whittaker models which one usually encounters in the literature, see for instance [13] p. 181, consist of quickly decreasing functions. I also need Whittaker models the functions in which grow exponentially.

5.2. Differential equation.

For $u = 0$ or $\frac{1}{2}$ the character ν^u of N is defined by

$$(5.2.1) \qquad \nu^u(n(x)) = e^{iux}.$$

Let U be a g-K-space contained in $H^\tau(s)$ for some $s \in \mathbb{C}$ or $U = E_1^\delta$. Let X be an intertwining operator from U into the functions in $C^\infty(G)$ which satisfy

$$(5.2.2) \qquad f(ng) = \nu^u(n) f(g) \quad \text{for all } n \in N,\ g \in G.$$

Again g and K act in $C^\infty(G)$ on the right. Then the Casimir operator ω acts in XU as multiplication by $\frac{1}{4}-s^2$. If $f \in XU$ has weight r then f is known as soon as we know f_1 and f_{-1} defined by

(5.2.3) $\qquad f_\varepsilon(y) = f(a(y)j_\varepsilon)$.

As $\underline{\omega} f = (\frac{1}{4}-s^2)f$ we get the differential equation

(5.2.4) $\qquad f''_\varepsilon(y) + (-u^2+\varepsilon r u y^{-1}+(\frac{1}{4}-s^2)y^{-2})f_\varepsilon(y) = 0.$

This equation has two linearly independent solutions. This means that f has to be an element of a four dimensional space. So we may expect at most four models for a given u and a given space U. This will turn out to be true in general; in the case $\tau = 0,1$ there are additional restrictions, due to the fact that $\tau \equiv -\tau(2)$.

5.3. Standard and logarithmic models.

In this subsection $u = 0$. A solution of (5.2.4) is $f_\varepsilon(y) = y^{\frac{1}{2}+s}$.

5.3.1. Definition. Let $\tau = 0,1$, $\delta = \pm 1$, $s \in \mathbb{C}$. The standard model $St^\delta_\tau(s)$ of $H^\delta_\tau(s)$ is defined by

(5.3.1) $\qquad St^\delta_\tau(s)\phi^\delta_r(s)(n(x)a(y)j_\varepsilon k(\theta)) = (\delta,\varepsilon)y^{\frac{1}{2}+s}e^{ir\theta}$

with

(5.3.2) $\qquad (\delta,\varepsilon) = \begin{cases} -1 & \text{if } \delta = \varepsilon = -1 \\ 1 & \text{else} \end{cases}$

Let $0 < \tau < 1$, $s \in \mathbb{C}$, $\eta = \pm 1$. The standard model $St^\eta_\tau(s)$ of $H_\tau(s)$ is given by

(5.3.3) $\qquad St^\eta_\tau(s)\phi_r(s)(n(x)a(y)j_\varepsilon k(\theta)) = \begin{cases} \sqrt{2}\, y^{\frac{1}{2}+s}e^{ir\theta} & \text{if } \varepsilon r \equiv \eta\tau(2) \\ 0 & \text{if } \varepsilon r \equiv -\eta\tau(2). \end{cases}$

It is easy to check that these models are $\underline{g}$-K-morphisms.
The factor $\sqrt{2}$ in (5.3.3) compensates for $St^\eta_\tau(s)$ being zero on one half of G.
The solution $f_\varepsilon(y) = y^{\frac{1}{2}-s}$ of (5.2.4) gives rise to models $St^\delta_\tau(-s)\iota^\delta_\tau(s)$. For general s one gets in this way a system of linearly independent models.
Standard models of D_b one obtains by the inclusions $D^\delta_b \subset H^\delta_\tau(\frac{1}{2}(b-1))$, $D_b \subset H_\tau(\frac{1}{2}(b-1))$. These are the only ones.

If $s = 0$ then (5.2.4) has also a solution $f_\varepsilon(y) = y^{\frac{1}{2}}\ln y$. If $\tau = 1$ there are no models of $H^\delta_1(0)$ corresponding to this solution.

5.3.2. Definition. Let $\delta = \pm 1$, $\tau = 0$. The logarithmic model L^δ_0 of $H^\delta_0(0)$ is given by

(5.3.4) $\qquad L^\delta_0\phi^\delta_r(0)(n(x)a(y)j_\varepsilon k(\theta)) = (\delta,\varepsilon)y^{\frac{1}{2}}(\ln y + a_r)e^{ir\theta}.$

Let $0 < \tau < 1$, $\eta = \pm 1$. The logarithmic model L^η_τ of $H_\tau(0)$ is given by

(5.3.5) $\quad L_\tau^\eta \phi_r(0)(n(x)a(y)j_\varepsilon k(\theta)) = \begin{cases} \sqrt{2}y^{\frac{1}{2}}(\ln y + a_{\varepsilon\eta r})e^{ir\theta} & \underline{\text{if}}\ \varepsilon r \equiv \eta\tau(2) \\ 0 & \underline{\text{if}}\ \varepsilon r \equiv -\eta\tau(2). \end{cases}$

<u>In both cases</u> a_r <u>is defined by</u>

(5.3.6) $\quad a_\tau = \Gamma'(\frac{1-\tau}{2})\Gamma(\frac{1-\tau}{2})^{-1}$

(5.3.7) $\quad (a_{r+2} - a_r)(r+1) = 2$ <u>for all</u> $r \equiv \tau(2)$.

The choice of a_τ makes that the following relations hold:

(5.3.8) $\quad \lim\limits_{s\to 0} \frac{1}{2s} (St_0^\delta(s) - St_0^\delta(-s)\iota_0^\delta(s))\psi = L_0^\delta\psi$ if $\tau = 0$, $\delta = \pm 1$

(5.3.9) $\quad \lim\limits_{s\to 0} \frac{1}{2s} (St_\tau^\eta(s) - St_\tau^\eta(-s)\iota_\tau^1(s))\psi = L_\tau^\eta\psi$ if $0 < \tau < 1$, $\eta = \pm 1$.

These limits converge uniformly on sets of the form $\{na(y)k | n \in N,\ k \in K,\ y_1 \leq y \leq y_2\}$ for ψ in the standard basis of $H_\tau^{(\delta)}$.
In the case $\tau = 1$, $s = 0$, the logarithmic case leads to the following definition.

5.3.3. <u>Definition</u>. <u>The logarithmic model</u> L_1^δ <u>of</u> E_1^δ <u>is given by</u>

(5.3.10) $\quad L_1^\delta\hat{\phi}_r^\delta(0)(n(x)a(y)j_\varepsilon k(\theta)) = (\delta,\varepsilon)y^{\frac{1}{2}} \ln y\ e^{ir\theta}$

(5.3.11) $\quad L_1^\delta\phi_r^\delta(0) = St_1^\delta(0)\phi_r^\delta(0)$.

Another logarithmic model of E_1^δ is $L_1^{-\delta}\iota_E^\delta$. There are also the standard models $St_1^\delta(0)p_E^\delta$ and $St_1^{-1}(0)p_E^{-\delta}\iota_E^\delta$. All other models in left-N-invariant functions are linear combinations of these four.
In the notations of (3.3.28)

(5.3.12) $\quad \lim\limits_{s\to 0} St_1^\delta(s) \oplus St_1^\delta(-s)\ \hat{\phi}_r^\delta(s) = L_1^\delta\hat{\phi}_r^\delta(0)$

$\quad \lim\limits_{s\to 0} St_1^\delta(s) \oplus St_1^\delta(-s)\ \phi_r^\delta(s) = L_1^\delta\phi_r^\delta(0)$;

the limits are uniform on the same sets as in (5.3.8-9).

5.4. <u>Whittaker models.</u>

For $u = \frac{1}{4}$ equation (5.2.4) is the Whittaker differential equation. The solutions $W_{\kappa,s}$ and $M_{\kappa,s}$ are discussed in subsection 13.2. The solution $W_{\kappa,s}(y)$ decreases quickly for $y \to \infty$, whereas in general $M_{\kappa,s}(y)$ increases exponentially. Still $M_{\kappa,s}$ gives rise to the easiest Whittaker model.

5.4.1. Definition. Let $s \in \mathbb{C}$, $s \notin \{-\frac{1}{2}, -1, -\frac{3}{2}, \ldots\}$. For $\tau = 0,1$, $\delta = \pm 1$ the Whittaker model $M_\tau^\delta(s)$ of $H_\tau^\delta(s)$ is given by

$$M_\tau^\delta(s)\phi_r^\delta(s)(n(x)a(y)j_\varepsilon k(\theta)) = (\delta,\varepsilon)e^{\frac{1}{2}ix}M_{\frac{1}{2}\varepsilon r,s}(y)e^{ir\theta}. \tag{5.4.1}$$

For $0 < \tau < 1$, $\eta = \pm 1$ the Whittaker model $M_\tau^\eta(s)$ of $H_\tau(s)$ is given by

$$M_\tau^\eta(s)\phi_r(s)(n(x)a(y)j_\varepsilon k(\theta)) = \begin{cases} \sqrt{2}e^{\frac{1}{2}ix}M_{\frac{1}{2}\varepsilon r,s}(y)e^{ir\theta} & \text{if } \varepsilon r \equiv \eta\tau(2) \\ 0 & \text{if } \varepsilon r \equiv -\eta\tau(2). \end{cases} \tag{5.4.2}$$

In general these Whittaker models do not satisfy the condition of moderate growth mentioned by Jacquet-Langlands, see [13] thm. 5.13, p. 181:

$$|f(a(y)k)| \ll y^a \quad \text{for} \quad y \to \infty, \text{ for some } a \in \mathbb{R} \tag{5.4.3}$$

for all f in the model. Whittaker models satisfying this condition are the following ones:

5.4.2. Definition. Let $s \in \mathbb{C}$ be general. For $\tau = 0,1$, $\delta = \pm 1$ the Whittaker model $W_\tau^\delta(s)$ of $H_\tau^\delta(s)$ is given by

$$W_\tau^\delta(s)\phi_r^\delta(s)(n(x)a(y)j_\varepsilon k(\theta)) = (\delta,\varepsilon)e^{\frac{1}{2}ix}\Gamma(\tfrac{1}{2}-s-\tfrac{1}{2}\varepsilon r)W_{\frac{1}{2}\varepsilon r,s}(y)e^{ir\theta}. \tag{5.4.4}$$

For $0 < \tau < 1$, $\eta = \pm 1$ the Whittaker model $W_\tau^\eta(s)$ of $H_\tau(s)$ is given by

$$W_\tau^\eta(s)\phi_r(s)(n(x)a(y)j_\varepsilon k(\theta)) = \begin{cases} \sqrt{2}e^{\frac{1}{2}ix}\Gamma(\tfrac{1}{2}-s-\tfrac{1}{2}\varepsilon r)W_{\frac{1}{2}\varepsilon r,s}(y)e^{ir\theta} & \text{if } \varepsilon r \equiv \eta\tau(2) \\ 0 & \text{if } \varepsilon r \equiv -\eta\tau(2). \end{cases} \tag{5.4.5}$$

To check that 5.4.1 and 5.4.2 indeed define g-K-operators, one needs the relations (13.2.10), (13.2.11).
For general $s \in \mathbb{C}$ and $0 < \tau < 1$, we have now defined all four possible Whittaker models of $H_\tau(s)$. For general $s \in \mathbb{C}$ and $\tau = 0,1$ we have also exhausted all possibilities; for here the action of j gives an additional condition relating f_1 and f_{-1} in 5.2. In the points $s = \frac{1}{2}(b-1)$, $b \equiv \pm\tau(2)$, the Γ-factor may give poles.

5.4.3. Lemma. If $\tau = 0,1$, $\delta = \pm 1$, $b \equiv \tau(2)$, $b \geq 1$, then $\operatorname{res}_{s \to \frac{1}{2}(b-1)} W_\tau^\delta(s)$ is the only Whittaker model of $H_\tau^\delta(\frac{1}{2}(b-1))$ satisfying (5.4.3).
If $\tau = 0,1$, $\delta = \pm 1$, $b \equiv \tau(2)$, $b < 1$, then $\operatorname{res}_{s \to \frac{1}{2}(b-1)} W_\tau^\delta(s)$ is the only Whittaker model of $H_\tau^\delta(\frac{1}{2}(b-1))$ satisfying (5.4.3); its restriction to D_b^δ is zero.
If $0 < \tau < 1$, $b \in Y_\tau$, then the only Whittaker models of $H_\tau(\frac{1}{2}(b-1))$ satisfying

(5.4.3) <u>are</u> $W^{\eta}_{\tau}(\frac{1}{2}(b-1))$, <u>obtained by holomorphic continuation, and</u> $\operatorname{res}_{s \to \frac{1}{2}(b-1)} W^{-\eta}_{\tau}(s)$, <u>with</u> $\eta = \pm 1$, $b \equiv \eta\tau(2)$; <u>the last one of these models is zero on</u> D_b.

<u>Proof</u>. Let $b \equiv \pm\tau(2)$. Any Whittaker model X of $H^{(\delta)}_{\tau}(\frac{1}{2}(b-1))$ satisfying (5.4.3) is of the form

(5.4.6) $$X\phi^{(\delta)}_{r}(\tfrac{1}{2}(b-1))(n(x)a(y)j_{\varepsilon}k(\theta)) = e^{\frac{1}{2}ix}c(\varepsilon,r)W_{\frac{1}{2}\varepsilon r,\frac{1}{2}(b-1)}(y)e^{ir\theta}.$$

Let $\delta = 1$ if $0 < \tau < 1$. By considering the action of j we see that

(5.4.7) $$c(-\varepsilon,r) = \delta c(\varepsilon,-r).$$

So we need only consider $\varepsilon = 1$. The action of $\underline{E}^{\pm}$ gives, with help of (13.2.11), the conditions:

(5.4.8) $$(b+r)c(1,r+2) = -2c(1,r)$$

(5.4.9) $$(b-r)c(1,r-2) = (b-r)(1-\tfrac{1}{2}(b+r))c(1,r).$$

(5.4.8) implies (5.4.9). From (5.4.8) follows:

(5.4.10) $$c(1,-b) = 0$$

(5.4.11) $c(1,r) = 0$ for all $r \equiv b(2)$, $r \leqslant -b$

(5.4.12) there is a constant c_1 such that
$$c(1,r) = c_1 \operatorname{res}_{s \to \frac{1}{2}(b-1)} \Gamma(\tfrac{1}{2}-s-\tfrac{1}{2}r) \text{ for all } r \equiv b(2).$$

(5.4.13) $(0 < \tau < 1$ only) there is a constant c_2 such that
$$c(1,r) = c_2\Gamma(1-\tfrac{1}{2}b-\tfrac{1}{2}r) \text{ for all } r \equiv -b(2).$$

Now it is easy to check that all possibilities for c indeed occur among the models given in the lemma.
The other assertions are easily checked.

5.4.4. <u>Definition</u>. <u>Let</u> $\tau = 0,1$, $b \equiv \tau(2)$, $b \geqslant 1$. <u>The Whittaker model</u> D^b <u>of</u> D_b <u>is obtained as</u> $\sqrt{2}(-1)^{\tau}(b-1)!\ \operatorname{res}_{s \to \frac{1}{2}(b-1)} W^{1}_{\tau}(s)$ <u>restricted to</u> $D^{1}_{b} \cong D_b$.
<u>Let</u> $0 < \tau < 1$, $b \in Y_{\tau}$. <u>The Whittaker model</u> D^b <u>of</u> D_b <u>is obtained as</u> $\Gamma(1-b)^{-1}W^{\eta}_{\tau}(\frac{1}{2}(b-1))$ <u>restricted to</u> D_b <u>with</u> $\eta = \pm 1$, $\eta b \equiv \tau(2)$.

5.4.5. <u>Lemma</u>. <u>For</u> $b \in Y_{\tau}$ <u>the model</u> D^b <u>of</u> D_b <u>is the only Whittaker model of</u> D_b <u>satisfying</u> (5.4.3). <u>If</u> $\tau = 0,1$, $b \in Y^{f}_{\tau}$, $\delta = \pm 1$, <u>then</u> D^{δ}_{b} <u>has no Whittaker models satisfying</u> (5.4.3).

<u>Proof</u>. In the same way as that of lemma 5.4.3.

Remark. The factors in definition 5.4.4 simplify formulas later on. For instance, for all $b \in Y_\tau$

(5.4.14) $$(D^b\phi_b[b])^+(z) = \sqrt{2}e^{\frac{1}{2}iz}$$
$$(D^b\phi_b[b])^-(z) = 0.$$

To check this use (13.2.9).

5.4.6. Remark. $-\sqrt{2}\ \operatorname{res}_{s \to 0} W_1^1(s)$ has the character of a D^b; I denote it by D^1.

5.5. Relations between the models.

A family of maps $P(s): H^\tau(s) \to C^\infty(G)$ I call holomorphic or meromorphic if for each holomorphic section ξ of H^τ and each $g \in G$ the map

(5.5.1) $$s \mapsto (P(s)\xi(s))(g)$$

is holomorphic or meromorphic. For each section ξ I denote by $P\xi$ the map $s \mapsto P(s)\xi(s)$ with values in $C^\infty(G)$.
The standard and Whittaker models give rise to this kind of maps between the sheaf of sections of H^τ and the sheaf of functions with values in $C^\infty(G)$.

5.5.1. Definition. The maps St, M and W are defined by

$$St(s) = \begin{pmatrix} St_\tau^1(s) & 0 \\ 0 & St_\tau^{-1}(s) \end{pmatrix}, \quad M(s) = \begin{pmatrix} M_\tau^1(s) & 0 \\ 0 & M_\tau^{-1}(s) \end{pmatrix},$$

$$W(s) = \begin{pmatrix} W_\tau^1(s) & 0 \\ 0 & W_\tau^{-1}(s) \end{pmatrix},$$

with respect to the decomposition (3.5.3).
St is holomorphic, W and M are meromorphic, as is easily checked.
Recall that in 3.5 the spaces D_b, $b \in Y_\tau$, have got a standard embedding in $H^\tau(\frac{1}{2}(b-1))$. The models D^b are supposed to act on this embedding.

5.5.2. Lemma. M is holomorphic outside $\{-\frac{1}{2}, -1, -\frac{3}{2}, \ldots\}$.
For each $b \in Y_\tau$ and for $b = \tau = 1$:

(5.5.2) $$M(\tfrac{1}{2}(b-1))|D_b = D^b.$$

Proof. The holomorphy is clear from definition 5.4.1. Formula (5.5.2) may be checked by a computation using (13.2.8). If one is content to consider only a generating element one may use (5.4.14).

Remark. In (5.5.2) one sees an advantage of the way D_b has been embedded in $H^{\tau}(\frac{1}{2}(b-1))$.

5.5.3. Lemma. W is holomorphic at general points s. It satisfies:

$$W\iota = W. \tag{5.5.3}$$

If $\tau = 0,1$, $b \equiv \tau(2)$, $b \geq 1$, then

$$\left(\operatorname*{res}_{s \to \frac{1}{2}(b-1)} W(s)\right)\Big| D_b = (-1)^{\tau}((b-1)!)^{-1} D^b, \tag{5.5.4}$$

$$\left(\operatorname*{res}_{s \to \frac{1}{2}(b-1)} W(s)\right)\Big| \tilde{D}_b = 0. \tag{5.5.5}$$

If $\tau = 0,1$, $b \in Y^f_\tau$, $\delta = \pm 1$, then

$$\left(\operatorname*{res}_{s \to \frac{1}{2}(b-1)} W(s)\right)\Big| D^{\delta}_b = 0. \tag{5.5.6}$$

If $0 < \tau < 1$, $b \in Y_\tau$ then

$$\lim_{s \to \frac{1}{2}(b-1)} W(s)|D_b = \Gamma(1-b)D^b; \quad \operatorname*{res}_{s \to \frac{1}{2}(b-1)} W(s)|\tilde{D}_b = 0. \tag{5.5.7}$$

$$\operatorname*{res}_{s \to \frac{1}{2}(b-1)} W(s) = \Gamma(1-b)D^{2-b} \operatorname*{res}_{s \to \frac{1}{2}(b-1)} \iota(s). \tag{5.5.8}$$

Proof. The holomorphy is clear from 5.4.2. Property (5.5.3) amounts to

$$W(-s)\iota(s) = W(s) \tag{5.5.9}$$

for all general $s \in \mathbb{C}$; this is easily checked. For (5.5.4) and (5.5.5) take $r \equiv b(2)$, $|r| \geq b$. Then

$$W(s) = \frac{1}{\sqrt{2}} (\phi^1_r(s) \pm \operatorname{sgn}(r)\phi^{-1}_r(s))(n(x)a(y)j_\varepsilon k(\theta)) \tag{5.5.10}$$

$$= \sum_{\delta = \pm 1} \frac{1}{\sqrt{2}} (\delta, \pm\operatorname{sgn} r)(\delta,\varepsilon) W^1_\tau(s)\phi^1_r(s)(n(x)a(y)j_\varepsilon k(\theta)).$$

Taking $\operatorname*{res}_{s \to \frac{1}{2}(b-1)}$ one obtains

$$\sum_{\delta = \pm 1} \frac{1}{\sqrt{2}} (\delta, \pm\varepsilon\operatorname{sgn} r) \frac{1}{\sqrt{2}(-1)^{\tau}(b-1)!} \; . \tag{5.5.11}$$

$$D^b\left(\frac{1}{\sqrt{2}} (\phi^1_r(\tfrac{1}{2}(b-1)) + \operatorname{sgn}(r)\phi^{-1}_r(\tfrac{1}{2}(b-1)))(n(x)a(y)j_\varepsilon k(\theta))\right).$$

But $D^b(\ldots)$ is only nonzero if $\varepsilon = \operatorname{sgn} r$. This gives (5.5.4) and (5.5.5).

(5.5.6) is clear from lemma 5.4.3.

(5.5.7) and (5.5.8) follow from lemma 5.4.3 and 5.4.4 and from (3.5.12). The assertion (5.5.7) should be understood in the following way: if ξ is a constant section of H^{τ}, i.e. a linear combination of the standard basis with constant coefficients, such that $\xi(\frac{1}{2}(b-1)) \in D_b$ then

$$(5.5.12) \qquad \lim_{s \to \frac{1}{2}(b-1)} W(s)\xi(s) = \Gamma(1-b)D^b\xi(\tfrac{1}{2}(b-1)).$$

6. Intertwining operators.

6.1. Introduction.

There are some intertwining operators between the models discussed in section 5. These operators turn up when one wants to compute Fourier coefficients of Eisenstein series or Poincaré series. See for instance [22], p. 319, [15] subsection 2.2, [27].

In this section I determine these intertwining operators explicitly. The computations are all known, at least in the case $\tau = 0$; see the references above. The other cases give no additional problems. The results are stated in subsection 6.8.

The intertwining operators go from a St- or M-model to a St- or W-model. The case St-St is the easiest one. The other cases are reduced to it. In section 14 I shall need convolution operators which are generalizations of the intertwining operators of this section. In subsection 6.4 I introduce these operators. This has the advantage that the computations in subsection 6.5 are more transparent. In particular the occurrence of the Bessel differential equation in the M-W-case turns out to be due to an eigenfunction equation for the Casimir operator.

The idea of the discussion in subsection 6.5 I have got from Niebur's computation of the Fourier coefficients of Poincaré series, [27].

6.2. The operator ${}_{\mu}V^{\varepsilon}(t)$.

Application of the relations in G, see 2.2, gives

(6.2.1) $$w^{-1} j_\varepsilon a(t^{-1}) n(x) a(y) j_\eta k(\theta) =$$

$$n(-\varepsilon t x(x^2+y^2)^{-1}) a(ty(x^2+y^2)^{-1}) j_{\varepsilon\eta} k(\theta - \tfrac{1}{2}\pi\varepsilon\eta + \eta \arctan xy^{-1})$$

6.2.1. Lemma. Let $t > 0$, $\varepsilon = \pm 1$. Suppose that the measurable function f on G satisfies

(6.2.2) $$|f(na(y)k)| \ll y^{\frac{1}{2}+\sigma} \quad \text{for } y \searrow 0$$

uniformly in $n \in N$, $k \in K$, for some $\sigma > 0$.

Then

(6.2.3) $$\int_N |f(w^{-1} j_\varepsilon a(t^{-1}) ng)| \, dn$$

converges; if $g \in Na(y)K$, then the integral is estimated by $t^{\frac{1}{2}+\sigma} y^{\frac{1}{2}-\sigma}$.

Proof. From (6.2.1) follows that we may estimate the integral by

(6.2.4) $$\int_{-\infty}^{+\infty} (ty(x^2+y^2)^{-1})^{\frac{1}{2}+\sigma} dx.$$

6.2.2. Definition. Let $u = 0$ or $\frac{1}{2}$, $\varepsilon = \pm 1$, $t > 0$. Suppose the measurable function f on G satisfies the estimate (6.2.2). Then the function ${}_uV^\varepsilon(t)f$ on G is defined by

(6.2.5) $${}_uV^\varepsilon(t)f(g) = \int_{-\infty}^{+\infty} e^{-iux} f(w^{-1} j_\varepsilon a(t^{-1})n(x)g)\,dx.$$

Remark that ${}_uV^\varepsilon(t)f$ is continuous for continuous f.

(6.2.6) $${}_uV^\varepsilon(t)f(n(x)g) = e^{iux}\,{}_uV^\varepsilon(t)f(g).$$

(6.2.7) $$|{}_uV^\varepsilon(t)f(a(y)k)| \ll t^{\frac{1}{2}+\sigma} y^{\frac{1}{2}-\sigma} \text{ for } y \to \infty$$

for t bounded and σ as in lemma 6.2.1.

6.3. ${}_uV^\varepsilon(t)$ as intertwining operator.

Let $s \in \mathbb{C}$, Re $s > 0$. The functions in $St(s)H^\tau(s)$ and in $M(s)H^\tau(s)$ satisfy the assumptions in lemma 6.2.1. For the case of $M(s)$ see definition 5.4.1 and (13.2.4). So ${}_uV^\varepsilon(t)$ acts on these models.

6.3.1. Lemma. ${}_uV^\varepsilon(t)$ acting on $St(s)H^\tau(s)$ or $M(s)H^\tau(s)$ commutes with the right g-K-action.

Proof. The K-action gives no problem. Let $\underline{X} \in \underline{g}_{\mathbb{R}}$ and f in the model. Then

(6.3.1) $$\underline{X}\,{}_uV^\varepsilon(t)f(g) =$$

$$\lim_{h \to 0} \int_{-\infty}^{+\infty} e^{-iux} \underline{X} f(w^{-1} j_\varepsilon a(t^{-1})n(x)g e^{h_1 \underline{X}})\,dx$$

with h_1 situated between 0 and h. Using (2.2.9) one sees that for h_1 bounded the estimate for the integrand may be made independent of h_1. So we may move the limit inside the integral.

6.3.2. Corollary. ${}_uV^\varepsilon(t)St(s)$ and ${}_uV^\varepsilon(t)M(s)$ are models of $H^\tau(s)$. If $u = 0$ it are standard models, if $u = \frac{1}{2}$ Whittaker models.

From the estimate (6.2.7) follows a restriction for the resulting model. If $u = 0$ it has to be built up from $St(-s)\iota(s)$; if $u = \frac{1}{2}$ and s is general it has to be built up from $W(s)$. Hence the following result:

6.3.3. Lemma. For $u, u_1 \in \{0, \frac{1}{2}\}$, $\varepsilon = \pm 1$, $t > 0$, s general, Re $s > 0$, there are 2×2 matrices ${}_u v^\varepsilon_{u_1}(t,s)$ such that

$$(6.3.2)\qquad \begin{aligned} {}_0V^{\varepsilon}(t)St(s) &= St(-s)\iota(s)\,{}_0v_0^{\varepsilon}(t,s)\\ {}_{\frac12}V^{\varepsilon}(t)St(s) &= W(s)\,{}_{\frac12}v_0^{\varepsilon}(t,s)\\ {}_0V^{\varepsilon}(t)M(s) &= St(-s)\iota(s)\,{}_0v_{\frac12}^{\varepsilon}(t,s)\\ {}_{\frac12}V^{\varepsilon}(t)M(s) &= W(s)\,{}_{\frac12}v_{\frac12}^{\varepsilon}(t,s). \end{aligned}$$

These matrices refer to the decomposition (3.5.3) of $H^{\tau}(s)$.
In the following subsections I determine the matrices ${}_u v_{u_1}^{\varepsilon}(t,s)$. Once that is done the case of nongeneral s can be handled easily.

6.4. Convolution operators.

In this subsection I give a generalization of the operators ${}_uV^{\varepsilon}(t)$.
NAwN and NAjwN are open subsets of G; they are components of the big cell NHwN in the Bruhat decomposition of G.

6.4.1. Definition. Let $u,u_1 \in \{0,\frac12\}$. ${}_uQ_{u_1}$ is the set of functions $q \in C^{\infty}(G)$ satisfying

$$(6.4.1)\qquad q(n(x)gn(x_1)) = e^{iux+iu_1x_1}q(g)$$

$$(6.4.2)\qquad |q| \in C_c(N\backslash G/N)$$

$$(6.4.3)\qquad \operatorname{supp} q \subset NAwN \cup NAjwN$$

A function $q \in {}_uQ_{u_1}$ is completely determined by two functions $q_1, q_{-1} \in C_c^{\infty}(0,\infty)$:

$$(6.4.4)\qquad q(a(y)j_{\varepsilon}w) = q_{\varepsilon}(y).$$

If $q \in {}_uQ_{u_1}$, then $\underline{\omega}q \in {}_uQ_{u_1}$. Application of (2.5.6) gives

$$(6.4.5)\qquad (\underline{\omega}q)_{\varepsilon}(y) = -y^2q_{\varepsilon}''(y) - \varepsilon uu_1yq_{\varepsilon}(y).$$

6.4.2. Definition. Let $q \in {}_uQ_{u_1}$. For functions f satisfying the assumption (6.2.2) and

$$(6.4.6)\qquad f(n(x)g) = e^{iu_1x}f(g)$$

define $C_qf = q*f$ by

$$(6.4.7)\qquad q*f(g) = \int_{G/N} q(x)f(x^{-1}g)\,d\dot{x}.$$

This definition makes sense, for by (2.4.11)

$$(6.4.8)\qquad \int_{G/N} q(x)f(x^{-1}g)d\dot{x} =$$

$$= \frac{1}{2\pi}\int_A \sum_{\varepsilon=\pm 1} \int_N q(naj_\varepsilon w)f(w^{-1}j_\varepsilon a^{-1}n^{-1}g)\alpha(a)^{-1}dadn$$

$$= \frac{1}{2\pi}\sum_{\varepsilon=\pm 1}\int_0^\infty q_\varepsilon(t)\,{}_uV^\varepsilon(t)f(g)t^{-2}dt.$$

If we take $q_{-\varepsilon} = 0$ and let q_ε tend to a delta-distribution at t_0 with respect to $t^{-2}dt$, then C_q tends to $\frac{1}{2\pi}\,{}_uV^\varepsilon(t_0)$.
For $q \in {}_uQ_{u_1}$ the operator C_q acts on the models $St(s)H^\tau(s)$ and $M(s)H^\tau(s)$ for Re $s > 0$. The discussion of subsection 6.3 can be repeated for this operator.

6.4.3. Proposition. *For* $u,u_1 \in \{0,\frac{1}{2}\}$, s *general*, Re $s > 0$, *there is for each* $q \in {}_uQ_{u_1}$ *a* 2×2 *matrix* $v(q,s)$ *such that*

$$(6.4.9)\qquad C_qSt(s) = St(-s)\iota(s)v(q,s) \qquad \textit{if } u = u_1 = 0$$

$$C_qSt(s) = W(s)v(q,s) \qquad \textit{if } u = \tfrac{1}{2},\ u_1 = 0$$

$$C_qM(s) = St(-s)\iota(s)v(q,s) \qquad \textit{if } u = 0,\ u_1 = \tfrac{1}{2}$$

$$C_qM(s) = W(s)v(q,s) \qquad \textit{if } u = u_1 = \tfrac{1}{2}.$$

6.5. General computations.

The suppositions are those of proposition 6.4.3. For each $f \in St(s)H^\tau(s)$ or $M(s)H^\tau(s)$

$$(6.5.1)\qquad \underline{\omega}f = (\tfrac{1}{4}-s^2)f.$$

But as $\underline{\omega}$ is invariant under left and right translations we get

$$(6.5.2)\qquad C_{\underline{\omega}q}f = (\tfrac{1}{4}-s^2)C_qf.$$

So

$$(6.5.3)\qquad v(\underline{\omega}q,s) = v(q,s)(\tfrac{1}{4}-s^2).$$

The coefficients of $v(.,s)$ are distributions on $(0,\infty)$ satisfying a differential equation. In (6.4.5) we see that this differential equation is elliptic.
So there are analytic functions F_1,F_{-1} on $(0,\infty)$ with 2×2 matrices as values, such that

(6.5.4) $$v(q,s) = \sum_{\varepsilon=\pm 1} \int_0^\infty q_\varepsilon(t) F_\varepsilon(t) t^{-2} dt$$

(6.5.5) $$t^2 F''_\varepsilon(t) + (\tfrac{1}{4} - s^2 + \varepsilon u u_1 t) F_\varepsilon(t) = 0$$

If we let q_1 or q_{-1} tend to a delta-distribution on $(0,\infty)$ as indicated in subsection 6.4, we get

(6.5.6) $$F_\varepsilon(t) = \frac{1}{2\pi} \, {}_u v^\varepsilon_{u_1}(t,s).$$

By (6.2.7) we obtain

(6.5.7) $$|F_\varepsilon(t)| \ll t^{\frac{1}{2} + \mathrm{Re}\, s} \quad \text{for } t \searrow 0.$$

This condition excludes one fundamental solution of (6.5.5). So there are 2×2 matrices ${}_u c^\varepsilon_{u_1}(s)$ such that

(6.5.8) $$F_\varepsilon(t) = \begin{cases} t^{\frac{1}{2}+s} \, {}_u c^\varepsilon_{u_1}(s) & \text{if } uu_1 = 0 \\ t^{\frac{1}{2}} J^\varepsilon_{2s}(t^{\frac{1}{2}}) \, {}_u c^\varepsilon_{u_1}(s) & \text{if } uu_1 = \frac{1}{4} \end{cases}$$

with the notation $J^1_{2s} = J_{2s}$ and $J^{-1}_{2s} = I_{2s}$, see (14.2.4).
Remark that for $t \searrow 0$:

(6.5.9) $$F_\varepsilon(t) \sim \begin{cases} t^{\frac{1}{2}+s} \, {}_u c^\varepsilon_{u_1}(s) & \text{if } uu_1 = 0 \\ 2^{-2s} \Gamma(2s+1)^{-1} t^{\frac{1}{2}+s} \, {}_u c^\varepsilon_{u_1}(s) & \text{if } uu_1 = \frac{1}{4} \end{cases}$$

6.5.1. Lemma. Let f satisfy

(6.5.10) $$f(n(x)g) = e^{iu_1 x} f(g)$$

(6.5.11) $$f(a(y)k) \sim y^{\frac{1}{2}+s} f_1(k) \quad \text{for} \quad y \searrow 0$$

uniformly in $k \in K$, then

(6.5.12) $$\lim_{t \searrow 0} t^{-\frac{1}{2}-s} \, {}_u V^\varepsilon(t) f(a(y) j_\eta k(\theta)) =$$

$$y^{\frac{1}{2}-s} \int_{-\infty}^{\infty} e^{-iuxy} (x^2+1)^{-\frac{1}{2}-s} f_1(j_{\varepsilon\eta} k(\theta - \tfrac{1}{2}\pi\varepsilon\eta + \eta \arctan x))\, dx$$

Proof. Immediate from (6.2.1) and definition 6.2.2.

6.5.2. Lemma. If $\phi \in H^\tau(s)$ then for $y \searrow 0$, uniformly in $k \in K$

(6.5.13) $\quad M(s)\phi(a(y)k) \sim y^{\frac{1}{2}+s}St(s)\phi(k)$

Proof. Immediate from definitions 5.3.1 and 5.4.1 and the power series expansion (13.2.2).

6.5.3. Lemma. ${}_0C_{\frac{1}{2}}(s) = {}_0C_0(s)$; ${}_{\frac{1}{2}}C_{\frac{1}{2}}(s) = 2^{+2s}\Gamma(2s+1)\,{}_{\frac{1}{2}}C_0(s)$.

Proof. Take $Y(s) = St(-s)\iota(s)$ if $u = 0$ and $Y(s) = W(s)$ if $u = \frac{1}{2}$. For $\phi \in H^\tau(s)$;

$$(6.5.14)\qquad \lim_{t\downarrow 0} t^{-\frac{1}{2}-s}\,{}_uV^\varepsilon(t)M(s)\phi(a(y)j_\eta k(\theta))$$

$$= \lim_{t\downarrow 0} t^{-\frac{1}{2}-s}Y(s)F_\varepsilon(t)\phi(a(y)j_\eta k(\theta))$$

$$= \begin{cases} 2\pi\, St(-s)\iota(s)\,{}_0C^\varepsilon_{\frac{1}{2}}(s)\phi(a(y)j_\eta k(\theta)) & \text{if } u = 0 \\ 2\pi\, W(s)2^{-2s}\Gamma(2s+1)^{-1}\,{}_{\frac{1}{2}}C^\varepsilon_{\frac{1}{2}}(s)\phi(a(y)j_\eta k(\theta)) & \text{if } u = \frac{1}{2}, \end{cases}$$

see (6.3.2), (6.5.6) and (6.5.9).

$$(6.5.15)\qquad \lim_{t\downarrow 0} t^{-\frac{1}{2}-s}\,{}_uV^\varepsilon(t)M(s)\phi(a(y)j_\eta k(\theta))$$

$$= y^{\frac{1}{2}-s}\int_{-\infty}^{+\infty} e^{-iuxy}(x^2+1)^{-\frac{1}{2}-s}St(s)\phi(j_{\varepsilon\eta}k(\theta-\tfrac{1}{2}\pi\varepsilon\eta+\eta\arctan x))dx$$

$$= {}_uV^\varepsilon(1)St(s)\phi(a(y)j_\eta k(\theta))$$

$$= Y(s)\,{}_uv^\varepsilon_0(1,s)\phi(a(y)j_\eta k(\theta))$$

$$= 2\pi Y(s)\,{}_uC^\varepsilon_0(s)\phi(a(y)j_\eta k(\theta)),$$

see lemma 6.5.1, lemma 6.5.2, definition 6.2.2, lemma 6.3.3, (6.5.6) and (6.5.9). Comparision of (6.5.14) and (6.5.15) proves the lemma.

6.6. The case $u = 0$.

For the case $u = u_1 = 0$ I need the following integral formula, valid for $\operatorname{Re} s > 0$, $r \in \mathbb{R}$:

$$(6.6.1)\qquad \int_{-\infty}^{+\infty} (x^2+1)^{-\frac{1}{2}-s}e^{ir\arctan x}dx =$$

$$\pi 2^{1-2s}\Gamma(2s)\Gamma(\tfrac{1}{2}+s+\tfrac{1}{2}r)^{-1}\Gamma(\tfrac{1}{2}+s-\tfrac{1}{2}r)^{-1}.$$

It can be easily derived from either one of the first two formulas on p. 9 of [25].

6.6.1. Lemma. If $\phi \in H^\tau(s)$ has weight r, then

$$(6.6.2)\qquad {}_0V^\varepsilon(t)St(s)\phi(a(y)j_\eta k(\theta)) = t^{\frac12+s}\pi 2^{+1-2s}\Gamma(2s)\Gamma(\tfrac12+s+\tfrac12 r)^{-1}\Gamma(\tfrac12+s-\tfrac12 r)^{-1} \cdot y^{\frac12-s}St(s)\phi(j_{\varepsilon\eta}k(\theta-\tfrac12\pi\varepsilon\eta)).$$

Proof. Use (6.2.1) and definition 6.2.1.

6.6.2. Definition. Let $s \in \mathbb{C}$ be general, $\varepsilon = \pm 1$.
$X^\varepsilon(s)\colon H^\tau(s) \to H^\tau(s)$ is given by the matrices

$$(6.6.3)\qquad e^{-\frac12\pi i\tau\varepsilon}\begin{pmatrix}1 & 0\\ 0 & \varepsilon\end{pmatrix} \quad \text{if } \tau = 0,1$$

$$\begin{pmatrix}1 & 0\\ 0 & c_{-1}(-s)\end{pmatrix} Y^\varepsilon \begin{pmatrix}e^{-\frac12\pi i\tau} & 0\\ 0 & e^{\frac12\pi i\tau}\end{pmatrix} \quad \text{if } 0 < \tau < 1;$$

with

$$(6.6.4)\qquad Y^1 = \begin{pmatrix}1 & 0\\ 0 & 1\end{pmatrix},\ Y^{-1} = \begin{pmatrix}0 & 1\\ 1 & 0\end{pmatrix}.$$

This defines an operator $X^\varepsilon\colon H^\tau \to H^\tau$ meromorphic over $\mathbb{C}$.

6.6.3. Proposition. Let $\varepsilon = \pm 1$, $t > 0$, $s \in \mathbb{C}$, s general, $\operatorname{Re} s > 0$. Then

$$(6.6.5)\qquad {}_0V^\varepsilon(t)St(s) = 2^{+1-2s}\Gamma(2s)\sin\pi(\tfrac12+s+\tfrac12\tau)t^{\frac12+s}St(-s)\iota(s)X^\varepsilon(s).$$

Proof. Of course definition 6.6.2 has been given in such a way that (6.6.5) is just a reformulation of (6.6.2).
Let $\tau = 0$ or 1, $\delta = \pm 1$, $r \equiv \tau(2)$. By lemma 6.6.1

$$(6.6.6)\qquad {}_0V^\varepsilon(t)St(s)\phi_r^\delta(s)(a(y)j_\eta k(\theta))$$

$$= t^{\frac12+s}\pi 2^{+1-2s}\Gamma(2s)\Gamma(\tfrac12+s+\tfrac12 r)^{-1}\Gamma(\tfrac12+s-\tfrac12 r)^{-1} \cdot y^{\frac12-s}(\delta,\varepsilon\eta)e^{ir\theta-\frac12\pi i\varepsilon\eta r}$$

$$= 2^{+1-2s}\Gamma(2s)\sin\pi(\tfrac12+s+\tfrac12\tau)t^{\frac12+s} \cdot (-1)^{\frac12(r-\tau)}\Gamma(\tfrac12-s-\tfrac12 r)\Gamma(\tfrac12+s-\tfrac12 r)^{-1}y^{\frac12-s}(\delta,\varepsilon\eta)e^{ir\theta}(-1)^{\frac12(\iota-\tau)}e^{-\frac12\pi i\varepsilon\eta}$$

$$= 2^{+1-2s}\Gamma(2s)\sin\pi(\tfrac12+s+\tfrac12\tau)t^{\frac12+s} \cdot (\delta,\varepsilon\eta)e^{-\frac12\pi i\varepsilon\eta\tau}\gamma_r(s)y^{\frac12-s}e^{ir\theta}$$

(6.6.7) $$St(-s)\iota(s)X^{\varepsilon}(s)\phi_r^{\delta}(s)(a(y)j_{\eta}k(\theta)) = e^{-\frac{1}{2}\pi i\tau\varepsilon}(\delta,\varepsilon)\gamma_r(s)(\delta(-1)^{\tau},\eta)y^{\frac{1}{2}-s}e^{ir\theta}$$

This proves the proposition in this case, for $e^{-\frac{1}{2}\pi i\tau\varepsilon\eta} = e^{-\frac{1}{2}\pi i\tau\varepsilon}((-1)^{\tau},\eta)$.
In the case $0 < \tau < 1$ take $r \equiv \alpha\tau(2)$, $\alpha = \pm 1$, $\delta = \pm 1$.

(6.6.8) $$\begin{aligned} &{}_0V^{\varepsilon}(t)St(s)\phi_r^{\delta}(s)(a(y)j_{\eta}k(\theta)) \\ &= t^{\frac{1}{2}+s}\pi 2^{+1-2s}\Gamma(2s)\Gamma(\tfrac{1}{2}+s+\tfrac{1}{2}\alpha r)^{-1}\Gamma(\tfrac{1}{2}+s-\tfrac{1}{2}\alpha r)^{-1} \\ &\quad \cdot y^{\frac{1}{2}-s}\sqrt{2}\,\delta_{\varepsilon\eta\alpha,\delta}\,e^{ir\theta-\frac{1}{2}\pi i r\varepsilon\eta} \\ &= 2^{+1-2s}\Gamma(2s)\sin\pi(\tfrac{1}{2}+s+\tfrac{1}{2}\tau)t^{\frac{1}{2}+s} \\ &\quad \cdot e^{-\frac{1}{2}\pi i\tau\delta}\sqrt{2}\,\delta_{\eta\alpha,\varepsilon\delta}\,\gamma_{\alpha r}(s)y^{\frac{1}{2}-s}e^{ir\theta} \end{aligned}$$

(6.6.9) $$\begin{aligned} &St(-s)\iota(s)X^{\varepsilon}(s)\phi_r^{\delta}(s)(a(y)j_{\eta}k(\theta)) \\ &= e^{-\frac{1}{2}\pi i\tau\delta}c_{\varepsilon\delta}(-s)\gamma_{\varepsilon\delta\alpha r}(s)\sqrt{2}\,\delta_{\eta\alpha,\varepsilon\delta}\,y^{\frac{1}{2}-s}e^{ir\theta} \end{aligned}$$

Use (3.3.15) to see that (6.6.8) and (6.6.9) give the equality in the proposition

6.6.4. <u>Corollary</u>. <u>In the cases</u> $u = u_1 = 0$ <u>and</u> $u = 0$, $u_1 = \frac{1}{2}$

(6.6.10) $${}_0C^{\varepsilon}_{u_1}(s) = \pi^{-1}2^{-2s}\Gamma(2s)\sin\pi(\tfrac{1}{2}+s+\tfrac{1}{2}\tau)X^{\varepsilon}(s)$$

<u>Proof</u>. Use lemma 6.5.3 and (6.5.6), (6.5.8).

6.7. <u>The case $u = \frac{1}{2}$</u>.

6.7.1. <u>Lemma</u>. <u>For</u> $\phi \in H^{\tau}(s)$, $k \in K$

(6.7.1) $$\lim_{y\downarrow 0} y^{s-\frac{1}{2}}W(s)\phi(a(y)k) = \Gamma(2s)St(-s)\iota(s)\phi(k).$$

<u>Proof</u>. Use definitions 5.3.1, 5.4.2, (13.2.5) and (13.2.2).

6.7.2. <u>Lemma</u>. <u>For</u> $\phi \in H^{\tau}(s)$, $k \in K$

(6.7.2) $$\lim_{y\downarrow 0} y^{s-\frac{1}{2}}\,{}_{\frac{1}{2}}V^{\varepsilon}(t)St(s)\phi(a(y)k) = {}_0V^{\varepsilon}(t)St(s)\phi(k).$$

<u>Proof</u>. Suppose that ϕ has weight r.

(6.7.3) $$\lim_{y\downarrow 0} y^{s-\frac12}\,{}_{\frac12}V^{\varepsilon}(t)St(s)\phi(a(y)j_\eta k(\theta))$$

$$= \lim_{y\downarrow 0} t^{\frac12+s}\int_{-\infty}^{\infty} e^{-\frac12 ixy}(x^2+1)^{-\frac12-s}e^{ir\eta\arctan x}dx$$

$$.\ St(s)\phi(j_{\varepsilon\eta}k(\theta-\tfrac12\pi\varepsilon\eta))$$

$$= t^{\frac12+s}\int_{-\infty}^{+\infty}(x^2+1)^{-\frac12-s}e^{ir\eta\arctan x}dx\ St(s)\phi(j_{\varepsilon\eta}k(\theta-\tfrac12\pi\varepsilon\eta))$$

$$= {}_0V^{\varepsilon}(t)St(s)\phi(j_\eta k(\theta)).$$

6.7.3. Proposition. Let $\varepsilon = \pm 1$, $t > 0$, $s \in \mathbb{C}$, s general, $\mathrm{Re}\, s > 0$. Then

(6.7.4) $${}_{\frac12}V^{\varepsilon}(t)St(s) = 2^{+1-2s}\sin\pi(\tfrac12+s+\tfrac12\tau)t^{\frac12+s}W(s)X^{\varepsilon}(s).$$

Proof. Combining (6.3.2) with the lemmas 6.7.1 and 6.7.2 we get:

$$\Gamma(2s)St(-s)\iota(s)\,{}_{\frac12}v_0^{\varepsilon}(t,s)$$

$$= {}_0V^{\varepsilon}(t)St(s).$$

The assertion now follows from proposition 6.6.3.

6.7.4. Corollary.

(6.7.5) $${}_{\frac12}c_0^{\varepsilon}(s) = \pi^{-1}2^{-2s}\sin\pi(\tfrac12+s+\tfrac12\tau)X^{\varepsilon}(s)$$

(6.7.6) $${}_{\frac12}c_{\frac12}^{\varepsilon}(s) = \pi^{-1}\Gamma(2s+1)\sin\pi(\tfrac12+s+\tfrac12\tau)X^{\varepsilon}(s)$$

Proof. Use lemma 6.5.3.

6.7.5. Corollary. For $\mathrm{Re}\, s > 0$, $\kappa \in \mathbb{R}$, $s \notin \{-\frac12-\kappa, -\frac{3}{2}-\kappa, \ldots\}$

(6.7.8) $$W_{\kappa,s}(y) = \pi^{-1}2^{2s-1}\Gamma(\tfrac12+s+\kappa)y^{\frac12-s}\int_{-\infty}^{+\omega} e^{-\frac12 ixy}(x^2+1)^{-\frac12-s}e^{2i\kappa\arctan x}dx$$

Proof. Let $r = 2\kappa$, take $\tau \in [0,1]$ such that $r \equiv \eta\tau$, $\eta = \pm 1$, $\eta = 1$ if $r \in \mathbb{Z}$. By the proposition

(6.7.9) $${}_{\frac12}V^1(1)St(s)\phi_r^{\eta}(s)(a(y)) =$$

$$= 2^{+1-2s}\sin\pi(\tfrac12+s+\tfrac12\tau)W(s)X^1(s)\phi_r^{\eta}(s)(a(y)).$$

On the other hand this equals

(6.7.10) $\int_{-\infty}^{+\infty} e^{-\frac{1}{2}ixy} St(s)\phi_r^{\eta}(s)(a(y^{-1}(x^2+1)^{-1})k(-\frac{1}{2}\pi+\arctan x))y\,dx.$

In the case $\tau = 0,1$

(6.7.11) $$2^{+1-2s}\sin\pi(\tfrac{1}{2}+s+\tfrac{1}{2}\tau)\Gamma(\tfrac{1}{2}-s-\tfrac{1}{2}r)W_{\frac{1}{2}r,s}(y)e^{-\frac{1}{2}\pi i\tau}$$
$$= \int_{-\infty}^{+\infty} e^{-\frac{1}{2}ixy}(1+x^2)^{-\frac{1}{2}-s}e^{ir\arctan x}dx\,y^{\frac{1}{2}-s}e^{-\frac{1}{2}\pi ir}.$$

The corollary follows easily.
The case $0 < \tau < 1$ is handled similarly.

6.8. Results.

Now we may gather the results proved in the previous subsections.
For the C_q I shall just state formulas for the $v(q,s)$.
Only the case $u = u_1 = \frac{1}{2}$ is needed later on, in section 14.

6.8.1. Proposition. Let $q \in {}_uQ_{u_1}$, $u,u_1 \in \{0,\frac{1}{2}\}$, let C_q be the operator defined in 6.4.2 and let $v(q,s)$ be the matrix defined in 6.4.3. Then for general s, $\mathrm{Re}\, s > 0$:

(6.8.1) $$v(q,s) = \sum_{\varepsilon=\pm 1}\int_0^{\infty} q(a(y)j_{\varepsilon}w)F_{\varepsilon}(y)y^{-2}dy$$

with

(6.8.2) $$F_{\varepsilon}(y) = \begin{cases} y^{\frac{1}{2}+s}\pi^{-1}2^{-2s}\Gamma(2s)\sin\pi(\frac{1}{2}+s+\frac{1}{2}\tau)X^{\varepsilon}(s) & \text{if } u = 0 \\ y^{\frac{1}{2}+s}\pi^{-1}2^{-2s}\sin\pi(\frac{1}{2}+s+\frac{1}{2}\tau)X^{\varepsilon}(s) & \text{if } u = \frac{1}{2},\ u_1 = 0 \\ y^{\frac{1}{2}}J^{\varepsilon}_{2s}(y^{\frac{1}{2}})\,\frac{1}{\pi}\,\Gamma(2s+1)\sin\pi(\frac{1}{2}+s+\frac{1}{2}\tau)X^{\varepsilon}(s) & \text{if } u = u_1 = \frac{1}{2} \end{cases}$$

X^{ε} has been defined in 6.6.2; for J^{ε}_{2s} see (14.2.4).

Proof. See (6.5.6), (6.5.8) and the corollaries 6.6.4 and 6.7.4.

For the ${}_uV^{\varepsilon}(t)$ I discuss the cases separately. In each case I also consider nongeneral s.

6.8.2. Proposition. Let $t > 0$, $\varepsilon = \pm 1$. Then ${}_0V^{\varepsilon}(t)St$ is holomorphic on $\mathrm{Re}\, s > 0$. For general $s \in \mathbb{C}$, $\mathrm{Re}\, s > 0$, it is given by

(6.8.3) $${}_0V^{\varepsilon}(t)St(s) = 2^{+1-2s}\Gamma(2s)\sin\pi(\tfrac{1}{2}+s+\tfrac{1}{2}\tau)t^{\frac{1}{2}+s}St(-s)\iota(s)X^{\varepsilon}(s).$$

For $b \in Y_\tau$, $b > 1$

$$(6.8.4) \qquad {}_0V^{\varepsilon}(t)St(\tfrac{1}{2}(b-1))|D_b+\tilde{D}_b = 0.$$

Proof. For general s see proposition 6.6.3. For $s \to \frac{1}{2}(b-1)$ one checks the behaviour of the right hand side of (6.8.3); use (3.5.11) and (3.5.12).

6.8.3. Proposition. Let $t > 0$, $\varepsilon = \pm 1$. Then ${}_{\frac{1}{2}}V^{\varepsilon}(t)St$ is holomorphic on $\mathrm{Re}\ s > 0$. For general s, $\mathrm{Re}\ s > 0$:

$$(6.8.5) \qquad {}_{\frac{1}{2}}V^{\varepsilon}(t)St(s) = 2^{+1-2s}\sin\pi(\tfrac{1}{2}+s+\tfrac{1}{2}\tau)t^{\frac{1}{2}+s}W(s)X^{\varepsilon}(s).$$

For $b \in Y_\tau$, $b > 1$:

$$(6.8.6) \qquad {}_{\frac{1}{2}}V^{\varepsilon}(t)St(\tfrac{1}{2}(b-1))|D_b = \begin{cases} \pi 2^{2-b}t^{\frac{1}{2}b}e^{-\frac{1}{2}\pi ib}\Gamma(b)^{-1}D^b & \text{if } \varepsilon = 1 \\ 0 & \text{if } \varepsilon = -1. \end{cases}$$

Proof. The holomorphy follows from the definition of ${}_{\frac{1}{2}}V^{\varepsilon}(t)$; one may also check that the right hand side of (6.8.5) may be holomorphically continued. (6.8.5) is given in proposition 6.7.3. To prove (6.8.6) consider the behaviour of the right hand side of (6.8.5) applied to a constant section ϕ of H^{τ} for which $\phi(\frac{1}{2}(b-1)) \in D_b$. Use (5.5.4), (5.5.5) and (5.5.7).

6.8.4. Proposition. Let $t > 0$, $\varepsilon = \pm 1$. Then ${}_0V^{\varepsilon}(t)M$ is holomorphic on $\mathrm{Re}\ s > 0$. For general s, $\mathrm{Re}\ s > 0$:

$$(6.8.7) \qquad {}_0V^{\varepsilon}(t)M(s) = 2^{+1-2s}t^{\frac{1}{2}+s}\Gamma(2s)\sin\pi(\tfrac{1}{2}+s+\tfrac{1}{2}\tau)St(-s)\iota(s)X^{\varepsilon}(s).$$

For $b \in Y_\tau$, $b > 1$

$$(6.8.8) \qquad {}_0V^{\varepsilon}(t)M(\tfrac{1}{2}(b-1))|D_b = 0.$$

Proof. (6.8.7) follows from lemma 6.3.3, (6.5.6) and (6.5.8).
If $\tau = 0,1$ then

$$(6.8.9) \qquad {}_0V^{\varepsilon}(t)M(\tfrac{1}{2}(b-1)) =$$

$$. \ 2^{2-b}t^{\frac{1}{2}}\Gamma(b-1)(-1)^{\frac{1}{2}(b+\tau)}\pi St(\tfrac{1}{2}(1-b)) \operatorname*{res}_{s\to\frac{1}{2}(b-1)} \iota(s)X^{\varepsilon}(\tfrac{1}{2}(b-1)).$$

Use (3.5.11) to get (6.8.8).
If $0 < \tau < 1$ then for $\eta = \pm 1$, $r \equiv \pm\tau(2)$, $b \equiv \eta\tau(2)$

$$(6.8.10) \qquad {}_0V^{\varepsilon}(t)M(\tfrac{1}{2}(b-1))\phi_r^{\eta}(\tfrac{1}{2}(b-1))$$

$$= 2^{-b}t^{\frac{1}{2}b}\Gamma(b-1)e^{-\frac{1}{2}\pi i\tau\eta}St(\tfrac{1}{2}(1-b)) \lim_{s\to\frac{1}{2}(b-1)} (\sin\pi(\tfrac{1}{2}+s+\tfrac{1}{2}\varepsilon\eta\tau)\iota(s))\phi_r^{\varepsilon\eta}(\tfrac{1}{2}(b-1))$$

If $\varepsilon = 1$, then the limit is $\sin\pi\frac{1}{2}(b+\eta\tau) \lim_{s\to\frac{1}{2}(b-1)} \iota(s)$, which has been defined on $\phi_r^\eta(\frac{1}{2}(b-1))$, see (3.5.9) and 3.3, and maps $\phi_r^\eta(\frac{1}{2}(b-1)) \in D_b$ onto 0, see (3.5.12). If $\varepsilon = -1$ then the limit equals $(-1)^{\frac{1}{2}(b-\eta\tau)}\pi \operatorname{res}_{s\to\frac{1}{2}(b-1)} \iota(s)$, which is zero on $\phi_r^\eta(\frac{1}{2}(b-1)) \in \tilde{D}_b$, see (3.5.12); remark that $Y^{-1}D_b = \tilde{D}_b$.

6.8.5. Proposition. Let $t > 0$, $\varepsilon = \pm 1$. Then ${}_{\frac{1}{2}}V^\varepsilon(t)M$ is holomorphic on $\operatorname{Re} s > 0$. For general s, $\operatorname{Re} s > 0$

(6.8.11) $${}_{\frac{1}{2}}V^\varepsilon(t)M(s) = 2t^{\frac{1}{2}}J^\varepsilon_{2s}(t^{\frac{1}{2}})\Gamma(2s+1)\sin\pi(\tfrac{1}{2}+s+\tfrac{1}{2}\tau)W(s)X^\varepsilon(s).$$

For $b \in Y_\tau$, $b > 1$:

(6.8.12) $${}_{\frac{1}{2}}V^\varepsilon(t)D^b = {}_{\frac{1}{2}}V^\varepsilon(t)M(\tfrac{1}{2}(b-1))\,|\,D_b = \begin{cases} 2\pi t^{\frac{1}{2}}J_{b-1}(t^{\frac{1}{2}})e^{-\frac{1}{2}\pi i b}D^b & \text{if } \varepsilon = 1 \\ 0 & \text{if } \varepsilon = -1. \end{cases}$$

Proof. In the same way as the previous propositions. Use the lemmas 5.5.2 and 5.5.3.

7. Fourier coefficients of automorphic models.

7.1. Introduction.

Classically, holomorphic modular forms have a Fourier series expansion

(7.1.1) $$\sum_{n=0}^{\infty} a_n e^{2\pi i n z}.$$

Real analytic modular forms also have a Fourier series expansion; see [36], §2. In this section "taking a Fourier coefficient" is considered as an intertwining operator between an automorphic model and a standard or Whittaker model. This operator is introduced in subsection 7.3. As the $\underline{g}$-K-space occurring in the definition of automorphic models is built up of irreducible ones in a simple way, this intertwining operator is determined by some scalars; these scalars I call Fourier coefficients. This I study in 7.4. In the study of Eisenstein series it is convenient to have some notation to handle the Fourier coefficients of order zero together; this notation I fix in subsection 7.5. The Maass-Selberg relation , needed for the continuation of Eisenstein series, I state in 7.6.
The idea of considering "taking a Fourier coefficient" as an intertwining operator is not new; it is at the background of the Jacquet-Langlands theory in [13]. I do not know a reference where it has been worked out explicitly. The contents of 7.5 and 7.6 is standard material; see [21], Ch. XIV, §14 , [37], §9.

7.2. Preliminaries.

In 4.2 for each cusp c a representative $g_c N g_c^{-1}$ has been chosen, and an element $g_c \in G_0$. I denote $N_c = g_c N g_c^{-1}$. The group $N_c \cap \Gamma$ is infinitely cyclic.
Choose g_c in such a way that

(7.2.1) $$g_c\, n(1)\, g_c^{-1} \text{ generates } N_c \cap \Gamma$$

(7.2.2) $$\widetilde{\overline{g_c^{-1}}} = g_c^{-1}.$$

(7.2.1) can always be reached by replacing g_c by $g_c a$ with $a \in A$. To satisfy (7.2.2) we choose the right element of $g_c Z_0$. The choice of g_c in the examples in 4.2.3 satisfies (7.2.1) and (7.2.2).
Condition (7.2.1) is a normalization which makes the formulas in the sequel easier. Condition (7.2.2) I only use to relate my computations to those of Roelcke. Denote $A = \overline{g_c^{-1}}$. Then $A^{-1}\infty$ is the parabolic point fixed by $\tilde{N}_c$. If $M \in Sl_2(\mathbb{R})$ such that

(7.2.3) $$AM = \begin{pmatrix} * & * \\ c & d \end{pmatrix},$$

then

(7.2.4) $$g_c^{-1}\tilde{M} = \widetilde{AM} = \widetilde{\widetilde{AM}}\, k(-2\pi \underline{w}(A,M))$$

and

(7.2.5) $\quad g_c^{-1}\widetilde{M}\,p(z) = \widetilde{AM}\,p(z)k(-2\pi\underline{w}(A,M)) = p(AM.z)k(-\arg_{(-\pi,\pi]}(cz+d)-2\pi\underline{w}(A,M))$

as may be checked using (2.2.12), (2.2.11), (2.2.9).

7.2.1. Definition. Λ is the set of pairs (c,n), with c a cusp and $n \in \mathbb{R}$ such that

(7.2.6) $\quad \chi(g_c\, n(1) g_c^{-1}) = e^{2\pi i n}.$

If $\alpha \in \Lambda$, then c_α and n_α are defined by

(7.2.7) $\quad \alpha = (c_\alpha, n_\alpha).$

$\Lambda^1 = \{\alpha \in \Lambda \mid n_\alpha \neq 0\}$; $\Lambda^0 = \{\alpha \in \Lambda \mid n_\alpha = 0\}$.

7.2.2. Notations. For $\alpha \in \Lambda$:

(7.2.8) $\quad g_\alpha = g_{c_\alpha},\ \nu_\alpha = \nu_{c_\alpha} = g_a\, n(1) g_\alpha^{-1},\ N_\alpha = N_{c_\alpha}.$

(7.2.9) $\quad \Gamma_\alpha = N_\alpha \cap \Gamma\ ;\ \Delta_\alpha = Z_0\Gamma_\alpha$

So Γ_α and Δ_α depend only on c_α. The normalizer of Γ_α in Γ is Δ_α; it is the direct product of Z_0 and Γ_α. The group Γ_α is infinitely cyclic, generated by ν_α. In the example of the theta group in 4.2.3:

(7.2.10) $\quad \chi_\theta(n(2)) = 1,\ \chi_\theta(w) = e^{\frac{1}{4}\pi i}$ (see (2.7.4)

So

(7.2.11) $\quad \chi_\theta(\nu_\infty) = 1,\ \chi_\theta(\nu_1) = \chi_\theta(wn(-2)) = e^{\frac{1}{4}\pi i}$

(7.2.12) $\quad \Lambda^0 = \{(\infty,0)\}$

$\Lambda^1 = \{(\infty,n) \mid n \in \mathbb{Z},\ n \neq 0\} \cup \{(1,n) \mid n \in \frac{1}{8} + \mathbb{Z}\}.$

The set Λ enumerates the possibilities for taking a Fourier coefficient; if $\alpha \in \Lambda$ then c_α indicates the cusp and n_α indicates the character of N.

7.3. The operator F_α.

Let f be a function satisfying

(7.3.1) $\quad f(\gamma g) = \chi(\gamma) f(g)$ for $\gamma \in \Gamma,\ g \in G$;

this property has been studied in 4.4.

For each cusp c and $m \in \mathbb{Z}$

(7.3.2) $\quad f(g_c\, n(m) g) = \chi(\nu_c^m) f(g_c g).$

This means, under some smoothness conditions on f:

(7.3.3) $$f(g_c n(x) g) = \sum_{\substack{\alpha\in\Lambda \\ c_\alpha = c}} e^{2\pi i n_\alpha x} \int_0^1 e^{-2\pi i n_\alpha x_1} f(g_c n(x_1) g)\, dx_1,$$

a Fourier series expansion. The "Fourier coefficient"

(7.3.4) $$f_\alpha(g) = \int_0^1 e^{-2\pi i n_\alpha x_1} f(g_\alpha n(x_1) g)\, dx_1$$

satisfies

$$f_\alpha(n(x)g) = f_\alpha(g)$$

if $\alpha \in \Lambda^0$; this is the left-N-behaviour of a standard model. If $\alpha \in \Lambda^1$, we do not get the left-N-behaviour of Whittaker models; some transformation is needed. Remark that

$$f_\alpha(g) = \int_{\Gamma_\alpha\backslash N_\alpha} \chi_\alpha(n)^{-1} f(ng_\alpha g)dn, \tag{7.3.5}$$

with the character χ_α of N_α defined by

$$\chi_\alpha(g_\alpha n(x) g_\alpha^{-1}) = e^{2\pi i n_\alpha x}. \tag{7.3.6}$$

The Haar measure on N_α is obtained from the Haar measure on N by conjugation with g_α. This leads to the following definitions:

7.3.1. Notations. For $\alpha \in \Lambda$:

$$\varepsilon(\alpha) = \begin{cases} 1 & \text{if } \alpha \in \Lambda^0 \\ \operatorname{sgn} n_\alpha & \text{if } \alpha \in \Lambda^1 \end{cases} \tag{7.3.7}$$

$$h_\alpha = \begin{cases} g_\alpha & \text{if } \alpha \in \Lambda^0 \\ g_\alpha a((4\pi|n_\alpha|)^{-1}) j_{\varepsilon(\alpha)} & \text{if } \alpha \in \Lambda^1 \end{cases} \tag{7.3.8}$$

7.3.2. Definition. Let $\alpha \in \Lambda$. For every $f \in C^\infty(G)$ satisfying (7.3.1) the function $F_\alpha f$ is defined by

$$F_\alpha f(g) = \int_{\Gamma_\alpha\backslash N_\alpha} \chi_\alpha(n)^{-1} f(nh_\alpha g)dn \tag{7.3.9}$$

For $f \in C^\infty(G)$ the Fourier series expansion in (7.3.3) converges absolutely. We get, for $k \in K$:

$$f(g_c n(x)a(y)k) = \sum_{\substack{\alpha\in\Lambda^1 \\ c_\alpha = c}} e^{2\pi i n_\alpha x} F_\alpha f(a(4\pi|n_\alpha|y)j_{\varepsilon(\alpha)}k) + F_{(c,0)} f(a(y)k) \quad \text{if } (c,0) \in \Lambda^0. \tag{7.3.10}$$

Further

$$F_\alpha f(n(x)g) = \begin{cases} F_\alpha f(g) & \text{if } \alpha \in \Lambda^0 \\ e^{\frac{1}{2}ix} F_\alpha f(g) & \text{if } \alpha \in \Lambda^1 \end{cases} \tag{7.3.11}$$

7.3.3. Proposition. Let α and f be as in 7.3.2. Then $F_\alpha f \in C^\infty(G)$. The operator F_α intertwines the g-K-action on the K-finite elements of $C^\infty(G)$ satisfying (7.3.1) and the K-finite elements of $C^\infty(G)$ satisfying (7.3.11).

Proof. The domain of integration in (7.3.9) is compact. Hence we can differentiate inside the integral.

7.4. Fourier coefficients of automorphic models

Let U be a g-K-space, contained in some $H^\tau(s)$, or $U = E_1^\delta$. For each automorphic model $T \in A(U)$ and $\alpha \in \Lambda$ we may consider the g-K-operator $F_\alpha T$. From (7.3.11) and proposition 7.3.3 follows that it is a standard or Whittaker model of U. Moreover, from condition (4.5.2) in the definition of automorphic models follows that the model $F_\alpha T$ satisfies

(7.4.1) $\quad |F_\alpha T\phi(n\, a(y)k)| \ll y^a \quad$ for $\quad y \to \infty$

for some $a \in \mathbb{R}$, uniformly in $n \in N$ and $k \in K$ for all $\phi \in U$. In the case $\alpha \in \Lambda^1$ condition (7.4.1) restricts the possibilities for $F_\alpha T$.

In the case $0 < \tau < 1$ there is another reduction of the possibilities: if $\phi \in U$ has weight $r \equiv \varepsilon\tau(2)$, then $T\phi$ lives on $G_0 j_\varepsilon$, see (4.4.4). From the definition of F_α follows that $F_\alpha T\phi$ lives on $G_0 j_{\varepsilon\varepsilon(\alpha)}$. For the case $\alpha \in \Lambda^0$ follows:

7.4.1. Lemma. Let $\alpha \in \Lambda^0$. If U is one of the following g-K-spaces and $T \in A(U)$, then $F_\alpha T$ is a linear combination of the models of U given in the following table:

U	models
$H_\tau^\delta(s)$, $\begin{cases} \tau=0,1, & s \text{ general}, s\neq 0, \delta = \pm 1 \\ \tau=0,1, & 2s+1 \in Y_\tau^f, \quad \delta = \pm 1 \end{cases}$	$St_\tau^\delta(s)$, $St_\tau^{\delta(-1)^\tau}(-s)\,\iota_\tau^\delta(s)$
$H_\tau^\delta(\frac{1}{2}(b-1))$, $\tau = 0,1$, $b \in Y_\tau$, $\delta = \pm 1$	$\begin{cases} St_\tau^\delta(\frac{1}{2}(b-1)) \\ St_\tau^{\delta(-1)^\tau}(\frac{1}{2}(1-b)) \operatorname{res}_{s\to\frac{1}{2}(b-1)} \iota_\tau^\delta(s) \end{cases}$
$H_0^\delta(0)$, $\delta = \pm 1$	$St_0^\delta(0)$, L_0^δ
$H_1^\delta(0)$, see D_1	
$H_\tau(s)$, $0 < \tau < 1$, $2s+1 \not\equiv -\tau(2)$, $s \neq 0$	$St_\tau^1(s)$, $St_\tau^1(-s)\,\iota_\tau^1(s)$
$H_\tau(\frac{1}{2}(b-1))$, $0 < \tau < 1$, $b \equiv -\tau(2)$	$\begin{cases} St_\tau^1(\frac{1}{2}(b-1)), \\ St_\tau^1(\frac{1}{2}(1-b)) \operatorname{res}_{s\to\frac{1}{2}(b-1)} \iota_\tau^1(s) \end{cases}$
$H_\tau(0)$, $0 < \tau < 1$	$St_\tau^1(0)$, L_τ^1
D_b^δ, $\tau = 0,1$, $b \in Y_\tau^f$, $\delta = \pm 1$	$St_\tau^\delta(\frac{1}{2}(b-1))\,\vert D_b^\delta$
D_b, $\tau = 0,1$, $b \in Y_\tau$ or $b = \tau = 1$	$\begin{cases} St(\frac{1}{2}(b-1))\,\vert D_b \\ St(\frac{1}{2}(b-1))\,\vert \tilde{D}_b \cong D_b \end{cases}$
D_b, $0 < \tau < 1$, $b \in Y_\tau$	$St_\tau^1(\frac{1}{2}(b-1))$

Remark that in the case $U = D_b$, $\tau = 0,1$, $b \in Y_\tau$ I use the embedding of D_b in $H^\tau(\frac{1}{2}(b-1))$; see (3.5.6), (3.5.7).

Proof. Use the results on standard and logarithmic models in subsection 5.3.

Remark. In subsection 7.5 I shall introduce notations to handle the case $\alpha \in \Lambda^0$ in another way.

7.4.2. Proposition. Let $\alpha \in \Lambda^1$ and let U be one of the following g-K-spaces. Then there is a linear form

$T \mapsto \alpha(U,T)$ such that

$F_\alpha T = \alpha(H^\delta_\tau(s),T) W^\delta_\tau(s)$ if $\tau = 0,1$, $\delta = \pm 1$, s general

$F_\alpha T = \alpha(H_\tau(s),T) W^{\varepsilon(\alpha)}_\tau(s)$ if $0 < \tau < 1$, s general or $2s+1 \equiv \varepsilon(\alpha)\tau(2)$

$F_\alpha T = \alpha(H^\delta_\tau(\frac{1}{2}(b-1)),T) \operatorname{res}_{s\to\frac{1}{2}(b-1)} W^\delta_\tau(s)$ if $\tau = 0,1$, $\delta = \pm 1$, $b \equiv \tau(2)$

$F_\alpha T = \alpha(H_\tau(\frac{1}{2}(b-1)),T) \operatorname{res}_{s\to\frac{1}{2}(b-1)} W^{\varepsilon(\alpha)}_\tau(s)$ if $0 < \tau < 1$, $b \equiv -\varepsilon(\alpha)\tau(2)$

$F_\alpha T = \alpha(D_b,T) D^b$ if $b \in Y_\tau$ or $b = \tau = 1$

$F_\alpha T = 0$ if $T \in A(D^\delta_b)$, $b \in Y^f_\tau$, $\delta = \pm 1$

Proof. Use the lemmas 5.4.3 and 5.4.5.

Remark. The numbers $\alpha(U,T)$ I call Fourier coefficients.

7.4.3. Proposition. Let $\alpha \in \Lambda^1$. If $\tau = 0,1$, $\delta = \pm 1$, $b \in Y_\tau$ then for all $T \in A(H^\delta_\tau(\frac{1}{2}(b-1)))$: $\alpha(H^\delta_\tau(\frac{1}{2}(b-1)),T) = \sqrt{2}\,(-1)^\tau (b-1)!\ \alpha(D_b, T|D^\delta_b)$.
If $0 < \tau < 1$, $b \in Y_\tau$, $b \equiv \varepsilon(\alpha)\tau(2)$ then for all $T \in A(H_\tau(\frac{1}{2}(b-1)))$
$\alpha(H^\delta_\tau(\frac{1}{2}(b-1)),T) = \Gamma(1-b)^{-1}\, \alpha(D_b, T|D_b)$.
If $0 < \tau < 1$, $b \in Y_\tau$, $b \equiv -\varepsilon(\alpha)\tau(2)$, then for all $T \in A(D_b)$: $\alpha(D_b,T) = 0$.

Proof. The first two assertions follow from definition 5.4.4. In the last case $F_\alpha T$ and D^b live on different components of G.

The Fourier expansion (7.3.10) may for $f \in TU$ be reformulated using the Fourier coefficients given in proposition 7.4.2. In propositions 4.5.6-8 a correspondence is given between spaces of automorphic models and spaces of automorphic forms. For the forms occurring in these propositions I now write down explicitly the Fourier series expansion. In this subsection I restrict myself to the "non-constant" part of the expansion. The "constant" term I study in the next subsection.
Fix a cusp c. Let, as in 7.2, $A = \overline{g_c}^{-1}$. Suppose $A = \begin{pmatrix} * & * \\ c & d \end{pmatrix}$. Then from (7.2.5) follows

(7.4.2) $$p(z) = g_c\, p(Az)\, k(-\arg_{(-\pi,\pi]}(cz+d)).$$

I use Roelcke's notation, see [36], (2.9):

(7.4.3) $$z_A = x_A + iy_A = Az.$$

Let $\tau = 0,1$, $\delta = \pm 1$, s general. Let $T \in A(H^\delta_\tau(s))$. Then the non-constant part of the Fourier series expansion of ${}^R(T\,\phi^\delta_\tau(s))^+(z)$ is

(7.4.4) $$e^{-i\tau \arg_{(-\pi,\pi]}(cz+d)} \sum_{\substack{\alpha\in\Lambda^1 \\ c_\alpha = c}} \alpha(H^\delta_\tau(s),T)(\delta,\varepsilon(\alpha)) \,.$$
$$.\ \Gamma(\tfrac{1}{2}-s-\tfrac{1}{2}\varepsilon(\alpha)\tau)\, W_{\frac{1}{2}\varepsilon(\alpha)\tau,s}(4\pi|n_\alpha|y_A) e^{2\pi i n_\alpha x_A} .$$

This may be compared with Roelcke's result (2.13), (2.10), [36].

Let $\tau = 0,1$, $b \equiv \tau(2)$, $b \geq 1$. Let $\psi = \frac{1}{\sqrt{2}}(\phi_b^1(\frac{1}{2}(b-1)) + \phi_b^{-1}(\frac{1}{2}(b-1)))$, as in (4.5.16). By proposition 4.5.7 each $T \in A(D_b)$ is determined by two holomorphic automorphic forms $(T\psi)^+$ and $(T\psi)^-$. For the non-constant part of the Fourier series expansion of $(T\psi)^+(z)$ we find

$$(7.4.5) \qquad \pi^{\frac{1}{2}b} 2^{b+\frac{1}{2}} (cz+d)^{-b} \sum_{\substack{\alpha\in\Lambda^1 \\ c_\alpha=c,\,\varepsilon(\alpha)=1}} n_\alpha^{\frac{1}{2}b}\, \alpha(D_b,T) e^{2\pi i n_\alpha z_A}$$

and for $(T\psi)^-(z)$

$$(7.4.6) \qquad \pi^{\frac{1}{2}b} 2^{b+\frac{1}{2}} (cz+d)^{-b} \sum_{\substack{\alpha\in\Lambda^1 \\ c_\alpha=c,\,\varepsilon(\alpha)=-1}} |n_\alpha|^{\frac{1}{2}b}\, \overline{\alpha(D_b,T)}\, e^{2\pi i |n_\alpha| z_A}.$$

In the case b=1 I use the notation

$$(7.4.7) \qquad \alpha(D_1,.) = -2^{-\frac{1}{2}} \alpha(H_1^{\pm 1}(0),.).$$

Let $0 < \tau < 1$, s general, $T \in A(H_\tau(s))$. The non-constant part of the Fourier series expansion of ${}^R(T\psi)^+(z)$ in proposition 4.5.6 is

$$(7.4.8) \qquad e^{-\tau i \arg_{(-\pi,\pi]}(cz+d)} \sqrt{2} \sum_{\substack{\alpha\in\Lambda^1 \\ c_\alpha=c}} \alpha(H_\tau(s),T)\Gamma(\tfrac{1}{2}-s-\tfrac{1}{2}\varepsilon(\alpha)\tau).$$
$$.W_{\frac{1}{2}\varepsilon(\alpha)\tau,s}(4\pi|n_\alpha|y_A)e^{2\pi i n_\alpha x_A}$$

If $0 < \tau < 1$, $b \equiv \pm\tau(2)$, then $T \in A(D_b)$ is determined by one holomorphic form $(T\psi)^\pm$, see proposition 4.5.8. The non-constant part of the Fourier series expansion is given by (7.4.5) or (7.4.6).

7.5. The constant term in the Fourier series expansion.

It will turn out to be convenient to consider the Fourier coefficients corresponding to Λ^0 as a vector. In the study of Eisenstein series this is a usual device.

7.5.1. Definition. E is the vector space over $\mathbb{C}$ with basis $\{\underline{e}_\lambda | \lambda \in \Lambda^0\}$. It has a scalar product $((.,.))$ for which this basis is orthonormal.

7.5.2. Notation. For functions $f \in C^\infty(G)$ satisfying (7.3.1) I define

$$(7.5.1) \qquad {}^*Ff = \sum_{\lambda\in\Lambda^0} F_\lambda f.\underline{e}_\lambda.$$

So $^*Ff$ is a function on G with values in E. If U is a g-K-space and $T \in A(U)$, then $^*FT$ is a standard model of U in the functions on G with values in E.

The results in lemma 7.4.1 I state in the following way:

7.5.3. Proposition. Let U be one of the following g-K-spaces. There exist linear maps $T \mapsto \underline{a}^\cdot(.,T)$ and $T \mapsto \underline{b}^\cdot(.,T)$ with values in E such that for all $T \in A(U)$:

$$*FT = \underline{a}_\tau^\delta(s,T)St_\tau^\delta(s) + \underline{b}_\tau^\delta(s,T)St_\tau^{\delta(-1)^\tau}(-s)\iota_\tau^\delta(s)$$

$\underline{\text{if}}$ $U = H_\tau^\delta(s), \tau=0,1,$ $\left\{\begin{matrix} \underline{s\ \text{general}},\ s \neq 0 \\ 2s+1 \in Y_\tau^f \end{matrix}\right\}$, $\delta = \pm 1$

$$*FT = \underline{a}_\tau^\delta(\tfrac{1}{2}(b-1),T)St_\tau^\delta(\tfrac{1}{2}(b-1)) + \underline{b}_\tau^\delta(\tfrac{1}{2}(b-1),T)St_\tau^{\delta(-1)^\tau}(-\tfrac{1}{2}(b-1)) \operatorname*{res}_{s\to\frac{1}{2}(b-1)} \iota_\tau^\delta(s)$$

$\underline{\text{if}}$ $U = H_\tau^\delta(\tfrac{1}{2}(b-1)),\ \tau = 0,1,\ \delta = \pm 1,\ b \in Y_\tau$

$$*FT = \underline{a}_0^\delta(0,T)St_0^\delta(0) + \underline{b}_0^\delta(0,T)L_0^1$$

$\underline{\text{if}}$ $U = H_0^\delta(0),\ \delta = \pm 1$

$$*FT = \underline{a}_\tau(s,T)St_\tau^1(s) + \underline{b}_\tau(s,T)St_\tau^1(-s)\iota_\tau^1(s)$$

$\underline{\text{if}}$ $U = H_\tau(s),\ 0 < \tau < 1,\ 2s+1 \not\equiv -\tau(2),\ s \neq 0$

$$*FT = \underline{a}_\tau(\tfrac{1}{2}(b-1),T)St_\tau^1(\tfrac{1}{2}(b-1)) + \underline{b}_\tau(\tfrac{1}{2}(b-1),T)St_\tau^1(\tfrac{1}{2}(1-b)) \operatorname*{res}_{s\to\frac{1}{2}(b-1)} \iota_\tau^1(s)$$

$\underline{\text{if}}$ $U = H_\tau(\tfrac{1}{2}(b-1)),\ 0 < \tau < 1,\ b \equiv -\tau(2)$

$$*FT = \underline{a}_\tau(0,T)St_\tau^1(0) + \underline{b}_\tau(0,T)L_\tau^1$$

$\underline{\text{if}}$ $U = H_\tau(0),\ 0 < \tau < 1$

$$*FT = \underline{a}_b(T)St(\tfrac{1}{2}(b-1))|D_b + \underline{b}_b(T)St(\tfrac{1}{2}(b-1))\tilde{\phi}$$

$\underline{\text{if}}$ $U = D_b,\ \tau = 0,1,\ b \in Y_\tau$ $\underline{\text{or}}$ $b = \tau = 1$ and $\tilde{\phi}\ \phi_r[b] = \tilde{\phi}_r[b]$

$$*FT = \underline{a}_b^\delta(T)St_\tau^\delta(\tfrac{1}{2}(b-1))|D_b^\delta$$

$\underline{\text{if}}$ $U = D_b^\delta,\ \tau = 0,1,\ b \in Y_\tau^f,\ \delta = \pm 1$

$$*FT = \underline{a}_b(T)St_\tau^1(\tfrac{1}{2}(b-1))$$

$\underline{\text{if}}$ $U = D_b,\ 0 < \tau < 1,\ b \in Y_\tau$

The constant terms in the Fourier series expansions in 7.4 may now be written down. I use the same notations as in 7.4. I suppose $\lambda = (c,0) \in \Lambda^0$; otherwise there is no constant term at the cusp c.

In the case of (7.4.4):

$$(7.5.2)\qquad e^{-i\tau \arg_{(-\pi,\pi]}(cz+d)}((\underline{a}_\tau^\delta(s,T)y_A^{\frac{1}{2}+s} + \underline{b}_\tau^\delta(s,T)\Gamma(\tfrac{1}{2}-s-\tfrac{1}{2}\tau)\Gamma(\tfrac{1}{2}+s-\tfrac{1}{2}\tau)^{-1}y_A^{\frac{1}{2}-s},\underline{e}_\lambda)).$$

In the case of (7.4.8):

$$(7.5.3)\qquad \sqrt{2}\, e^{-i\tau \arg_{(-\pi,\pi]}(cz+d)}((\underline{a}_\tau(s,T)y_A^{\frac{1}{2}+s} + \underline{b}_\tau(s,T)\Gamma(\tfrac{1}{2}-s-\tfrac{1}{2}\tau)\Gamma(\tfrac{1}{2}+s-\tfrac{1}{2}\tau)^{-1}y_A^{\frac{1}{2}-s},\underline{e}_\lambda)).$$

In the cases (7.4.5) and (7.4.6):

$$(7.5.4)\qquad \sqrt{2}\,(cz+d)^{-b}((\underline{a}_b(T),\underline{e}_\lambda))$$

$$(7.5.5)\qquad \sqrt{2}\,(cz+d)^{-b}\overline{((\underline{b}_b(T),\underline{e}_\lambda))}$$

If $0 < \tau < 1$, $b \equiv \pm\tau(2)$ then the constant term of the Fourier series expansion of $(T\ \phi_b[b])^{\pm}(z)$ is given by

$$(7.5.6)\qquad \sqrt{2}\,(cz+d)^{-b}\begin{cases}((\underline{a}_b(T),\underline{e}_\lambda)) & \text{if } b\equiv\tau(2)\\ \overline{((\underline{a}_b(T),\underline{e}_\lambda))} & \text{if } b\equiv-\tau(2)\end{cases}$$

7.5.4. The bundle E_2 .

The remainder of subsection 7.5 is a reformulation of the results of proposition 7.5.3. The ungainly list in this proposition is concealed with help of the language of bundles. This will be convenient when we study Eisenstein series.

The standard models of $H_\tau^{(\delta)}(s)$ for $s\in\mathbb{C}$ form a bundle the structure of which we now study.

Let τ be fixed, and if $\tau=0$ or 1, let $\delta=\pm1$ be fixed. For each general $s\in\mathbb{C}$, $s\neq0$, we have a linear map

$$(7.5.7)\qquad A(H_\tau^{(\delta)}(s))\to E^2 \;:\; T\mapsto(\underline{a}_\tau^{(\delta)}(s,T),\underline{b}_\tau^{(\delta)}(s,T)).$$

This map describes $T\mapsto{}^*FT$ with help of the formulas in proposition 7.5.3. The extension to other s may be described in the following way.

7.5.5. Definition.

$$U^g=\begin{cases}\mathbb{C}\backslash\{\tfrac12(b-1)\,|\,b\in Y_\tau \text{ or } b=1\} & \underline{\text{if}}\ \tau=0\ \underline{\text{or}}\ 1\\ \mathbb{C}\backslash\{\tfrac12(b-1)\,|\,b\equiv-\tau(2)\text{ or } b=1\} & \underline{\text{if}}\ 0<\tau<1\end{cases}$$

For $b\in Y_\tau\cup\{1\}$ if $\tau=0,1$ and for $b\equiv-\tau(2)$ or $b=1$ if $0<\tau<1$:

$$U_b=\{s\in\mathbb{C}\,|\,|s-\tfrac12(b-1)|<\rho\},\ \rho\ \underline{\text{small}},\ \rho>0.$$

These sets form an open covering of $\mathbb{C}$. All triple intersections are empty. Moreover the U_b intersect only U^g and not each other. A bundle over $\mathbb{C}$ may be defined with help of this covering. By E^2 I mean $E\times E$.

7.5.6. Definition. The bundle $E_2=E^2\times\mathbb{C}$ over $\mathbb{C}$ is defined by the following local homeomorphisms:

$$\Omega^g\;:\;E^2\times U^g\to E_2\;:\;(\underline{a},\underline{b};s)\to(\underline{a},\underline{b};s)$$

For $\tau\neq1$:

$$\Omega_0\;:\;E^2\times U_0\to E_2\;:\;(\underline{a},\underline{b};s)\to\begin{cases}(\underline{a},\underline{b};0) & \underline{\text{if}}\ s=0\\ (\underline{a}+\frac{1}{2s}\underline{b},\ \frac{-1}{2s}\underline{b};s) & \underline{\text{if}}\ s\neq0\end{cases}$$

For $\tau=1$, $\delta=\pm1$

$$\Omega_0\;:\;E^2\times U_0\to E_2\;:\;(\underline{a},\underline{b};s)\to\begin{cases}(\underline{a},\underline{b};0) & \underline{\text{if}}\ s=0\\ (\underline{a}+2s\underline{b},\ \underline{a};s) & \underline{\text{if}}\ s\neq0\end{cases}$$

For $b\in Y_\tau$ if $\tau=0,1$ and for $b\equiv-\tau(2)$ if $0<\tau<1$:

$$\Omega_b\;:\;E^2\times U_b\to E_2\;:\;(\underline{a},\underline{b};s)\to\begin{cases}(\underline{a},\underline{b};\tfrac12(b-1)) & \underline{\text{if}}\ s=\tfrac12(b-1)\\ (\underline{a},(s-\tfrac12(b-1))\underline{b};s) & \underline{\text{if}}\ s\neq\tfrac12(b-1)\end{cases}$$

One sees that E_2 is a bundle with holomorphic transition matrices. The connection with the formulas in proposition 7.5.3 becomes clear after defining the following map from E_2 into standard models with values in E.

7.5.7. Definition. If $\tau = 0,1$, $\delta = \pm 1$, then St_τ^δ is defined by

$$St_\tau^\delta(\underline{a},\underline{b};s) = \begin{cases} \underline{a}\ St_\tau^\delta(s) + \underline{b}\ St_\tau^{\delta(-1)^\tau}(-s)\iota_\tau^\delta(s) & \text{if } s \in U_g \\ \underline{a}\ St_\tau^\delta(\tfrac12(b-1)) + \underline{b}\ St_\tau^{\delta(-1)^\tau}(\tfrac12(1-b))\ \operatorname*{res}_{s\to\frac12(b-1)} \iota_\tau^\delta(s) & \text{if } s=\tfrac12(b-1),\ b\in Y_\tau \\ \underline{a}\ St_0^\delta(0) + \underline{b}\ L_0^\delta & \text{if } s=0,\ \tau=0 \end{cases}$$

If $0 < \tau < 1$ then

$$St_\tau(\underline{a},\underline{b};s) = \begin{cases} \underline{a}\ St_\tau^1(s) + \underline{b}\ St_\tau^1(-s)\iota_\tau^1(s) & \text{if } s \in U_g \\ \underline{a}\ St_\tau^1(\tfrac12(b-1)) + \underline{b}\ St_\tau^1(\tfrac12(1-b))\ \operatorname*{res}_{s\to\frac12(b-1)} \iota_\tau^1(s) & \text{if } s=\tfrac12(b-1),\ b\equiv-\tau\ (2) \\ \underline{a}\ St_\tau^1(0) + \underline{b}\ L_\tau^1 & \text{if } s = 0 \end{cases}$$

The definition of $St_\tau^{(\delta)}$ has been given in such a way that for all $T \in A(H_\tau^\delta(s))$, if not $\tau = 1$ and $s = 0$:

$$(7.5.8) \qquad {}^*FT = St_\tau^{(\delta)}(\underline{a}_\tau^{(\delta)}(s,T),\underline{b}_\tau^{(\delta)}(s,T);s).$$

7.5.8. Definition. Let $\tau = 1$, $\delta = \pm 1$. For $s \in U_0$, $s \neq 0$, $\underline{a},\underline{b} \in E$ the model $\widehat{St}_1^\delta(\underline{a},\underline{b};s)$ of $H_1^\delta(s)\oplus H_1^\delta(-s)$ is given by

$$(7.5.9) \qquad \widehat{St}_1^\delta(\underline{a},\underline{b};s) = (\underline{a}\ St_1^\delta(s) + \underline{b}\ St_1^{-\delta}(-s)\iota_1^\delta(s)) \oplus$$
$$\oplus\ (\underline{b}\ St_1^\delta(-s) + \underline{a}\ St_1^{-\delta}(s)\iota_1^\delta(-s));$$

for $s = 0$ the model $\widehat{St}_1^\delta(\underline{a},\underline{b};0)$ of E_1^δ is given by

$$(7.5.10) \qquad \widehat{St}_1^\delta(\underline{a},\underline{b};0) = \underline{a}(L_1^\delta + L_1^{-\delta}\iota_E^\delta) + \underline{b}(St_1^\delta(0)p_E^\delta + St_1^{-\delta}(0)p_E^{-\delta}\iota_E^\delta).$$

Let $s \in U_0$, $s \neq 0$; then for each $T \in A(H_\tau^\delta(s))$ there is $\tilde T \in A(H_\tau^\delta(-s))$ such that ${}^R(T\phi_1^\delta(s))^+ = {}^R(\tilde T\phi_1^\delta(-s))^+$; see proposition 4.5.6. From (7.5.2) follows that

$$(7.5.11) \qquad \underline{a}(-s,\tilde T) = \Gamma(-s)\Gamma(s)^{-1}\ \underline{b}(s,T)$$
$$\underline{b}(-s,\tilde T) = \Gamma(-s)\Gamma(s)^{-1}\ \underline{a}(s,T).$$

Definition 7.5.8. has been given in such a way that

$$(7.5.12) \qquad {}^*F(T\oplus\Gamma(s)\Gamma(-s)^{-1}\ \tilde T) = \widehat{St}_1^\delta(\underline{a}(s,T),\underline{b}(s,T);s).$$

7.5.9. Proposition. Let ψ be a constant section of H_τ^δ and let $\underline{q}$ be a continuous section of E_2 on some subset U of $\mathbb{C}$, not containing 0 in the case $\tau = 1$.
Then the function $s \mapsto St_\tau^{(\delta)}(\underline{q}(s))\psi(g)$ is continuous on U, uniformly for g in sets of the form $N\{a(y)\,|\,y_1 \leq y \leq y_2\}K$.
If $\tau = 1$, $\underline{q}$ is a continuous section of E_2 on U_0 and ψ is a section ϕ_r^δ or $\hat\phi_r^\delta$ as given in (3.3.28), then the function $s \mapsto \widehat{St}_\tau^\delta(\underline{q}(s))\psi(g)$ is continuous on U_0, uniform on the same subsets of G.
The proof comes down to straightforward checking that the definitions are right.

7.5.10. Definition. $N \subset E_2$ is defined by

$$N(s) = \{(\underline{a}_\tau^{(\delta)}(s,T),\underline{b}_\tau^{(\delta)}(s,T);s)\,|\,T \in A(H_\tau^{(\delta)}(s))\},$$

unless $s = 0$, $\tau = 1$; and if $\tau = 1$:

$$N(0) = \{\underline{n} \in E_2(0) \mid \underline{\text{there is a}}\ T \in A(E_1^{\delta})\ \underline{\text{for which}}\ {}^*FT = \widehat{\underline{St}}_1^{\delta}(\underline{n})\}.$$

In general $N(s)$ consists of those points of $E_2(s)$ which actually occur as Fourier coefficients of automorphic models. In the case $s = 0$, $\tau = 1$ we only consider automorphic models of E_1^{δ} with a Fourier coefficient of order zero of a prescribed form.

7.6. Maass-Selberg relation.

This relation restricts the Λ^0-Fourier coefficients of automorphic models of $H_\tau^{(\delta)}(s)$. The idea of the Maass-Selberg relation in a simple situation (weight zero, modular group) is explained on p. 331-333 of [22]. I do not give a proof but use the results of Roelcke in [37], §9.

7.6.1. Proposition (Maass-Selberg relation). Let $s \in \mathbb{C}$, $\tau \in [0,1]$, $\delta = \pm 1$ if $\tau = 0,1$, $s \neq 0$ if $\tau = 1$. Then for each $T \in A(H_\tau^{(\delta)}(s))$ and each $\hat{T} \in A(H_\tau^{(\delta)}(\bar{s}))$:

$$((\underline{a}_\tau^{(\delta)}(s,T), \underline{b}_\tau^{(\delta)}(\bar{s},\tilde{T}))) = ((\underline{b}_\tau^{(\delta)}(s,T), \underline{a}_\tau^{\delta}(\bar{s},\tilde{T}))). \tag{7.6.1}$$

Proof. Let $T \in H_\tau^{(\delta)}(s)$, $r \equiv \tau(2)$ such that $\phi_r^{(\delta)}(s)$ generates $H_\tau^{(\delta)}(s)$. Take $f = {}^R(T\phi_r^{(\delta)}(s))^+$. We have seen in (4.5.15) that f is an automorphic form in the sense of Roelcke.

Take $\lambda \in \Lambda^0$ and use the notation $A = \overline{g_\lambda^{-1}}$ and $z_A = Az$, as in 7.4. The constant term in the Fourier series expansion of f at the cusp c_λ is

$$e^{-ir\arg(cz+d)}(\underline{\underline{b}}_\lambda y_A^{\frac12+s} + \underline{\underline{c}}_\lambda y_A^{\frac12-s}) \quad \text{if } s \neq 0 \tag{7.6.2}$$

$$e^{-ir\arg(cz+d)}(\underline{\underline{b}}_\lambda + \underline{\underline{c}}_\lambda \ln y_A)\, y_A^{\frac12} \quad \text{if } s = 0$$

with

$$\sum_{\lambda\in\Lambda^0} \underline{\underline{b}}_\lambda \underline{e}_\lambda = \begin{cases} \underline{a}_\tau^{(\delta)}(s,T) & \text{if } s \neq 0 \\ \underline{a}_\tau^{(\delta)}(0,T) + a_r\, \underline{b}_0^{(\delta)}(0,T) & \text{if } s = 0,\ \tau \neq 1 \end{cases} \tag{7.6.3}$$

$$\sum_{\lambda\in\Lambda^0} \underline{\underline{c}}_\lambda \underline{e}_\lambda = \begin{cases} \underline{b}_\tau^{(\delta)}(s,T)\, \dfrac{\Gamma(\frac12-s-\frac12 r)}{\Gamma(\frac12+s-\frac12 r)} & \text{if } s \in U^g \\ \underline{b}_\tau^{(\delta)}(s,T)\, \dfrac{(-1)^{\frac12(b+r)}}{\Gamma(\frac12(b+1))\Gamma(\frac12(b-r))} & \text{if } b = 2s+1 \in Y_\tau,\ b \equiv -\tau(2) \\ \underline{b}_\tau^{(\delta)}(0,T) & \text{if } s = 0,\ \tau \neq 1. \end{cases}$$

This follows from proposition 7.5.3. The condition that $\phi_r^{(\delta)}(s)$ generates $H_\tau^{(\delta)}(s)$ implies that $|r| \leq b-2$ if $\tau = 0,1$ and $b = 2s+1 \in Y_\tau$ and that $r \geq 2-b$ if $0 < \tau < 1$ and $b = 2s+1 \equiv -\tau(2)$.

The $\underline{\underline{b}}_\lambda$ and $\underline{\underline{c}}_\lambda$ are the coefficients in (2.12) and (2.16) of [36] with the r of Roelcke equal to is.

For $\hat{T} \in A(H_\tau^{(\delta)}(\bar{s}))$ we define $\hat{\underline{\underline{b}}}_\lambda$ and $\hat{\underline{\underline{c}}}_\lambda$ similarly. Formula (9.7) of [37] states:

$$\sum_{\lambda\in\Lambda^0} (\bar{\underline{\underline{c}}}_\lambda \hat{\underline{\underline{b}}}_\lambda - \bar{\underline{\underline{b}}}_\lambda \hat{\underline{\underline{c}}}_\lambda) = 0. \tag{7.6.4}$$

Comparison with (7.6.3) gives (7.6.1).

Remark. The Maass-Selberg relation also applies to the case $\tau = 1$, $s = 0$. Indeed, let $T, T_1 \in A(E_1^\delta)$ and suppose

$$(7.6.5) \qquad {}^*FT = \underline{a}\, L_1^\delta + \underline{b}\, L_1^{-\delta} \iota_E^\delta + \underline{c}\, St_1^\delta(0) p_E^\delta + \underline{d}\, St_1^{-\delta}(0) p_E^{-\delta} \iota_E^\delta,$$

and for T_1 the same with $\underline{a}_1, \underline{b}_1, \underline{c}_1, \underline{d}_1 \in \mathcal{E}$.
Applying Roelcke's formula (9.7) to ${}^R(T\hat{\phi}_1^\delta(0))^+$ and ${}^R(T\hat{\phi}_1^\delta(0))^-$ we get

$$(7.6.6) \qquad \begin{aligned} &((\underline{a},\underline{c}_1)) + ((\underline{b},\underline{d}_1)) = ((\underline{c},\underline{a}_1)) + ((\underline{d},\underline{b}_1)) \\ &((\underline{a},d_1)) + (\underline{b},\underline{c}_1)) + 2\Gamma'(1)((\underline{a},\underline{b}_1)) = ((\underline{d},\underline{a}_1)) + ((\underline{c},\underline{b}_1)) + \\ &\qquad + 2\Gamma'(1)((\underline{b},\underline{a}_1)). \end{aligned}$$

7.6.2. Reformulation. The Maass-Selberg relation may be stated in the language of the bundle $\mathcal{E}_2$. Define a duality between $\mathcal{E}_2(s)$ and $\mathcal{E}_2(\bar{s})$ by

$$(7.6.7) \qquad (((\underline{a},\underline{b};s),(\hat{\underline{a}},\hat{\underline{b}};\bar{s}))) = ((\underline{a},\hat{\underline{b}})) - ((\underline{b},\hat{\underline{a}})).$$

This duality is not continuous in s: if $\underline{q}$ is a section of $\mathcal{E}_2$ over $U \subset \mathbb{C}$ and $\hat{\underline{q}}$ a section over $\bar{U}$, then the function

$$(7.6.8) \qquad s \mapsto ((\underline{q}(s),\hat{\underline{q}}(\bar{s})))$$

is not necessarily continuous at the points in $\mathbb{C}\setminus U^g$. But the zero set is closed. The Maass-Selberg relation may be written as

$$(7.6.9) \qquad ((N(s),N(\bar{s}))) = 0,$$

at least valid if not $\tau = 1$ and $s = 0$. Comparison of (7.5.10), definition 7.5.10 and (7.6.6) shows that it holds for $\tau = 1$, $s = 0$ as well.

8. The operators Θ_α.

8.1. Introduction.

In the theory of holomorphic automorphic forms the Fourier coefficients of cusp forms may be obtained by taking the Petersson scalar product with Poincaré series; see e.g. [23], p. 286. This may be generalized. In this section I study an operator Θ_α for each $\alpha \in \Lambda$. They turn left-N-equivariant functions into left-Γ-equivariant ones. The definition is given in subsection 8.2. In 8.3 I give an adjointness relation for the operators F_α and Θ_α. In the proof of the sum formula it is essential to have an expression for $F_\alpha \Theta_\beta$, with $\alpha,\beta \in \Lambda^0$. In Kuznetsov's proof of the sum formula, [15], it occurs as a formula for the Petersson scalar product of two Poincaré series in the sense of Selberg, [15], §2.2. In [1] I use a result of Niebur, [27], giving the Fourier coefficients of real analytic Poincaré series of weight zero. The expression for $F_\alpha \Theta_\beta$ is given in subsection 8.5. In this expression Kloosterman sums occur. These sums are generalizations of the Kloosterman sums

(8.1.1) $$\sum_{x \bmod c} e^{2\pi i c^{-1}(nx+m\tilde{x})},$$

with $x\tilde{x} \equiv 1(c)$. These last sums are connected with the modular group and $\tau = 0$. In subsection 8.4 I discuss Kloosterman sums for general Γ and τ.

8.2. Definition of Θ_α.

In this subsection $\alpha \in \Lambda$ is fixed. Take $u = 0$ if $\alpha \in \Lambda^0$ and $u = \frac{1}{2}$ if $\alpha \in \Lambda^1$. Consider a function f on G satisfying

(8.2.1) $$f(n(x)g) = e^{iux} f(g).$$

Then the function f_α, given by

(8.2.2) $$f_\alpha(g) = f(h_\alpha^{-1} g),$$

with h_α as in 7.3.1, satisfies

(8.2.3) $$f_\alpha(ng) = \chi_\alpha(n) f_\alpha(g) \qquad \text{for all } n \in N_\alpha,$$

see (7.3.6).

So, in particular, f_α is left-Γ_α-χ-equivariant. One might try to define a left-Γ-χ-equivariant function by

(8.2.4) $$\sum_{\gamma \in \Gamma_\alpha \backslash \Gamma} f_\alpha(\gamma g)\chi(\gamma)^{-1},$$

but this will never converge as the infinite center Z_0 is contained in Γ. We have to make f_α left-Δ_α-χ-equivariant.
All functions we want to consider are of type τ, i.e. they transform on the right according to the characters

(8.2.5) $\quad k(\pi 1) \mapsto e^{\pi i \tau 1}$ and $k(\pi 1) \mapsto e^{-\pi i \tau 1}$

of Z_0. Hence they transform on the left according to the same characters of Z_0.

8.2.1. Definition. Let $\varepsilon = \pm 1$. For each function f on G of type τ the function $P_\varepsilon f$ on G is defined by

(8.2.6) $\quad P_\varepsilon f = f \quad$ if $\tau = 0,1$

(8.2.7) $$\left.\begin{aligned} & f = P_1 f + P_2 f \\ & P_\varepsilon f(k(\pi 1)g) = e^{\pi i \tau \varepsilon 1} P_\varepsilon f(g) \text{ for all } k(\pi 1) \in Z_0 \end{aligned}\right\} \text{ if } 0 < \tau < 1.$$

Now if f satisfies (8.2.1) then the function

(8.2.8) $\quad g \mapsto P_{\varepsilon(\alpha)} f(h_\alpha^{-1} g)$

is left-Δ_α-χ-equivariant; see 7.3.1 for the definition of $\varepsilon(\alpha)$.

8.2.2. Definition. Let $\alpha \in \Lambda$ be given and let the function f on G of type τ satisfy (8.2.1). Then

(8.2.9) $$\Theta_\alpha f(g) = \sum_{\gamma \in \Delta_\alpha \backslash \Gamma} \chi(\gamma)^{-1} P_{\varepsilon(\alpha)} f(h_\alpha^{-1} \gamma g),$$

if this sum converges absolutely, uniformly on compact subsets of G.
If $\Theta_\alpha f$ is well defined, then it satisfies

(8.2.10) $\quad \Theta_\alpha f(\gamma g) = \chi(\gamma) \Theta_\alpha f(g) \quad$ for all $\gamma \in \Gamma$.

8.2.3. Proposition. Let α and f be as in 8.2.2. Suppose that

(8.2.11) $\quad |f(a(y)k)| \ll y^{1+\varepsilon} \quad$ for $y \downarrow 0$

for some $\varepsilon > 0$, uniformly in $k \in K$.
Then $\Theta_\alpha f$ is well defined.
If c is a cusp, $c \neq c_\alpha$, then

(8.2.12) $\quad |\Theta_\alpha f(g_c n a(y) k| \ll 1 \quad$ for $y \to \infty$

uniformly in $n \in N$, $k \in K$.

(8.2.13) $\quad |\Theta_\alpha f(g_\alpha n a(y) k) - P_{\varepsilon(\alpha)} f(h_\alpha^{-i} g_\alpha a(y) k| \ll 1 \quad$ for $y \to \infty$

uniformly in $n \in N$, $k \in K$.
The constants in the estimates (8.2.12), (8.2.13) contain as a factor the constant in (8.2.11) describing the dependence on f.

Proof. To prove this result I translate the situation to the upper half plane. From (7.2.5) follows that

$$(8.2.14) \qquad |P_{\varepsilon(\alpha)} f(h_\alpha^{-1} \gamma p(z)k)| \ll (\operatorname{Im} A\bar{\gamma} z)^{1+\varepsilon} \text{ for } \operatorname{Im} A\bar{\gamma} z \searrow 0.$$

Here $A = \overline{g_\alpha^{-1}}$. So the series in (8.2.9) is absolutely estimated by

$$(8.2.15) \qquad \sum_{M \in \overline{\Delta_\alpha} \backslash \overline{\Gamma}} (\operatorname{Im} AMz)^{1+\varepsilon}.$$

This is an Eisenstein series of weight zero on the upper half plane, in the domain of convergence. The convergence, uniform on compact sets, follows from Satz 1 and Satz 2 of [30] or from theorem 3C of [23], p. 276.
If a compact set C in h is given then we can find $y_0 > 0$ such that $\operatorname{Im} AMz \leq y_0$ for all $M \in \overline{\Gamma}$ and all $z \in C$. This proves that $\Theta_\alpha f$ is well defined.
Moreover if we take $\operatorname{Im} z \geq y_1$ for some y_1 then $\operatorname{Im}(AM\bar{g}_c z) \leq y_2$ for all those z, for all $M \in \overline{\Gamma}$ for some $y_2 > 0$, if $c \neq c_\alpha$. Hence boundedness of

$$(8.2.16) \qquad \sum_{M \in \overline{\Delta_\alpha} \backslash \overline{\Gamma}} (\operatorname{Im} AM\bar{g}_c z)$$

for those z gives (8.2.12). The same argument works for $c = c_\alpha$, if we delete $M \in \bar{\Delta}_\alpha$.
The boundedness of the sum in (8.2.16) follows again from [30] or from results on Eisenstein series, e.g. [37], lemma 10.2. This gives (8.2.12) and almost (8.2.13). In the last case the sum over $\gamma \in \Delta_\alpha \backslash \Gamma$ with $\gamma \notin \Delta_\alpha$ is bounded.
Remark that Petersson, Lehner and Roelcke at the cited places do not sum over $\overline{\Delta}_\alpha \backslash \overline{\Gamma}$ but over $\overline{\Gamma}_\alpha \backslash \overline{\Gamma}$. As (8.2.15) gives an Eisenstein series of weight zero, this difference does not matter. To complete the proof of (8.2.13) remark that we know that

$$(8.2.17) \qquad |\Theta_\alpha f(g_\alpha n a(y)k) - P_{\varepsilon(\alpha)} f(h_\alpha^{-1} g_\alpha n a(y)k|$$

is bounded , for only the term with $\gamma \in \Delta_\alpha$ is missing.

Proposition 8.2.3 implies that Θ_α is well defined on a lot of $\underline{g}$-K-spaces. If $\alpha \in \Lambda^0$ and $\operatorname{Re} s > \frac{1}{2}$ then Θ_α is well defined on $St(s)H^\tau(s)$; in this way one obtains Eisenstein series. If $\alpha \in \Lambda^1$, and $\operatorname{Re} s > \frac{1}{2}$ then $\Theta_\alpha M(s)H^\tau(s)$ consist of real analytic Poincaré series, with in general exponential growth; see [26], [27]. I work out the example of holomorphic Poincaré series. Let $\alpha \in \Lambda^1$, $b \in Y_\tau$, $b > 2$.

Then Θ_α is well defined on D^bD_b. I shall now compute $D^b\phi_b[b]$.
From (5.4.14) follows

(8.2.18) $$D^b\phi_b[b](p(z)k(\theta)) = \sqrt{2}\, y^{\frac{1}{2}b} e^{\frac{1}{2}iz} e^{ib\theta}$$

$$D^b\phi_b[b](p(z)jk(\theta)) = 0$$

Clearly $P_{\varepsilon(\alpha)}D^b\phi_b\ [b] = 0$ if $b \not\equiv \varepsilon(\alpha)\tau(2)$. So in the case $0 < \tau < 1$ I assume $b \equiv \varepsilon(\alpha)\tau(2)$.
From (8.2.18) and the definition of h_α follows that $\Theta_\alpha D^b\phi_b\,[b]\,(p(z)j_{-\varepsilon(\alpha)}) = 0$.
With help of (7.2.5) we compute

(8.2.19) $$\Theta_\alpha D^b\phi_b[b](p(z)j_{\varepsilon(\alpha)}) =$$

$$\sqrt{2}(4\pi|n_\alpha|)^{\frac{1}{2}b} y^{\frac{1}{2}b} \sum_{M\in\bar{\Delta}_\alpha\backslash\bar{\Gamma}} \chi(\tilde{M})^{-1} e^{-ib\varepsilon(\alpha)(\arg_{(-\pi,\pi]}(c_Mz+d_M)+2\pi \underline{w}(A,M))}$$

$$\cdot\ |c_Mz+d_M|^{-b} e^{2\pi i n_\alpha \mathrm{Re}\, AMz - 2\pi|n_\alpha|\mathrm{Im}\, AMz}.$$

Here I use the notation

(8.2.20) $$AM = \begin{pmatrix} * & * \\ c & d \end{pmatrix}.$$

In the case $\varepsilon(\alpha) = 1$ we get, with v^+ and v^- as in (4.4.3),

(8.2.21) $$(\Theta_\alpha D^b\phi_b[b])^+(z) =$$

$$\sqrt{2}(4\pi n_\alpha)^{\frac{1}{2}b} \sum_{M\in\bar{\Delta}_\alpha\backslash\bar{\Gamma}} v^+(M)^{-1}(c_Mz+d_M)^{-b} e^{-2\pi i b\underline{w}(A,M)} e^{2\pi i n_\alpha AMz}.$$

In the case $\varepsilon(\alpha) = -1$:

(8.2.22) $$(\Theta_\alpha D^b\phi_b[b])^-(z) =$$

$$\sqrt{2}(4\pi|n_\alpha|)^{\frac{1}{2}b} \sum_{M\in\bar{\Delta}_\alpha\backslash\bar{\Gamma}} v^-(M)^{-1}(c_Mz+d_M)^{-b} e^{-2\pi i b\underline{w}(A,M)} e^{2\pi i|n_\alpha|AMz}.$$

These are indeed holomorphic Poincaré series; compare with [28], p. 172. Use that $v^\pm(M)^{-1}e^{-2\pi i b\underline{w}(A,M)} = v^\pm(A)v^\pm(AM)^{-1}$.

8.2.4. Proposition. Let $\alpha \in A$. Let V be contained in $C^\infty(G)$ and let it be a g-K-space with respect to the right action. Suppose all element of V satisfy (8.2.1) and the estimate (8.2.11). Then Θ_α maps V into $C^\infty(G)$ and intertwines the g-K-actions.

Proof. The point one may worry about is the differentiability and the g-action.

But as the convergence of the series for Θ_α is uniform on compact sets, $\Theta_\alpha f$ is differentiable and derivatives may be taken inside the sum.

8.3. Adjointness of Θ_α and F_α.

Before discussing any adjointness relation I have to fix the scalar products.

8.3.1. Scalar product for N-equivariant functions.
Θ_α acts on functions f on G satisfying

$$(8.3.1) \qquad f(n(x)g) = e^{iux} f(g)$$

with u as in 8.2. For measurable functions f, f_1 with this property:

$$(8.3.2) \qquad \langle f, f_1 \rangle = \begin{cases} \int_{Z_0 N\backslash G} f(g)\overline{f_1(g)}\,d\dot{g} & \text{if } \tau = 0,1 \\ \sum_{\varepsilon=\pm 1} \int_{Z_0 N\backslash G} P_\varepsilon f(g)\overline{P_\varepsilon f_1(g)}\,d\dot{g} & \text{if } 0 < \tau < 1 \end{cases}$$

provided $P_1 f.\overline{P_1 f_1}$ and $P_{-1} f.\overline{P_{-1} f_1}$ are integrable on $Z_0 N\backslash G$ with respect to $d\dot{g}$. This is for instance the case if

$$(8.3.3) \qquad |f(a(y)k)\overline{f_1(a(y)k)}| \ll \begin{cases} y^{1+\varepsilon} & \text{for } y \downarrow 0 \\ y^{1-\varepsilon} & \text{for } y \to \infty \end{cases}$$

uniformly in $k \in K$, for some $\varepsilon > 0$. It is also the case if $f, f_1 \in {}^uS_\tau$, see 3.6.

8.3.2. Scalar product for Γ-equivariant functions.
On the other hand we have functions ϕ on G satisfying

$$(8.3.4) \qquad \phi(\gamma g) = \chi(\gamma)\phi(g) \quad \text{for all } \gamma \in \Gamma.$$

The χ-equivariance implies the right Z_0-behaviour on the left, so we need not separate the cases $\tau = 0,1$ and $0 < \tau < 1$. For measurable ϕ and ϕ_1 satisfying (8.3.4):

$$(8.3.5) \qquad \langle \phi, \phi_1 \rangle = \int_{\Gamma\backslash G} \phi(g)\overline{\phi_1(g)}\,dg,$$

provided this integral converges absolutely.
Let us translate this to the upper half plane. Each $\gamma \in \Gamma$ may be written as $\gamma = \zeta\tilde{\tilde{\gamma}}$ with $\zeta \in Z_0$. Suppose $\bar{\gamma} = \begin{pmatrix} * & * \\ c & d \end{pmatrix}$; then

$$(8.3.6) \qquad \gamma p(z)k = p(\bar{\gamma}.z)\, k(-\arg_{(-\pi,\pi]}(cz+d))k.$$

So by (2.4.4)

(8.3.7) $$\langle \phi,\phi_1\rangle = \frac{1}{2\pi}\int_{\Gamma\backslash h}\sum_{\varepsilon=\pm 1}\int_0^{\pi}\phi(p(z)j_\varepsilon k(\theta))\overline{\phi_1(p(z)j_\varepsilon k(\theta))}\,d\theta\, y^{-2}dydx.$$

For ϕ with weight r and ϕ_1 with weight r_1 we get

(8.3.8) $$\langle \phi,\phi_1\rangle = \frac{1}{2\pi}\int_0^{\pi} e^{i(r-r_1)\theta}\,d\theta.$$

$$\cdot\int_{\Gamma\backslash h}\left({}^R\phi^+(z)\overline{{}^R\phi_1^+(z)} + \overline{{}^R\phi^-(z)}\,{}^R\phi_1^-(z)\right)y^{-2}dxdy.$$

For $\tau = 0,1$ we get

(8.3.9) $$\langle \phi,\phi_1\rangle = \tfrac{1}{2}\delta_{r,r_1}\left(({}^R\phi^+,{}^R\phi_1^+) + ({}^R\phi_1^-,{}^R\phi^-)\right)$$

in Roelcke's notation, [36], p. 302. If $0<\tau<1$ condition (8.3.4) implies that ϕ and ϕ_1 live on only one component of G.

(8.3.10) $$\langle \phi,\phi_1\rangle = \begin{cases}\tfrac{1}{2}\delta_{r,r_1}({}^R\phi^+,{}^R\phi_1^+) & \text{if } r\equiv\tau(2)\\[1ex] \tfrac{1}{2}\delta_{r,r_1}({}^R\phi_1^-,{}^R\phi^-) & \text{if } r\equiv-\tau(2)\end{cases}$$

If we use the functions $\phi^\pm,\phi_1^\pm$ given by (2.3.2), we may express $\langle\phi,\phi_1\rangle$ with help of the Petersson scalar product; cf. [23], p. 36 or p. 284.
The scalar product is well defined if for each cusp c

(8.3.11) $$|\phi(g_c na(y)k)\overline{\phi_1(g_c na(y)k)}| \ll y^{1-\varepsilon} \quad \text{for } y\to\infty$$

uniformly in $n\in N$, $k\in K$, for some $\varepsilon>0$.

8.3.3. <u>Proposition. (Adjointness relation)</u>. <u>Let</u> $\alpha\in\Lambda$.
<u>Let</u> f <u>be a function on</u> G, <u>satisfying</u> (8.3.1),

(8.3.12) $$|f(a(y)k)| \ll y^a \quad \text{for } y\searrow 0$$

<u>uniformly in</u> k, <u>for some</u> $a>1$, <u>and</u>

(8.3.13) $$|f(a(y)k)| \ll y^b \quad \text{for } y\to\infty$$

<u>uniformly in</u> k, <u>for some</u> $b\in\mathbb{R}$.
<u>Let</u> ϕ <u>be a function on</u> G, <u>satisfying</u> (8.3.4) <u>and</u>

(8.3.14) $$|\phi(g_c na(y)k)| \ll y^d \quad \text{for } y\to\infty$$

<u>uniformly in</u> $n\in N$, $k\in K$ <u>for some</u> $d\in\mathbb{R}$, <u>for each cusp</u> c.
<u>Suppose that</u> $b+d<1$.
<u>Then</u> $\langle\Theta_\alpha f,\phi\rangle$ <u>and</u> $\langle f,F_\alpha\phi\rangle$ <u>are well defined, and</u>

(8.3.15) $\qquad \langle \Theta_\alpha f,\phi\rangle = v_\alpha \langle f, F_\alpha \phi\rangle$

with

(8.3.16) $\qquad v_\alpha = 1$ if $\alpha \in \Lambda^0$, $v_\alpha = 4\pi |n_\alpha|$ if $\alpha \in \Lambda^1$.

Proof. From proposition 8.2.3, (8.3.3) and (8.3.11) follows that the integrals defining both scalar products converge. Moreover, if we replace everything in the definitions of $\Theta_\alpha f$ and $F_\alpha\phi$ by its absolute value we still have convergence. So we may desintegrate and reintegrate in the following way. Suppose that $P_{\varepsilon(\alpha)} f = f$:

(8.3.17) $\qquad \langle \Theta_\alpha f,\phi\rangle =$

$$\sum_{\gamma\in\Delta_\alpha\backslash\Gamma} \int_{\Gamma\backslash G} f(h_\alpha^{-1}\gamma g)\overline{\chi(\gamma)\phi(g)}\,dg =$$

$$\int_{\Delta_\alpha\backslash G} f(h_\alpha^{-1} g)\overline{\phi(g)}\,dg =$$

$$\int_{Z_0N_\alpha\backslash G} \int_{\Delta_\alpha\backslash Z_0N} f(h_\alpha^{-1} ng)\overline{\phi(ng)}\,dn\,d\dot g =$$

$$\int_{Z_0N\backslash G} \int_{\Gamma_\alpha\backslash N_\alpha} f(h_\alpha^{-1} ng_\alpha g)\overline{\phi(ng_\alpha g)}\,dn\,d\dot g =$$

$$\int_0^\infty \sum_{\varepsilon=\pm 1} \int_{Z_0\backslash K_0} \int_0^1 f(a(v_\alpha)j_{\varepsilon(\alpha)} n(x)a(y)j_\varepsilon k)$$

$$\cdot\ \overline{\phi(g_\alpha n(x)a(y)j_\varepsilon k)}\,y^{-2}\,dx\,dy\,dk =$$

$$\int_0^\infty \sum_{\varepsilon=\pm 1} \int_{Z_0\backslash K_0} \int_0^1 f(a(y)j_\varepsilon k)e^{2\pi i n_\alpha x}\overline{\phi(g_\alpha n(x)a(v_\alpha^{-1}y)j_{\varepsilon\varepsilon(\alpha)}k)}$$

$$\cdot\ v_\alpha y^{-2}\,dx\,dy\,dk =$$

$$v_\alpha \int_{Z_0N_0\backslash G} f(g) \int_{\Gamma_\alpha\backslash N_\alpha} \chi_\alpha(n)\overline{\phi(nh_\alpha g)}\,dn\,d\dot g =$$

$$v_\alpha\langle f, F_\alpha\phi\rangle\ .$$

If $P_{\varepsilon(\alpha)} f = 0$ we have $\langle \Theta_\alpha f,\phi\rangle = 0$, but also $F_\alpha\phi = P_{\varepsilon(\alpha)}F_\alpha\phi$, so $\langle f, F_\alpha\phi\rangle = 0$ as well.

8.4. Kloosterman sums.

Let α and $\beta \in \Lambda$ be given. From the Bruhat decomposition follows that G_0 is the disjoint union of $g_\alpha NH_0 g_\beta^{-1}$ and $g_\alpha Nw^{-1}H_0 Ng_\beta^{-1}$. As Γ is a discrete group:

(8.4.1) $$g_\alpha NH_0 g_\beta^{-1} \cap \Gamma = \begin{cases} \Delta_\alpha & \text{if } c_\alpha = c_\beta \\ \emptyset & \text{otherwise.} \end{cases}$$

Indeed, suppose $g = g_\alpha n a \zeta g_\beta^{-1} \in \Gamma$, with $n \in N$, $a \in A$, $\zeta \in Z_0$; then $gg_\beta n(1) g_\beta^{-1} g^{-1} = g_\alpha n(\alpha(a)) g_\alpha^{-1}$. This should be an element of Γ_α, so $\alpha(a) \in \mathbb{Z}$. But also $g^{-1} = g_\beta n_1 a^{-1} \zeta_1 g_\alpha^{-1} \in \Gamma$, so $\alpha(a)^{-1} \in \mathbb{Z}$. Hence $a = 1$.
The Bruhat decomposition is unique; therefore we may define :

8.4.1. Definition. Let $\alpha, \beta \in \Lambda$. For each $g \in g_\alpha Nw^{-1}H_0 Ng_\beta^{-1}$ the numbers $x'_{\alpha\beta}(g), x''_{\alpha\beta}(g) \in \mathbb{R}$, $c_{\alpha\beta}(g) > 0$ and $l_{\alpha\beta}(g) \in \mathbb{Z}$ are defined by

(8.4.2) $$g_\alpha^{-1} g g_\beta = n(x'_{\alpha\beta}(g)) w^{-1} k(\pi l_{\alpha\beta}(g)) a(c_{\alpha\beta}(g)^2) n(x''_{\alpha\beta}(g)).$$

Remark that

(8.4.3) $$\overline{g_\alpha^{-1} g g_\beta} = (-1)^{l_{\alpha\beta}(g)} \begin{pmatrix} c_{\alpha\beta}(g) x'_{\alpha\beta}(g) & c_{\alpha\beta}(g) x'_{\alpha\beta}(g) x''_{\alpha\beta}(g) - c_{\alpha\beta}(g)^{-1} \\ c_{\alpha\beta}(g) & c_{\alpha\beta}(g) x''_{\alpha\beta}(g) \end{pmatrix}$$

Right- or left-translation with $\zeta = k(\pi 1) \in Z_0$ leaves $x'_{\alpha\beta}$, $x''_{\alpha\beta}$ and $c_{\alpha\beta}$ invariant, and

(8.4.4) $$l_{\alpha\beta}(g\zeta) = l_{\alpha\beta}(\zeta g) = l_{\alpha\beta}(g) + 1.$$

Left translation with elements of Γ_α changes only $x'_{\alpha\beta}$ by an integer; similarly right translation with elements of Γ_β only changes $x''_{\alpha\beta}$. So $c_{\alpha\beta}$ is constant on double cosets $\Delta_\alpha \gamma \Delta_\beta$.

8.4.2. Definition. Let $\alpha, \beta \in \Lambda$.
$C_{\alpha\beta} = \{c_{\alpha\beta}(\gamma) \mid \gamma \in \Gamma \cap g_\alpha Nw^{-1}H_0 Ng_\beta^{-1}\}$.

Example. Let $\tilde{\Gamma} = Sl_2(\mathbb{Z})$. Then $c_\alpha = c_\beta = \infty$.
For $\gamma = \begin{pmatrix} a & b \\ c & d \end{pmatrix}$ we find, if $c \neq 0$:

(8.4.5) $$x'_{\alpha\beta}(\gamma) = ac^{-1}; \; x''_{\alpha\beta}(\gamma) = dc^{-1}; \; c_{\alpha\beta}(\gamma) = |c|;$$

$$l_{\alpha\beta}(\gamma) = \begin{cases} 0 & \text{if } c > 0 \\ 1 & \text{if } c < 0. \end{cases}$$

We find $C_{\alpha\beta} = \mathbb{N}$ and for each $c \in \mathbb{N}$:

$$(8.4.6) \qquad \{\gamma \in \Gamma \mid c_{\alpha\beta}(\gamma) = c\} = \bigcup_{\substack{0 \leq d \leq c-1 \\ (c,d)=1}} \Delta_\alpha \widetilde{\begin{pmatrix} a & b \\ c & d \end{pmatrix}} \Delta_\beta;$$

here $\begin{pmatrix} a & b \\ c & d \end{pmatrix} \in Sl_2(\mathbb{Z})$. We see that $c_{\alpha\beta}$ has the same value on only a finite number of double cosets. This is true in general:

8.4.3. Proposition. *Let $\alpha,\beta \in \Lambda$ and $c \in C_{\alpha\beta}$. There is a finite number of double cosets $\Delta_\alpha\gamma\Delta_\beta$ with $c_{\alpha\beta}(\gamma) = c$.*

This result follows from [23], p. 271. It will also appear as a corollary to proposition 8.5.1.

8.4.4. Definition. *Let $\tau \in [0,1]$, a discrete group Γ and a character χ as discussed in 4.2 and 4.3 be fixed. Then for $\alpha,\beta \in \Lambda$ and $c \in C_{\alpha\beta}$ the Kloosterman sum $S(\alpha,\beta;c)$ is defined by*

$$(8.4.7) \qquad S(\alpha,\beta;c) = \sum_{\substack{\gamma \in \Delta_\alpha\backslash\Gamma/\Delta_\beta \\ c_{\alpha\beta}(\gamma) = c}} e^{2\pi i(n_\alpha x'_{\alpha\beta}(\gamma)+n_\beta x''_{\alpha\beta}(\gamma))} \chi(\gamma)^{-1} e^{\pi i \tau l_{\alpha\beta}(\gamma)}$$

Remark that one always can choose γ in its double coset such that $l_{\alpha\beta}(\gamma) = 0$.

Example. Let again $\overline{\Gamma} = Sl_2(\mathbb{Z})$. Take $\alpha = (\infty,n)$, $\beta = (\infty,m)$.
Then for $c \geq 1$

$$(8.4.8) \qquad S(\alpha,\beta;c) = \sum_{\substack{0 \leq d \leq c-1 \\ (c,d)=1}} e^{2\pi i c^{-1}(na+md)}\, v^{+}\begin{pmatrix} a & b \\ c & d \end{pmatrix}^{-1}.$$

see (8.4.5) and (4.4.3). For each $d \in [0,c-1]$ with $(c,d) = 1$ one has chosen $\begin{pmatrix} a & b \\ c & d \end{pmatrix} \in Sl_2(\mathbb{Z})$.

8.4.5. Proposition. *Let $\alpha,\beta \in \Lambda$. Then $C_{\alpha\beta} = C_{\beta\alpha}$.*
For each $c \in C_{\alpha\beta}$

$$(8.4.9) \qquad \overline{S(\alpha,\beta;c)} = e^{-\pi i\tau} S(\beta,\alpha;c).$$

Proof. Remark that for $\gamma \in g_\alpha N w^{-1} H_0 N g_\beta^{-1} \cap \Gamma$,

$$(8.4.10) \qquad \gamma^{-1} = g_\beta n(-x''_{\alpha\beta}(\gamma)) w^{-1} k(\pi(1-l_{\alpha\beta}(\gamma)) a(c_{\alpha\beta}(\gamma)^2) n(-x'_{\alpha\beta}(\gamma)) g_\alpha^{-1}.$$

This implies the equality of $C_{\alpha\beta}$ and $C_{\beta\alpha}$. Further,

(8.4.11) $\overline{S(\alpha,\beta;c)} =$

$$\sum_{\substack{\gamma\in\Delta_\alpha\backslash\Gamma/\Delta_\beta \\ c_{\alpha\beta}(\gamma) = c}} e^{2\pi i(-n_\alpha x'_{\alpha\beta}(\gamma)-n_\beta x''_{\alpha\beta}(\gamma))}\chi(\gamma)e^{-\pi i\tau 1_{\alpha\beta}(\gamma)} =$$

$$\sum_{\substack{\gamma^{-1}\in\Delta_\beta\backslash\Gamma/\Delta_\alpha \\ c_{\beta\alpha}(\gamma^{-1}) = c}} e^{2\pi i(n_\beta x'_{\beta\alpha}(\gamma^{-1})+n_\alpha x''_{\beta\alpha}(\gamma^{-1}))}\chi(\gamma^{-1})^{-1}e^{\pi i\tau(1_{\beta\alpha}(\gamma^{-1})-1)}$$

$$= e^{-\pi i\tau}S(\beta,\alpha;c).$$

8.5. The composition $F_\beta\Theta_\alpha$.

Let $\alpha,\beta\in\Lambda$ be fixed. Let f be a function on G satisfying the conditions (8.2.1) and (8.2.11). Then $\Theta_\alpha f$ and $F_\beta\Theta_\alpha f$ are well defined; everything still converges if we take absolute values everywhere. So in computing $F_\beta\Theta_\alpha f$ the order of integration and summation may be changed.

(8.5.1) $F_\beta\Theta_\alpha f(g) =$

$$\int_{\Gamma_\beta\backslash N_\beta}\overline{\chi_\beta(n)}\sum_{\gamma\in\Delta_\alpha\backslash\Gamma}\overline{\chi(\gamma)}P_{\varepsilon(\alpha)}f(h_\alpha^{-1}\gamma nh_\beta g)dn$$

$$= \sum_{\gamma\in\Delta_\alpha\backslash\Gamma}\overline{\chi(\gamma)}\int_{\Gamma_\beta\backslash N_\beta}\overline{\chi_\beta(n)}P_{\varepsilon(\alpha)}f(h_\alpha^{-1}\gamma nh_\beta g)dn.$$

This sum is split in two parts, according to $g_\alpha^{-1}G_0g_\beta = NH_0 \mathbin{\underline{\cup}} Nw^{-1}H_0N$. The first part only occurs if $c_\alpha = c_\beta$; its contribution is

(8.5.2) $$\delta_{c_\alpha,c_\beta}\int_{\Gamma_\alpha\backslash N_\alpha}\overline{\chi_\beta(n)}P_{\varepsilon(\alpha)}f(h_\alpha^{-1}nh_\beta g)dn$$

$$= \delta_{c_\alpha c_\beta}\int_{\Gamma_\alpha\backslash N_\alpha}\overline{\chi_\beta(n)}\chi_\alpha(n)dn\ P_{\varepsilon(\alpha)}f(h_\alpha^{-1}h_\beta g)$$

$$= \delta_{c_\alpha c_\beta}\delta_{n_\beta n_\alpha}P_{\varepsilon(\alpha)}f(h_\alpha^{-1}h_\beta g)$$

$$= \delta_{\alpha\beta}P_{\varepsilon(\alpha)}f(g).$$

The contribution of the other part is

(8.5.3) $$\sum_{\substack{\gamma\in\Delta_\alpha\backslash\Gamma/\Gamma_\beta\\ \gamma\notin g_\alpha NH_0 g_\beta^{-1}}}\ \sum_{\delta\in\Gamma_\beta} \overline{\chi(\gamma)}\,\overline{\chi(\delta)} \int_{\Gamma_\beta\backslash N_\beta} \overline{\chi_\beta(n)}\, P_{\varepsilon(\alpha)} f(h_\alpha^{-1}\gamma\delta n h_\beta g)\,dn$$

$$= \sum_{\substack{\gamma\in\Delta_\alpha\backslash\Gamma/\Delta_\beta\\ \gamma\notin g_\alpha NH_0 g_\beta^{-1}}} \overline{\chi(\gamma)} \int_{N_\beta} \overline{\chi_\beta(n)}\, P_{\varepsilon(\alpha)} f(h_\alpha^{-1}\gamma n h_\beta g)\,dn$$

$$= \sum_{c\in C_{\alpha\beta}}\ \sum_{\substack{\gamma\in\Delta_\alpha\backslash\Gamma/\Delta_\beta\\ c_{\alpha\beta}(\gamma)=c}} \overline{\chi(\gamma)} \int_{-\infty}^{+\infty} e^{-2\pi i n_\beta x}\, P_{\varepsilon(\alpha)} f(h_\alpha^{-1}\gamma g_\beta n(x) g_\beta^{-1} h_\beta g)\,dx$$

$$= \sum_{c\in C_{\alpha\beta}}\ \sum_{\substack{\gamma\in\Delta_\alpha\backslash\Gamma/\Delta_\beta\\ c_{\alpha\beta}(\gamma)=c}} \chi(\gamma)^{-1} e^{2\pi i(n_\beta x''_{\alpha\beta}(\gamma)+n_\alpha x'_{\alpha\beta}(\gamma))}$$

$$\cdot\, e^{\pi i\tau l_{\alpha\beta}(\gamma)} \int_{-\infty}^{+\infty} e^{2\pi i n_\beta x}\, P_{\varepsilon(\alpha)} f(h_\alpha^{-1} g_\alpha w^{-1} a(c_{\alpha\beta}(\gamma)^2) n(x) g_\beta^{-1} h_\beta g)\,dx$$

$$= \sum_{c\in C_{\alpha\beta}} S(\alpha,\beta;c) \int_{-\infty}^{+\infty} e^{2\pi i n_\beta \varepsilon(\beta) v_\beta^{-1} x}$$

$$\cdot\, P_{\varepsilon(\alpha)} f(j_{\varepsilon(\alpha)} a(v_\alpha) w^{-1} a(c^2) n(\varepsilon(\beta) v_\beta^{-1} x) a(v_\beta^{-1}) j_{\varepsilon(\beta)} g)\ \frac{dx}{v_\beta}$$

$$= v_\beta^{-1} \sum_{c\in C_{\alpha\beta}} S(\alpha,\beta;c) e^{\frac{1}{2}\pi i\tau(\varepsilon(\alpha)-1)}$$

$$\cdot \int_{-\infty}^{+\infty} e^{i(2\pi|n_\beta|/v_\beta)x}\, P_{\varepsilon(\alpha)} f\Big(w^{-1} j_{\varepsilon(\alpha)\varepsilon(\beta)}\, a\Big(\frac{c^2}{v_\alpha v_\beta}\Big)\, n(x) g\Big)\,dx$$

$$= v_\beta^{-1} \sum_{c\in C_{\alpha\beta}} S(\alpha,\beta;c) e^{\frac{1}{2}\pi i\tau(\varepsilon(\alpha)-1)}\ {}_{u(\beta)}V^{\varepsilon(\alpha)\varepsilon(\beta)}(v_\alpha v_\beta c^{-2}) P_{\varepsilon(\alpha)} f(g).$$

So we get:

8.5.1. Proposition. Let $\alpha,\beta \in \Lambda$. Then on functions satisfying (8.2.1) and (8.2.11):

$$(8.5.4)\qquad F_\beta\Theta_\alpha = \delta_{\alpha\beta}P_{\varepsilon(\alpha)}$$

$$+\ v_\beta^{-1}e^{\frac{1}{2}\pi i\tau(\varepsilon(\alpha)-1)} \sum_{c\in C_{\alpha\beta}} S(\alpha,\beta;c)\,{}_{u(\beta)}V^{\varepsilon(\alpha)\varepsilon(\beta)}(v_\alpha v_\beta c^{-2})P_{\varepsilon(\alpha)}$$

<u>with</u>

$$(8.5.5)\qquad u(\beta) = 2\pi|n_\beta|v_\beta^{-1} = \begin{cases} 0 & \text{if } \beta\in\Lambda^0 \\ \frac{1}{2} & \text{if } \beta\in\Lambda^1. \end{cases}$$

<u>The series, and the integral defining</u> ${}_{u(\beta)}V^{\varepsilon(\alpha)\varepsilon(\beta)}(v_\alpha v_\beta c^{-2})$ <u>are absolutely convergent for a given function satisfying</u> (8.2.1) <u>and</u> (8.2.11).

The absolute convergence follows from proposition 8.2.3 and the definition of F_β.

8.5.2. <u>Corollary</u>. <u>Let</u> $\alpha,\beta\in\Lambda$ <u>and</u> $c\in C_{\alpha\beta}$. <u>The number of cosets</u> $\Delta_\alpha\gamma\Delta_\beta$ <u>with</u> $c_{\alpha\beta}(\gamma) = c$ <u>is finite.</u>

<u>Proof</u>. $\Delta_\alpha\gamma\Delta_\beta$ only depends on the cusps c_α and c_β, so take $\alpha,\beta\in\Lambda^0$, $\chi = 1$, $\tau = 0$ and take f positive of weight 0, satisfying (8.2.1) and (8.2.11). Then ${}_{u(\beta)}V^{\varepsilon(\alpha)\varepsilon(\beta)}(c^{-2})P_{\varepsilon(\alpha)}f$ may be supposed to be nonzero. The sum in (8.5.4) only converges if thereare a finite number of cosets.

8.5.3. <u>Proposition</u>. <u>Let</u> $\alpha,\beta\in\Lambda$, <u>let</u> $\varepsilon > 0$. <u>Then</u>

$$(8.5.6)\qquad \sum_{c\in C_{\alpha\beta}} |S(\alpha,\beta;c)|\,c^{-2-\varepsilon} < \infty.$$

<u>Proof</u>. For each $c\in C_{\alpha\beta}$ it is clear that

$$(8.5.7)\qquad |S(\alpha,\beta;c)| \leqslant \#\,\{\gamma\in\Lambda_\alpha\backslash\Gamma/\Delta_\beta \mid c_{\alpha\beta}(\gamma) = c\}.$$

The number in the right hand side of (8.5.7) depends only on Γ and the cusps c_α and c_β. By taking χ trivial we may arrange that there are $\lambda,\mu\in\Lambda^0$ such that $c_\lambda = c_\alpha$, $c_\mu = c_\beta$. Now apply proposition 8.5.1 to $F_\mu\Theta_\lambda St(\frac{1}{2}+\frac{1}{2}\varepsilon)\phi_0^0(\frac{1}{2}+\frac{1}{2}\varepsilon)$. We get the convergence of

$$(8.5.8)\qquad \sum_{c\in C_{\alpha\beta}} \#\,\{\gamma\in\Delta_\alpha\backslash\Gamma/\Delta_\beta \mid c_{\alpha\beta}(\gamma) = c\}\ 2^{-2-\varepsilon}\Gamma(1+\varepsilon)$$

$$.\ \sin\pi(1+\tfrac{1}{2}\varepsilon)c^{-2-\varepsilon}St(-\tfrac{1}{2}-\tfrac{1}{2}\varepsilon)\iota(\tfrac{1}{2}+\tfrac{1}{2}\varepsilon)\phi_0^0(\tfrac{1}{2}+\tfrac{1}{2}\varepsilon)(g).$$

This proves proposition 8.5.3.

9. Eisenstein models.

9.1. Introduction.

The theory of Eisenstein series is well known in the case $\tau = 0$, $\chi = 1$; see e.g. [14], [6], [21], Ch. XIV. Much more general algebraic groups are discussed in [11], [22]; this covers our cases $\tau = 0,1$, $\chi = 1$. Roelcke treats the general case in [37], except that he does not go into the proof of the analytic continuation. As I want to prove the sum formula completely, I have to go into the theory of Eisenstein series deeply. I shall use the method of Selberg as may be found in [11], [22].

In this section I discuss Eisenstein series in the domain of convergence. I give the definition of the Eisenstein model in 9.2. The connection with Roelcke's Eisenstein series I discuss in 9.3. The Fourier coefficients of the Eisenstein model may be computed with help of propositions 8.5.1 and 6.8.2-3. This I do in 9.4 and 9.5.
Everything in this section is well known. I just state it in the language I have developed in the previous sections.

9.2 Definition of the Eisenstein model.

The operators Θ_λ with $\lambda \in \Lambda^0$, see definition 8.2.2., enable us to construct automorphic models. The space E, defined in 7.5.1, is used to handle all $\lambda \in \Lambda^0$ together.

9.2.1. Definition. If f is a function on G with values in E for which $\Theta_\lambda((f,\underline{e}_\lambda))$ is well defined for all $\lambda \in \Lambda^0$, then the function $^*\Theta f$ on G is defined by

$$^*\Theta f = \sum_{\lambda \in \Lambda^0} \Theta_\lambda((f,\underline{e}_\lambda)). \tag{9.2.1}$$

Now take $s \in \mathbb{C}$, $\mathrm{Re}\, s > \frac{1}{2}$. From proposition 8.2.3 and the properties of standard models follows that Θ_λ is well defined on $\mathrm{St}(s)\, H^\tau(s)$ for all $\lambda \in \Lambda^0$.

9.2.2. Definition. $^*H^\tau = H^\tau \otimes_{\mathbb{C}} E$.

The models St, W and M act in a natural way on $^*H^\tau$. St(s), W(s) and M(s) map $^*H^\tau(s)$ into the functions on G with values in E.

9.2.3. Definition. Let $\mathrm{Re}\, s > \frac{1}{2}$. The Eisenstein model $\underline{E}(s)$ of $^*H^\tau(s)$ is given by

$$\underline{E}(s) = {}^*\Theta\, \mathrm{St}(s). \tag{9.2.2}$$

$\underline{E}(s)$ maps $^*H^\tau(s)$ into the functions on G.

9.2.4. Notation. For each $\underline{e} \in E$ the map from $H^\tau(s)$ into the functions on G given by

$$(9.2.3) \qquad \psi \mapsto \underline{E}(s)\,\psi \otimes \underline{e}$$

is denoted $\underline{E}(s)\underline{e}$.

9.2.5. Proposition. Let Re $s > \frac{1}{2}$. For each $e \in E$ the map $E(s)\underline{e}$ is an automorphic model of $H^{\tau}(s)$.

Remark. I call $\underline{E}(s)\underline{e}$ an Eisenstein model of $H^{\tau}(s)$.

Remark. We may view $\underline{E}(s)$ as a linear map $E \to A(H^{\tau}(s))$.

Proof of 9.2.5. We have already seen that $\underline{E}(s)\underline{e}$ is well defined. It is clear that functions in $\underline{E}(s)\underline{e}\ H^{\tau}(s)$ satisfy the Γ-equivariance (4.5.1). From proposition 8.2.3 and the properties of St(s) follows that the growth condition (4.5.2) holds. Proposition 8.2.4 ensures that $\underline{E}(s)\underline{e}$ intertwines the $\underline{g}$-K-actions.

9.2.6. Proposition. Let ψ be a constant section of $*H^{\tau}$. For each $g \in G$ the function $s \mapsto \underline{E}(s)\,\psi(g)$ is holomorphic on Re $s > \frac{1}{2}$. If C is a compact subset of G and α is a bounded functional on C, then

$$(9.2.4) \qquad s \mapsto \alpha(\underline{E}(s)\psi)$$

is holomorphic on Re $s > \frac{1}{2}$.

Remark. The last assertion is important, for it implies the holomorphy of the Fourier coefficients.

Proof. Take ψ a constant section of H^{τ}, take $\lambda \in \Lambda^{0}$ and consider Θ_{λ} St(s) ψ. From definition 5.3.1 follows that $\frac{d}{ds}$ St(s) $\psi(s)$ satisfies condition (8.2.11) uniformly in s in any compact set contained in $\{z \in \mathbb{C} \mid \text{Re } z > \frac{1}{2}\}$.

$$(9.2.5) \qquad \sum_{\gamma} \chi(\gamma)^{-1}\, h^{-1}(P_1\, St(s+h) - P_1\, St(s))\, \psi(g_{\lambda}^{-1}\, \gamma\, g)$$
$$= \sum_{\gamma} \chi(\gamma)^{-1}\, P_1 \frac{d}{ds}\, St(s+h_{\gamma g})\, \psi(g_{\lambda}^{-1}\, \gamma\, g)$$

with $|h_{\gamma g}| \leq |h|$. So for small h the series in (9.2.5) converge absolutely, uniformly for g in compact sets, uniformly in h. So the limit for $h \to 0$ exists and may be computed term by term. This means that Θ_{λ} St(s) $\psi(g)$ is holomorphic in s. As everything is uniform for g in compact sets, the second assertion follows as well.

9.3. Eisenstein series on the upper half plane.

In this subsection I show that under the correspondence given in (2.3.5) the functions in $\underline{E}(s)\ *H^{\tau}(s)$ with fixed weight give rise to Eisenstein series as considered by Roelcke in §10 of [37].

Let Re $s > \frac{1}{2}$, $\lambda \in \Lambda^0$, $\delta = \pm 1$, $\phi_r^\delta(s) \in H^\tau(s)$. If $0 < \tau < 1$ then take $\delta = 1$, for $\underline{E}(s)\underline{e}_\lambda$ is zero on the other component of $H^\tau(s)$. Put $f = \underline{E}(s)\underline{e}_{-\lambda}\ \phi_r^\delta(s)$.
From proposition 9.2.5 and (4.5.8) follows that ${}^R f^+$ and ${}^{\bar R} f^-$ are automorphic forms in the sense of Roelcke.
Let $\eta = \pm 1$.

(9.3.1)
$$f(p(z)j_\eta) = \Theta_\lambda\ St(s)\ \phi_r^\delta(s)\ (p(z)j_\eta)$$
$$= \sum_{\gamma \in \Delta_\lambda \backslash \Gamma} \chi(\gamma)^{-1}\ P_1\ St(s)\ \phi_r^\delta(s)\ (g_\lambda^{-1}\ \gamma\ p(z)j_\eta)$$
$$= \sum_{M \in \bar\Delta_\lambda \backslash \bar\Gamma} \chi(\tilde M)^{-1}\ St(s)\ \phi_r^\delta(s)\ (p(AMz)\ k(-\arg_{(-\pi,\pi]}(cz+d)).$$
$$.\ k(-2\pi \underline{w}(A,M))j_\eta)$$

in the notation of (7.2.5), $A = \overline{g_\lambda^{-1}}$,
$$= \sum_{M \in \bar\Delta_\lambda \backslash \bar\Gamma} \chi(\tilde M)^{-1}\ e^{-ir\eta \arg_{(-\pi,\pi]}(cz+d) - 2\pi i r\eta \underline{w}(A,M)}\ .$$
$$.\ St(s)\ \phi_r^\delta(s)\ (p(AMz)j_\eta)$$
$$= \sum_{M \in \bar\Delta_\lambda \backslash \bar\Gamma} v^+(M)^{-1}\ \sigma_{\eta r}(A,M)^{-1}\ j_{AM}(z,\eta r)^{-1}\ .$$
$$.\ \operatorname{Im}(AMz)^{\frac{1}{2}+s}\ St(s)\ \phi_r^\delta(s)\ (j_\eta)$$

see (4.4.3) and (1.10), (1.6) of [36].
If $\tau = 0,1$ we get

(9.3.2)
$${}^R f^+(z) = f(p(z))$$
$$= E(z,\tfrac{1}{2}+s,1;\overline{g_\lambda^{-1}},\bar\Gamma,r,\mathbb{C},v^+)$$

see (10.1) of [37], and

(9.3.3)
$${}^R f^-(z) = \overline{f(p(z)j)}$$
$$= \sum_{M \in \bar\Delta_\lambda \backslash \bar\Gamma} v^-(M)^{-1}\ \sigma_r(A,M)^{-1}\ j_{AM}(z,r)^{-1}(\operatorname{Im} AMz)^{\frac{1}{2}+\bar s}\ (\delta,-1)$$
$$= \delta\ E(z,\tfrac{1}{2}+\bar s,1;\overline{g_\lambda^{-1}},\bar\Gamma,r,\mathbb{C},v^-).$$

Remark that Roelcke sums over $\bar\Gamma_\lambda \backslash \bar\Gamma$ and puts in a factor $\frac{1}{2}$.
In the case $0 < \tau < 1$ we have $r \equiv \pm\tau(2)$; take $s_1 = s$ if $r \equiv \tau(2)$ and $s_1 = \bar s$ if $r \equiv -\tau(2)$. We get

(9.3.4) $\qquad {}^{R}f^{\pm}(z) = \sqrt{2}\, E(z,\tfrac{1}{2}+s_1,1;\overline{g_\lambda^{-1},\bar{\Gamma},r,\mathbb{C},v^{\pm}}).$

9.4. Fourier coefficients of Eisenstein models, order zero.

In this section I study $*F\, \underline{E}(s)$. Let us first take s general, Re $s > \frac{1}{2}$. Clearly $*F\, \underline{E}(s)$ is a standard model of $*H^{\tau}$ with values in E. From proposition 7.5.3. follows that there are linear maps $A^{\pm 1}(s)$ and $B^{\pm 1}(s) : E \to E$, such that with respect to the decomposition induced by (3.5.3):

(9.4.1) $$*F\, \underline{E}(s) = St(s)\begin{pmatrix} A^{1}(s) & 0 \\ 0 & A^{-1}(s)\end{pmatrix} + St(-s)\,\iota(s)\begin{pmatrix} B^{1}(s) & 0 \\ 0 & B^{-1}(s)\end{pmatrix}.$$

If $0 < \tau < 1$ then $A^{-1}(s) = B^{-1}(s) = 0$.
We compute the maps $A^{\pm 1}(s)$, $B^{\pm 1}(s)$ by taking a matrix coefficient and applying propostion 8.5.1.
Let $\lambda, \mu \in \Lambda^{0}$, $\psi \in H^{\tau}(s)$.

(9.4.2) $$((*F\, \underline{E}(s)\psi \otimes \underline{e}_\lambda, \underline{e}_\mu)) =$$

$$F_\mu\, \Theta_\lambda\, St(s)\,\psi = \delta_{\lambda\mu}\, P_1\, St(s)\,\psi$$

$$+ \sum_{c \in C_{\lambda\mu}} S(\lambda,\mu;c)\; {}_0V^{1}(c^{-2})\, P_1\, St(s)\,\psi.$$

9.4.1. Definition. Let $\varepsilon = \pm 1$. The map $\pi_\varepsilon : H^{\tau} \to H^{\tau}$ is given by the maps $\pi_\varepsilon(s) : H^{\tau}(s) \to H^{\tau}(s)$:

(9.4.3) $$\pi_\varepsilon(s) = 1 \text{ if } \tau = 0,1$$

$$\pi_1(s) = \begin{pmatrix} 1 & 0 \\ 0 & 0\end{pmatrix},\; \pi_{-1}(s) = \begin{pmatrix} 0 & 0 \\ 0 & 1\end{pmatrix} \text{ if } 0 < \tau < 1.$$

As the matrix for $\pi_\varepsilon(s)$ does not depend on s I often write π_ε instead of $\pi_\varepsilon(s)$. Clearly $P_\varepsilon\, St = St\, \pi_\varepsilon$, so we get for the expression in (9.4.2):

(9.4.4) $$St(s)\,\delta_{\lambda\mu}\,\pi_1\,\psi +$$

$$2^{+1-2s}\Gamma(2s)\sin\pi(\tfrac{1}{2}+s+\tfrac{1}{2}\tau) \sum_{c \in C_{\lambda\mu}} S(\lambda,\mu;c)\, c^{-1-2s}\,.$$

$$.\; St(-s)\,\iota(s)\,X^{1}(s)\,\pi_1\,\psi,$$

see proposition 6.8.2.

9.4.2. Definition. For Re $s > \frac{1}{2}$ the linear map $\underline{M}(s) : E \to E$ is defined by

(9.4.5) $$((\underline{M}(s)\underline{e}_\lambda, \underline{e}_\mu)) = 2^{+1-2s}\,\Gamma(2s)\sin\pi(\tfrac{1}{2}+s+\tfrac{1}{2}\tau)\,.$$

$$.e^{-\frac{1}{2}\pi i\tau} \sum_{c \in C_{\lambda\mu}} S(\lambda,\mu;c)c^{-1-2s}$$

From proposition 8.5.3 follows that $\underline{M}(s)$ is holomorphic on Re $s > \frac{1}{2}$.
Definition 6.6.2 implies $X^1(s)\,\pi_1 = e^{-\frac{1}{2}\pi i\tau}\,\pi_1$. So from (9.4.4) follows that the maps $A^{\pm 1}(s)$ and $B^{\pm 1}(s)$ in (9.4.1) are given by

(9.4.6) $$A^1(s) = I,\ A^{-1}(s) = \begin{cases} I & \text{if } \tau = 0,1 \\ 0 & \text{if } 0 < \tau < 1 \end{cases}$$

$$B^1(s) = \underline{M}(s),\ B^{-1}(s) = \begin{cases} \underline{M}(s) & \text{if } \tau = 0,1 \\ 0 & \text{if } 0 < \tau < 1 \end{cases}$$

So we have proved:

9.4.3. Proposition. For s general, Re $s > \frac{1}{2}$:

(9.4.7) $$*F\ \underline{E}(s) = St(s)\,\pi_1 + St(-s)\,\iota(s)\,\underline{M}(s)\,\pi_1.$$

Remark that $St(-s)\,\iota(s)\,\underline{M}(s)\,\pi_1$ acts on $*H^\tau(s) = H^\tau(s) \otimes E$; $St(-s)$, $\iota(s)$ and π_1 act on the first factor, $\underline{M}(s)$ on the second one.

9.4.4. Proposition. For all s with Re $s > \frac{1}{2}$

(9.4.8) $$\underline{M}(s)^* = \underline{M}(\bar{s}),$$

the star denotes the adjoint with respect to $((\cdot,\cdot))$.

Proof. It is immediat from the definition of $\underline{M}(s)$ and proposition 8.4.5 that (9.4.8) holds for general s. As $\underline{M}$ is continuous it holds for all s with Re $s > \frac{1}{2}$.

Remark. Proposition 9.4.4 follows also from the Maass-Selberg relation, proposition 7.6.1. Indeed, if s is general and if we embed $H_\tau^{(\delta)}(s)$ in $H^\tau(s)$ in the standard way, then for all $\underline{e} \in E$

(9.4.9) $$\underline{a}_\tau^{(\delta)}(s,\ \underline{E}(s)\underline{e} \mid H_\tau^{(\delta)}(s)) = \underline{e}$$

$$\underline{b}_\tau^{(\delta)}(s,\ \underline{E}(s)\underline{e} \mid H_\tau^{(\delta)}(s)) = \underline{M}(s)\underline{e},$$

by proposition 9.4.3. Now (7.6.1) is equivalent to (9.4.8) for general s.

9.4.5. Reformulation. I reformulate proposition 9.4.3 in the language of the bundle E_2 introduced in 7.5.4. Let $V = \{s \in \mathbb{C} \mid s \text{ general, Re } s > \frac{1}{2}\}$. Remark that $V \subset U^g$.
Define the map $\underline{N} : V \times E \to E_2$ by

(9.4.10) $$\underline{N}(s,\underline{e}) = (\underline{e},\underline{M}(s)\underline{e};s).$$

This map is holomorphic in s and linear in $\underline{e}$.

Proposition 9.4.3 states that for $s \in V$, $\underline{e} \in E$:

$$(9.4.11) \qquad {}^*F\ \underline{E}(s)\underline{e} \mid H_\tau^{(\delta)}(s) = \underline{St}_\tau^{(\delta)}\ \underline{N}(s,\underline{e}).$$

From prososition 9.2.6 follows that $s \mapsto {}^*F\ \underline{E}(s)\ \psi(s) \otimes \underline{e}(g)$ is holomorphic on $\mathrm{Re}\ s > \frac{1}{2}$ for constant sections ψ of $H_\tau^{(\delta)}$. If we prove that $\underline{N}(s,\underline{e})$ may be extended holomorphically to nongeneral s with $\mathrm{Re}\ s > \frac{1}{2}$, then proposition 7.5.9 implies that (9.4.11) has to be valid for all s with $\mathrm{Re}\ s > \frac{1}{2}$.
Let $s_0 \in \mathbb{C}$, $\mathrm{Re}\ s_0 > \frac{1}{2}$, $s_0 \notin V$. If $s_0 \in U^g$ then from the holomorphy of $\underline{M}(s)$ follows that $\underline{N}$ may be extended to s_0. The only case left is $s_0 = \frac{1}{2}(b-1)$, $b > 2$, $b \equiv -\tau(2)$.
For $s \in U_b$, $s \neq s_0$:

$$(9.4.12) \qquad \Omega_b^{-1}\ \underline{N}(s,\underline{e}) = (\underline{e}, (s-s_0)^{-1}\ \underline{M}(s)\underline{e};\ s).$$

From the occurrence of the factor $\sin \pi(\frac{1}{2}+s+\frac{1}{2}\tau)$ in (9.4.5) follows that we may extend $\underline{N}$ by defining

$$(9.4.13) \qquad \underline{N}(\tfrac{1}{2}(b-1),\underline{e}) = \Omega_b(\underline{e},\ \underline{M}'(\tfrac{1}{2}(b-1));\ \tfrac{1}{2}(b-1)).$$

9.4.6. <u>Proposition</u>. <u>The map</u> $\underline{N}$ <u>given in</u> (9.4.10) <u>has a holomorphic extension to all</u> s <u>with</u> $\mathrm{Re}\ s > \frac{1}{2}$. <u>For all</u> s <u>with</u> $\mathrm{Re}\ s > \frac{1}{2}$ <u>and all</u> $\underline{e} \in E$

$$(9.4.14) \qquad {}^*F\ \underline{E}(s)\underline{e} \mid H_\tau^{(\delta)}(s) = \underline{St}_\tau^{(\delta)}\ \underline{N}(s,\underline{e}).$$

9.4.7. <u>Corollary</u>. <u>If</u> $b \equiv -\tau(2)$, $b > 2$, <u>then</u>

$$(9.4.15) \qquad {}^*F\ \underline{E}(\tfrac{1}{2}(b-1)) = St(\tfrac{1}{2}(b-1))\ \pi_1 + \underline{M}'(\tfrac{1}{2}(b-1))\ St(\tfrac{1}{2}(1-b)) \operatorname*{res}_{s\to\frac{1}{2}(b-1)} \iota(s))\ \pi_1$$

<u>If</u> $0 < \tau < 1$, $b \equiv \tau(2)$, $b > 2$, <u>then</u>

$$(9.4.16) \qquad {}^*F\ \underline{E}(\tfrac{1}{2}(b-1)) = St(\tfrac{1}{2}(b-1))\ \pi_1 + \underline{M}(\tfrac{1}{2}(b-1))\ St(\tfrac{1}{2}(1-b))\ \iota(\tfrac{1}{2}(b-1))\ \pi_1.$$

Remark that in the last case $\iota(s)\pi_1$ is holomorphic at $s = \frac{1}{2}(b-1)$.

9.5. <u>Fourier coefficients of Eisenstein models, non-zero order</u>.

The idea is the same as in the previous subsection.
Let s be general, $\mathrm{Re}\ s > \frac{1}{2}$, $\alpha \in \Lambda^1$, $\lambda \in \Lambda^0$. From propositions 8.5.1 and 6.8.3 and definition 6.6.2 follows that

$$(9.5.1) \qquad F_\alpha\ \underline{E}(s)\underline{e}_\lambda = (4\pi|n_\alpha|)^{s-\frac{1}{2}}\ 2^{+1-2s}\ \sin \pi(\tfrac{1}{2}+s+\tfrac{1}{2}\tau).$$
$$\sum_{c \in \mathcal{C}_{\lambda\alpha}} S(\lambda,\alpha;c)c^{-1-2s}\ W(s)\ X^{\varepsilon(\alpha)}(s)\ \pi_1$$

9.5.1. Definition. For $\alpha \in \Lambda^1$ and Re $s > \frac{1}{2}$

$$(9.5.2)\qquad \underline{f}_\alpha(s) = (4\pi|n_\alpha|)^{s-\frac{1}{2}}\, 2^{+1-2s} \sin\pi(\tfrac{1}{2}+s+\tfrac{1}{2}\tau)e^{-\frac{1}{2}\pi i\tau}$$

$$\cdot \sum_{\mu\in\Lambda^0}\ \sum_{c\in C_{\alpha\mu}} S(\alpha,\mu;c)c^{-1-2s}\ \underline{e}_\mu .$$

From proposition 8.5.3 follows that $\underline{f}_\alpha(s)$ is holomorphic on Re $s > \frac{1}{2}$. Using proposition 8.4.5 we get from (9.5.1):

9.5.2. Proposition. Let $\alpha \in \Lambda^1$, Re $s > \frac{1}{2}$, s general, $\underline{e} \in E$, $\phi \in H^\tau(s)$. Then

$$(9.5.3)\qquad F_\alpha \underline{E}(s)\phi \otimes \underline{e} = ((\underline{e}, f_\alpha(\bar{s})))W(s)X^{\varepsilon(\alpha)}(s)\pi_1\phi$$

9.5.3. Corollary. Let $\alpha \in \Lambda^1$, s general, Re $s > \frac{1}{2}$, $\underline{e} \in E$.
If $\tau = 0,1$ then

$$(9.5.4)\qquad \alpha(H_\tau^\delta(s), \underline{E}(s)\underline{e}\,|\,H_\tau^\delta(s)) =$$

$$(\delta,\varepsilon(\alpha))e^{-\frac{1}{2}\pi i\tau\varepsilon(\alpha)}((\underline{e},\underline{f}_\alpha(\bar{s}))).$$

If $0 < \tau < 1$ then

$$(9.5.5)\qquad \alpha(H_\tau(s), \underline{E}(s)\underline{e}\,|\,H_\tau(s)) =$$

$$c_{\varepsilon(\alpha)}(-s)e^{-\frac{1}{2}\pi i\tau}((\underline{e},\underline{f}_\alpha(\bar{s}))).$$

Proof. Use proposition 7.4.2 and definition 6.6.2.

9.5.4. Corollary. Let $\alpha \in \Lambda^1$, $b \in Y_\tau$, $b > 2$, $\underline{e} \in E$.
If $\tau = 0,1$ then

$$(9.5.6)\qquad \alpha(H_\tau^\delta(\tfrac{1}{2}(b-1)), \underline{E}(\tfrac{1}{2}(b-1))\underline{e}\,|\,H_\tau^\delta(\tfrac{1}{2}(b-1))) =$$

$$(\delta,\varepsilon(\alpha))e^{-\frac{1}{2}\pi i\tau\varepsilon(\alpha)}((\underline{e},\underline{f}'_\alpha(\tfrac{1}{2}(b-1)))).$$

If $0 < \tau < 1$, $b \equiv -\varepsilon(\alpha)\tau(2)$, then

$$(9.5.7)\qquad \alpha(H_\tau(\tfrac{1}{2}(b-1)), \underline{E}(\tfrac{1}{2}(b-1))\underline{e}\,|\,H_\tau(\tfrac{1}{2}(b-1))) =$$

$$\begin{cases} 0 & \text{if } \varepsilon(\alpha) = -1 \\ e^{-\frac{1}{2}\pi i\tau}((\underline{e},\underline{f}'_\alpha(\tfrac{1}{2}(b-1)))) & \text{if } \varepsilon(\alpha) = 1 \end{cases}$$

If $0 < \tau < 1$, $b \equiv \varepsilon(\alpha)\tau(2)$, then

(9.5.8) $\alpha(H_\tau(\frac{1}{2}(b-1)), \underline{E}(\frac{1}{2}(b-1))\underline{e}|H_\tau(\frac{1}{2}(b-1))) =$

$$\begin{cases} e^{-\frac{1}{2}\pi i\tau}((\underline{e},\underline{f}_\alpha(\frac{1}{2}(b-1)))) & \underline{\text{if}}\ \varepsilon(\alpha) = 1 \\ \pi^{-1}\sin\pi b\ e^{-\frac{1}{2}\pi i\tau}((\underline{e},\underline{f}'_\alpha(\frac{1}{2}(b-1)))) & \underline{\text{if}}\ \varepsilon(\alpha) = -1 \end{cases} .$$

Proof. The idea is to use the holomorphy of $s \mapsto F_\alpha \underline{E}(s)\phi(s) \otimes \underline{e}(g)$ for constant sections ϕ of $H_\tau^{(\delta)}(s)$. This holomorphy follows from proposition 9.2.6.
I work out the case $\tau = 0,1$. The following expression is holomorphic at $s = \frac{1}{2}(b-1)$:

(9.5.9) $(\delta,\varepsilon(\alpha))e^{-\frac{1}{2}\pi i\tau\varepsilon(\alpha)}((\underline{e},\underline{f}_\alpha(\overline{s})))W_\tau^\delta(s)\phi(s)(g)$

$$= (\delta,\varepsilon(\alpha))e^{-\frac{1}{2}\pi i\tau\varepsilon(\alpha)}((\underline{e},(\overline{s}-\tfrac{1}{2}(b-1))^{-1}\underline{f}_\alpha(\overline{s})))$$

$$.\ (s-\tfrac{1}{2}(b-1))W_\tau^\delta(s)\phi(s)(g).$$

As $\underline{f}_\alpha(\frac{1}{2}(b-1)) = 0$ the limit of (9.5.9) for $s \to \frac{1}{2}(b-1)$ is

(9.5.10) $(\delta,\varepsilon(\alpha))e^{-\frac{1}{2}\pi i\tau\varepsilon(\alpha)}((\underline{e},\underline{f}'_\alpha(\frac{1}{2}(b-1))))(\operatorname{res}_{s\to\frac{1}{2}(b-1)} W_\tau^\delta(s))\phi(\frac{1}{2}(b-1)).$

This proves (9.5.6).

9.5.5. A well known example.
To get some confidence in the computations just made, I consider the case $\overline{\Gamma} = Sl_2(\mathbb{Z})$, $\chi = 1$, $\tau = 0$. In this case Λ^0 has only one element, so $E = \mathbb{C}$ and $\underline{M}(s)$ is a scalar.

(9.5.11) $\underline{M}(s) = 2^{1-2s}\Gamma(2s)\sin\pi(\frac{1}{2}+s) \sum_{c=1}^{\infty} \sum_{\substack{1\leq d\leq c \\ (c,d)=1}} c^{-1-2s}$

$$= 2^{1-2s}\Gamma(2s)\sin\pi(\tfrac{1}{2}+s)\zeta(2s)\zeta(2s+1)^{-1}.$$

If $\alpha = (\omega,n) \subset \Lambda^1$, then

(9.5.12) $\underline{f}_\alpha(s) = (4\pi|n|)^{s-\frac{1}{2}}2^{1-2s}\sin\pi(\frac{1}{2}+s) \sum_{c=1}^{\infty} \sum_{\substack{1\leq d\leq c \\ (c,d)=1}} e^{2\pi i c^{-1}an} c^{-1-2s}$

with $ad \equiv 1(c)$

$$= (4\pi|n|)^{s-\frac{1}{2}}2^{1-2s}\sin\pi(\tfrac{1}{2}+s)\sigma_{-2s}(|n|)\zeta(2s+1)^{-1}$$

$$\sigma_a(1) = \sum_{d|1} d^a.$$

So, for $\delta = \pm 1$, from (7.4.4) and (7.5.2) follows

(9.5.13) $\quad {}^{R}(\underline{E}(s)\phi_0^\delta(s))^+(z) =$

$$\underline{a}_0^\delta(s,\underline{E}(s)|H_0^\delta(s))y^{\frac{1}{2}+s} + \underline{b}_0^\delta(s,\underline{E}(s)|H_0^\delta(s))\Gamma(\tfrac{1}{2}-s)\Gamma(\tfrac{1}{2}+s)^{-1}y^{\frac{1}{2}-s}$$

$$+ \sum_{n\neq 0} (\infty,n)(H_0^\delta(s),\underline{E}(s)|H_0^\delta(s))(\delta,\mathrm{sgn}\ n)\Gamma(\tfrac{1}{2}-s)W_{0,s}(4\pi|n|y)e^{2\pi inx}$$

$$= y^{\frac{1}{2}+s} + \pi^{\frac{1}{2}}\Gamma(s)\zeta(2s)\Gamma(s+\tfrac{1}{2})^{-1}\zeta(2s+1)^{-1}y^{\frac{1}{2}-s}$$

$$+ \pi^{s+\frac{1}{2}}\Gamma(s+\tfrac{1}{2})^{-1}\zeta(2s+1)^{-1} \sum_{n\neq 0} |n|^{s-\frac{1}{2}}\sigma_{-2s}(|n|)W_{0,s}(4\pi|n|y)e^{2\pi inx}$$

On the other hand

(9.5.14) $\quad {}^{R}(\underline{E}(s)\phi_0^\delta(s))^+(z) = \underline{E}(s)\phi_0^\delta(s)(p(z))$

$$= \Theta_{(\infty,0)}\mathrm{St}(s)\phi_0^\delta(s)(p(z))$$

$$= \sum_{\gamma\in\Delta_{(\infty,0)}\backslash\Gamma} \mathrm{St}_0^\delta(s)\phi_0^\delta(s)(\gamma p(z))$$

$$= \sum_{\gamma\in\overline{\Delta_{(\infty,0)}}\backslash\overline{\Gamma}} (\mathrm{Im}\ \gamma z)^{\frac{1}{2}+s}$$

This may be compared with (3.4) in [15]; use that $W_{0,s}(y) = \pi^{-\frac{1}{2}}y^{\frac{1}{2}}K_s(\tfrac{1}{2}y)$.

10. $L^2(\Gamma\backslash G, \chi)$

10.1. Introduction

Now we start the study of the Hilbert space $L^2(\Gamma\backslash G, \chi)$ of Γ-equivariant functions which are square integrable. G acts on it by right translations. The spectral decomposition of this representation will give one side of the sum formula. $L^2(\Gamma\backslash G,\chi)$ is defined in 10.2. In 10.3 I consider automorphic models the elements of which are situated in $L^2(\Gamma\backslash G, \chi)$. These square integrable automorphic models correspond to square integrable automorphic forms on the upper half plane, as I show in 10.4. The most intrigueing part of $L^2(\Gamma\backslash G, \chi)$ consists of the cusp forms, defined in 10.5. The representation of G is completely reducible in this cuspidal part. To prove this I study in 10.4 a convolution operator. The material in this section is known for rather general algebraic groups and Γ-invariant functions; see e.g. Langlands [22] or Harish Chandra [11]. The case I need is given by Roelcke [36],[37] in the language of functions on the upper half plane. I have reformulated it in the language of automorphic models. Moreover, the estimates in 10.5 have to be discussed anyhow for the analytical continuation of Eisenstein series.

10.2. Definitions

10.2.1. Definition. $C_c^\infty(\Gamma\backslash G, \chi)$ consists of the functions $f \in C^\infty(G)$ satisfying

(10.2.1) $f(\gamma g) = \chi(\gamma) f(g)$ for $\gamma \in \Gamma$, $g \in G$

(10.2.2) f is right-K-finite, of type τ

(10.2.3) $|f| \in C_c(\Gamma\backslash G)$.

The scalar product defined in (8.3.5) converges for elements of $C_c^\infty(\Gamma\backslash G, \chi)$. Under the pointwise action $C_c^\infty(\Gamma\backslash G, \chi)$ is a $\underline{g}$-K-space with unitary structure.

10.2.2 Definition. $L^2(\Gamma\backslash G, \chi)$ is the Hilbert space obtained by completion of $C_c(\Gamma\backslash G, \chi)$. If there is no danger of confusion, I shall write L^2 instead of $L^2(\Gamma\backslash G,\chi)$.

Remark. After choice of a fundamental domain F for $\Gamma\backslash G$ one may view L^2 as the space of equivalence classes of measurable functions of F, square integrable with respect to the Haar measure on G. With help of relation (10.2.1) these functions may be defined on G. The right translate by an element of G of such a function again represents an element of L^2. In this way we get a representation $\underline{r}$ of G in L^2; this representation is unitary.

For $k \in K$ and f representing an element of L^2 we have $(k.f)(g) = f(gk)$. This pointwise action corresponds to the automorphism $\underline{r}(k)$ of L^2.

The Lie algebra $\underline{g}$ acts in the space of C^∞-vectors in L^2:

(10.2.4) $\underline{r}(\underline{X})f = \lim_{h\to 0} h^{-1}(\underline{r}(e^{h\underline{X}})f-f)$ for $\underline{X} \in \underline{g}_{\mathbf{R}}$.

If the C^∞-function f represents an element of L^2, then one may ask whether this element is a C^∞-vector and whether the pointwise differentiation corresponds to the differentiation in L^2.

I shall in the sequel often use the same symbol for an element of L^2 and functions representing this element.

10.2.3. Lemma. Let $f \in C^\infty(G)$ represent a C^∞-vector in $L^2(\Gamma\backslash G, \chi)$. Then $\underline{X}f$ represents $\underline{r}(\underline{X})f$ for all $\underline{X} \in \underline{g}$.

Proof. Let $g_0 \in G$ and let C be a compact neighbourhood of 1 in G such that $C \cap \Gamma = \{1\}$. One may choose a fundamental domain F of $\Gamma\backslash G$ such that $Cg_0 \subset F$. For $\phi \in C_c^\infty(\Gamma\backslash G, \chi)$ with support contained in ΓCg_0 I want to prove that for $\underline{X} \in \underline{g}_{\mathbb{R}}$:

$$\langle \underline{r}(\underline{X})f, \phi\rangle = \langle \underline{X}f, \phi\rangle. \tag{10.2.5}$$

But

$$\langle \underline{r}(\underline{X})f, \phi\rangle = \lim_{h\to 0} h^{-1} \int_F (f(ge^{h\underline{X}}) - f(g))\overline{\phi(g)}dg \tag{10.2.6}$$

$$= \lim_{h\to 0} h^{-1} \int_{Cg_0} (f(ge^{h\underline{X}}) - f(g))\overline{\phi(g)}dg$$

The limit may be taken inside the integral and the lemma follows.

10.2.4. Lemma. Let $\underline{X} \in \underline{g}_{\mathbb{R}}$. Suppose that $f \subset C^\infty(G)$ represents an element of $L^2(\Gamma\backslash G, \chi)$ and also $\underline{X}f \in L^2(\Gamma\backslash G, \chi)$. Then $\underline{r}(\underline{X})f$ is well defined and is represented by $\underline{X}f$.

Proof. Take Cg_0 and ϕ as in the proof of lemma 10.2.3. It is sufficient to consider

$$\lim_{h\to 0} \langle h^{-1}(\underline{r}(e^{h\underline{X}})f - f), \phi\rangle \tag{10.2.7}$$

$$= \lim_{h\to 0} h^{-1} \int_{Cg_0} (f(ge^{h\underline{X}}) - f(g))\overline{\phi(g)}dg.$$

Taking the limit inside the integral proves the lemma.

10.2.5. Lemma. If a K-finite eigenfunction f of ω represents an element of $L^2(\Gamma\backslash G, \chi)$, then this element is a C^∞-vector in $L^2(\Gamma\backslash G, \chi)$.

10.2.6. Proposition (Harish-Chandra). If the function f on G is a K-finite eigenfunction of ω then there are $\alpha_1, \alpha_{-1} \in C_c^\infty(G_0)$ such that

$$f * \alpha_{\underline{+1}}(j_{\underline{+1}}g_0) = f(j_{\underline{+1}}g_0) \qquad \text{for all } g_0 \in G_0. \tag{10.2.8}$$

Proposition 10.2.6 is a direct consequence of theorem 1 on p. 18 of [10].

Proof of lemma 10.2.5. Let

$$f * \alpha_{\underline{+1}}(j_{\underline{+1}}g_0) = \int_{G_0} f(j_{\underline{+1}}x^{-1})\alpha_{\underline{+1}}(xg_0)dx \tag{10.2.9}$$

and differentiate under the integral sign.

10.3. Square integrable automorphic models

Let V be a $\underline{r}(G)$-irreducible subspace of L^2. By proposition 3.4.5 there is an operator $T\colon U \to V_K$ intertwining the $\underline{g}$-K-action in U, one of the spaces discussed in 3.4.1 - 3.4.3, and the $\underline{r}(\underline{g})$ - $\underline{r}(K)$-action in the space of $\underline{r}(K)$-finite elements

V_K of V. Each element of V_K is an eigenfunction of $\underline{r}(\omega)$; if it is a weight element it is weakly an eigenfunction of an elliptic operator, so it is represented by a C^∞-function. From lemma 10.2.3 follows that the $\underline{r}(\underline{g})$-$\underline{r}$(K)-action on V_K coincides with the pointwise $\underline{g}$-K-action. So T satisfies all conditions in definition 4.5.1 of an automorphic model, except possibly condition (4.5.2). In subsection 10.5 we shall prove (4.5.2) as well.

10.3.1. Definition. Let U be a $\underline{g}$-K-space contained in $H^\tau(s)$ for some $s \in \mathbb{C}$. An automorphic model T of U is square integrable if $TU \subset L^2(\Gamma\backslash G, \chi)$.

If U has non-zero square integrable automorphic models then it is built up from the spaces discussed in 3:4.1-3, as it has a unitary structure.

Let U be irreducible and let T be a square integrable automorphic model of U. By lemma 10.2.4 the elements of TU are C^∞-vectors in L^2 and the $\underline{g}$-K-action and the $\underline{r}(\underline{g})$-$\underline{r}$(K)-action coincide. The reasoning in the proof of theorem 7 on p. 199 of [21] shows that the weight vectors in TU are analytic vectors in L^2. Theorem 1 on p. 99 of [21] applies to this situation. So the closure V of TU in L^2 is $\underline{r}(G_0)$-invariant. As $jTU = TU$ the closure V is also $\underline{r}(G)$-invariant. It is even $\underline{r}(G)$-irreducible. Indeed, suppose that W is a $\underline{r}(G)$-invariant subspace of V. As W is the closure of the space W_K of $\underline{r}$(K)-finite elements, we have to show that $W_K = \{0\}$ or TU. Suppose not; then there is an element $f \in W_K$ of some weight r with $f \notin TU$. But f is the limit of elements of TU, which may be supposed to be all of weight r. But then these approximating elements are multiples of the same element. So $f \notin TU$ is impossible.

10.3.2. Definition. Let U be one of the spaces in 3.4.1-3.4.3. $A^{sq}(U)$ consists of the square integrable automorphic models of U.

So $A^{sq}(U)$ is a linear subspace of $A(U)$.

10.3.3. Proposition. Let U be one of the spaces in 3.4.1-3. For each $T \in A^{sq}(U)$, $T \neq 0$, the closure of TU in $L^2(\Gamma\backslash G, \chi)$ is an $\underline{r}(G)$-irreducible subspace of $L^2(\Gamma\backslash G, \chi)$. All $\underline{r}(G)$-irreducible subspaces of $L^2(\Gamma\backslash G, \chi)$ arise in this way. Two non-zero elements of $A^{sq}(U)$ give the same $\underline{r}(G)$-irreducible subspace if and only if they are linearly dependent.

Proof. All assertions have been proved, except the growth condition (4.5.2). This will be proved in corollary 10.5.8.

10.3.4. Proposition. Let U be one of the spaces in 3.4.1-3. For $T, T_1 \in A^{sq}(U)$ the expression

$$(10.3.1) \qquad \langle T, T_1\rangle = \|u\|^{-2}\langle Tu, T_1u\rangle \qquad u \neq 0$$

does not depend on $u \in U$. It defines a positive definite scalar product on $A^{sq}(U)$.

Proof. The independence may be checked in three steps. 1. Replace u by λu. 2. Replace a weight vector u by the weight vector $\underline{E}^{\pm} u$ or ju. 3. Express a general u in an orthonormal basis consisting of weight vectors.

10.4. Square integrable automorphic forms

Let $T \in A^{sq}(U)$ and let $\phi \in U$ have weight r. Then the functions ${}^{R}(T\phi)^{\pm}$ are square integrable automorphic forms in the sense of Roelcke; see [36], p. 302. If F_1 is a fundamental domain for $\bar{\Gamma}\backslash h$, then

(10.4.1) $$F = \{p(z)k(\theta)j_\varepsilon \mid z \in F_1,\ 0 \leq \theta < \pi,\ \varepsilon = \pm 1\}$$

is a fundamental domain for $\Gamma\backslash G$.

(10.4.2) $$\langle T\phi, T\phi\rangle = \int_{\Gamma\backslash G} |T\phi(g)|^2 dg$$
$$= \int_{F_1} \int_0^\pi \sum_{\varepsilon=\pm 1} |T\phi(p(z)k(\theta)j_\varepsilon)|^2 \frac{1}{2\pi}\, d\theta\; y^{-2}dxdy$$
$$= \tfrac{1}{2} \sum_{\pm 1} \int_{F_1} |{}^{R}(T\phi)^{\pm}(z)|^2\, y^{-2}dxdy.$$

So, indeed, square integrability of $T\phi$ on $\Gamma\backslash G$ amounts to square integrability of ${}^{R}(T\phi)^{+}$ and ${}^{R}(T\phi)^{-}$ on $\bar{\Gamma}\backslash h$.

10.4.1. Proposition. Let $\mathrm{Re}\, s = 0$ or $0 < |s| < \frac{1}{2}(1-\tau)$, s general. The map in proposition 4.5.6 restricted to $A^{sq}(H_\tau^{(\delta)}(s))$ gives a bijection onto the space of square integrable automorphic forms in $F_{\tau,\frac{1}{4}-s^2}(\bar{\Gamma}, ¢, v^{+})$. Furthermore, for $T, T_1 \in A^{sq}(H_\tau^{(\delta)}(s))$:

(10.4.3) $$\langle T, T_1\rangle = \begin{cases} \tfrac{1}{2}\varepsilon_\tau \langle {}^{R}(T\psi)^{+}, {}^{R}(T_1\psi)^{+}\rangle & \text{if } \mathrm{Re}\, s = 0 \\ \tfrac{1}{2}\varepsilon_\tau \dfrac{\Gamma(\frac{1}{2}+s-\frac{1}{2}\tau)}{\Gamma(\frac{1}{2}-s-\frac{1}{2}\tau)} \langle {}^{R}(T\psi)^{+}, {}^{R}(T_1\psi)^{+}\rangle & \text{if } 0 < |s| < \frac{1}{2}(1-\tau) \end{cases}$$

with $\varepsilon_\tau = 2$ if $\tau = 0, 1$ and $\varepsilon_\tau = 1$ if $0 < \tau < 1$.

Proof. A square integrable automorphic form on h is obtained as ${}^{R}(T\psi)^{+}$, with $T \in A(H_\tau^{(\delta)}(s))$. From (10.4.2) follows that $T\psi \in L^2$. By lemma 10.2.5 and repeated application of lemma 10.2.3 we see that $T \in A^{sq}(H_\tau^{(\delta)}(s))$. The formula in (10.4.3) follows from (4.5.17), (4.5.18) and (4.4.5).

Now let $U = D_b$ for some $b \in Y_\tau$. For $\phi \in D_b$ of weight r and $T \in A^{sq}(D_b)$ we get

(10.4.4) $$\langle T\phi, T\phi\rangle = \tfrac{1}{2} \sum_{\varepsilon=\pm 1} \int_{F_1} |T(\phi)^{\pm}(z)|^2\, y^{r-2}\, dxdy$$
$$= \tfrac{1}{2}\|(T\phi)^{+}\|_{P,r}^2 + \tfrac{1}{2}\|(T\phi)^{-}\|_{P,r}^2.$$

Here $\|.\|_{P,r}$ denotes the norm with respect to the Petersson scalar pr duct, defined in weight r by

(10.4.5) $$\langle f, g\rangle_{P,r} = \int_{\bar{\Gamma}\backslash h} f(z)\overline{g(z)}y^{r-2}dxdy,$$

see [23], p. 284. As in the case of proposition 10.4.1 we may prove:

10.4.2. Proposition. Let $\tau = 0, 1$, $b \in Y_\tau$ or $b = \tau = 1$. The map in proposition 4.5.7 maps $A^{sq}(D_b)$ onto the space of pairs of square integrable automorphic forms in $\{\bar{\Gamma}, -b, v^{+}\} \oplus \{\bar{\Gamma}, -b, v^{-}\}$. For $T, T_1 \in A^{sq}(D_b)$

(10.4.6) $\langle T, T_1\rangle = \frac{1}{2}(b-1)!(\langle(T\psi)^+, (T_1\psi)^+\rangle_{P,b} + \langle(T_1\psi)^-, (T\psi)^-\rangle_{P,b})$

10.4.3. Proposition. Let $0 < \tau < 1$, $b \equiv \pm\tau(2)$. The map in proposition 4.5.8 maps $A^{sq}(D_b)$ onto the space of square integrable automorphic forms in $\{\overline{\Gamma}, -b, v^{\pm}\}$. For $T, T_1 \in A^{sq}(D_b)$

(10.4.7) $$\langle T, T_1\rangle = \begin{cases} \frac{1}{2}\Gamma(b)\langle(T\psi)^+, (T_1\psi)^+\rangle_{P,b} & \text{if } b \equiv \tau(2) \\ \frac{1}{2}\Gamma(b)\langle(T_1\psi)^-, (T\psi)^-\rangle_{P,b} & \text{if } b \equiv -\tau(2) \end{cases}$$

10.5. Estimates

For any reasonable function f on G and $\phi \in C_c^\infty(G)$ we may consider the convolution

(10.5.1) $f * \phi(g) = \int_G f(gx^{-1})\phi(x)dx.$

This is well defined for instance for $f \in C^\infty(G)$, and also for f representing an element of $L^2(\Gamma\backslash G, \chi)$. In the last case

(10.5.2) $f * \phi = \underline{r}(\check{\phi})f,$

with $\check{\phi}(x) = \phi(x^{-1})$.

In this subsection $f * \phi(g)$ will be estimated. This is needed to complete the proof of proposition 10.3.3. Another consequence is the complete reducibility of the cuspidal part of L^2, discussed in subsection 10.6. The estimates will also be used in section 11 to obtain results necessary for the analytic continuation of Eisenstein series.

It will be sufficient to consider $\phi \in C_c^\infty(G_0)$, extended to G by defining ϕ to be zero on $G_0 j$.

Now let f be locally square integrable on G satisfying (10.2.1), and let $\phi \in C_c^\infty(G_0)$. Then $f * \phi$ is well defined and

(10.5.3) $f * \phi(g) = \int_G f(x)\phi(x^{-1}g)dx.$

We shall estimate $f * \phi(g)$ for g in a socalled Siegel domain S. I take

(10.5.4) $S = \{g_c(n(x)a(y)k(\theta)j_\varepsilon \mid |x| \leq \frac{1}{2},\ y \geq y_0,\ 0 \leq \theta \leq \pi,\ \varepsilon = \pm 1\}$

with c a fixed cusp, $y_0 > 0$.

The following estimates may be compared with §4 Ch. XII of [21] and p. 327-330 of [22]. If we define

(10.5.5) $J_\phi(g, g_1) = \sum_{\gamma \in \Delta_c} \overline{\chi(\gamma)}\ \phi(g_1^{-1}\gamma g),$

then

(10.5.6) $f * \phi(g) = \int_{\Delta_c\backslash G} f(g_1)J_\phi(g, g_1)dg_1.$

By the formula of Poisson:

(10.5.7) $J_\phi(g, g_1) = \sum_{\substack{\alpha\in\Lambda \\ c_\alpha = c}} \hat{\phi}_{g,g_1}(n_\alpha)$

with

(10.5.8) $\hat{\phi}_{g,g_1}(u) = \int_{-\infty}^{+\infty} e^{-2\pi iut} \sum_{m\in\mathbb{Z}} e^{-\pi i\tau m}\ \phi(g_1^{-1}k(\pi m)g_c n(t)g_c^{-1}g)dt$

When estimating $J_\phi(g, g_1)$ we may restrict ourselves to $g_1 = g_c n(x_1)a(y_1)k(\theta_1)j_{\varepsilon_1}$

with $|x_1| \leq \frac{1}{2}$, $0 \leq \theta_1 < \pi$, $y_1 > 0$, $\varepsilon_1 = \pm 1$. We write $g = g_c n(x)a(y)k(\theta)j_\varepsilon$, and know that $|x| \leq \frac{1}{2}$, $y \geq y_0$, $0 \leq \theta \leq \pi$, $\varepsilon = \pm 1$, for $g \in S$.

10.5.1. Lemma. There is a compact set C in G such that

(10.5.9) $\quad g^{-1} g_c a(y) \in C$, and

(10.5.10) $\quad a(y)^{-1} g_c^{-1} g_1 \in C$ whenever $\phi(g_1^{-1} ng) \neq 0$ for some $n \in N_c Z_0$.

Proof. (cf [21], p. 236). Take $C_1 = \{n(x_2)k(\theta_2)j_{\varepsilon_2} \mid |x_2| \leq \frac{1}{2}y_0^{-1}, 0 \leq \theta_2 \leq \pi, \varepsilon_2 = \pm 1\}$, then C_1 is compact and $g^{-1} g_c a(y) \in C_1^{-1}$.

Let $C_2 = \mathrm{supp}(\phi)$ and let $g_1^{-1} ng \in C_2$ for some $n \in N_c Z_0$. So $n = g_c n(-t) g_c^{-1} k(-\pi m)$ with $t \in \mathbb{R}$, $m \in \mathbb{Z}$. Now $g^{-1} n^{-1} g_1 \in C_2^{-1}$ and

(10.5.11) $\quad g^{-1} n^{-1} g_1 = g^{-1} g_c a(y).\ n((t+x_1)y^{-1}).\ a(y^{-1}y_1).\ k(\theta_1 + \pi m).\ j_{\varepsilon_1}$.

As $g^{-1} g_c a(y) \in C_1^{-1}$, and as G is homeomorphic to the direct product $N \times A \times K_0 \times \{1, j\}$ we may conclude that

(10.5.12) $\quad |t + x_1| \leq y u_0$ $\qquad$ for some $u_0 > 0$

(10.5.13) $\quad y y_2 \leq y_1 \leq y y_3$ $\qquad$ for some y_2, y_3 with $0 < y_2 < y_3$

(10.5.14) $\quad m \in C_3$ $\qquad$ for some finite set $C_3 \subset \mathbb{Z}$.

This implies that

(10.5.15) $\quad a(y)^{-1} g_c^{-1} g_1 = n(x_1 y^{-1}) a(y^{-1} y_1) k(\theta_1) j_\varepsilon$

is contained in a compact set C_4, for $|x_1 y^{-1}| \leq \frac{1}{2} y_0^{-1}$, $y_2 \leq y^{-1} y_1 \leq y_3$, $0 \leq \theta_1 \leq \pi$.

Take $C = C_4 \cup C_1^{-1}$; this proves the lemma.

Remark that

$$|\hat{\phi}_{g,g_1}(u)| \leq \int_{-\infty}^{+\infty} \sum_{m\in\mathbb{Z}} |\phi(g_1^{-1} k(\pi m) g_c n(t) g_c^{-1} g)|\,dt.$$

From (10.5.12-14) follows:

10.5.2. Lemma. For g, g_1 satisfying the conditions mentioned before lemma 10.5.1:

(10.5.16) $\quad |\hat{\phi}_{g,g_1}(u)| \ll y$;

(10.5.17) $\quad$ if $\hat{\phi}_{g,g_1}(u) \neq 0$ then $y_2 \leq y^{-1} y_1 \leq y_3$.

10.5.3. Lemma. For each integer $d \geq 2$ and $u \neq 0$:

(10.5.18) $\quad |\hat{\phi}_{g,g_1}(u)| \ll y^{1-d} |u|^{-d}$.

Proof. (cf. [21], p. 237).

(10.5.19) $\quad \hat{\phi}_{g,g_1}(u) =$

$$\sum_{m\in C_3} y \int_{-\infty}^{+\infty} e^{-2\pi i u y t} e^{\pi i \tau m} \phi(g_1^{-1} k(-\pi m) g_c n(yt) g_c^{-1} g)\,dt$$

$$= y \sum_{m\in C_3} e^{\pi i \tau m} \int_{-\infty}^{+\infty} e^{-2\pi i u y t} \phi(g_1^{-1} g_c a(y).k(-\pi m) n(t).a(y)^{-1} g_c^{-1} g)\,dt$$

As $g_1^{-1} g_c a(y)$ and $a(y)^{-1} g_c^{-1} g$ vary in compact sets, integration by parts gives the lemma.

10.5.4. Lemma. Let $0 < y_2 \leq y_3$ and let $y \geq y_0$. If f represents an element of $L^2(\Gamma\backslash G, \chi)$, then

(10.5.20) $$\int_{\Delta_c\backslash G,\ y_2 \leq y^{-1}y_1 \leq y_3} |f(g_1)|dg_1 \ll \|f\|_2 .$$

If f satisfies

(10.5.21) $|f(g_c na(y_2)k)| \leq C_f y_2^d$ for $y_2 \geq y_0$

$n \in N$, $k \in K$ for some $d \in \mathbb{R}$ and some $C_f \geq 0$, then

(10.5.22) $$\int_{\Delta_c\backslash G,\ y_2 \leq y^{-1}y_1 \leq y_3} |f(g_1)|dg_1 \leq C_f C_5 y^{d-1}$$

with C_5 not depending on f.

Proof. The integral over $\Delta_c\backslash G$, $y_2 \leq y^{-1}y_1 \leq y_3$ may be taken over the compact set $\{g_c n(x_1)a(y_1)k(\theta_1)j_{\varepsilon_1} \mid |x_1| \leq \frac{1}{2},\ yy_2 \leq y_1 \leq yy_3,\ 0 \leq \theta_1 \leq \pi,\ \varepsilon_1 = \pm 1\}$. This set is contained in a finite number of fundamental domains of $\Gamma\backslash G$, independent of $y \geq y_0$. So (10.5.20) is clear, for vol $(\Gamma\backslash G) < \infty$, and hence 1 is square integrable on $\Gamma\backslash G$. To obtain (10.5.22) one may write out the integral and use (10.5.21).

10.5.5. Proposition. Let f be an automorphic form. Let $\phi \in C_c^\infty(G)$. Let c be a cusp. If $(c, 0) \notin \Lambda$ or if $F_{(c,0)}f = 0$ then

(10.5.23) $|f * \phi(g_c na(y)k)| \ll y^{-d_1}$ for $y \to \infty$

uniformly in $n \in N$ and $k \in K$, for each $d_1 \geq 2$. If $(c, 0) \in \Lambda^0$ then

(10.5.24) $|f * \phi(g_c na(y)k) - F_{(c,0)}f * \phi(na(y)k)| \ll y^{-d_1}$ for $y \to \infty$

uniformly in $n \in N$ and $k \in K$, for each $d_1 \geq 2$. The constants in the estimates contain the number C_f in (10.5.21) as a factor describing the dependence on f.

Proof. By the Γ-invariance it is sufficient to restrict $g = g_c na(y)k$ to a Siegel domain. If $(c, 0) \in \Lambda^0$ then in (10.5.7) the term $\hat{\phi}_{g,g_1}(0)$ occurs. It contributes to $f * \phi(g_c na(y)k)$:

(10.5.25) $$\int_{\Delta_c\backslash G} f(g_1)\, \hat{\phi}_{gg_1}(0)dg_1 =$$

$$\int_{Z_0N_c\backslash G} \int_{\Gamma_c\backslash N_c} f(n_1g_1) \sum_{m\in C_3} e^{\pi i \tau m} \int_{N_c} \phi(g_1^{-1}k(-\pi m)n_1^{-1}ng)dndn_1dg_1$$

$$= \int_{Z_0N_c\backslash G} F_{(c,0)}f(g_c^{-1}g_1) \sum_{m\in C_3} e^{\pi i \tau m} \int_{N_c} \phi(g_1^{-1}k(-\pi m)n^{-1}g)dndg_1$$

$$= F_{(c,0)}\, f * \phi(g_c^{-1}g).$$

The remainder of the sum in (10.5.7) is estimated by y^{-d_2} for each $d_2 \geq 2$, see lemma 10.5.3. Using lemma 10.5.4 the estimates follow.

10.5.6. Proposition. There are Siegel domains S_c, as indicated in (10.5.4), for all cusps c such that $G = \bigcup_{c\ \text{cusp}} \Gamma S_c$.

Proof. From the discussion of Petersson in §4 of [30] follows that

(10.5.26) $$\mathfrak{h} = \bigcup_{c\ \text{cusp}} \bar{\Gamma}\, \bar{g}_c\{x+iy \mid |x| \leq \tfrac{1}{2},\ y \geq y_c\}$$

for suitably chosen $y_c > 0$. The proposition follows.

10.5.7. Proposition. Let $\phi \in C_c^\infty(G_0)$. Then for each $f \in L^2(\Gamma\backslash G, \chi)$ and for each cusp c

(10.5.27) $|\underline{r}(\phi)f(g_c na(y)k)| \ll y\|f\|_2$ for $y \to \infty$

uniformly for $n \in N$, $k \in K$.

If $(c, 0) \notin \Lambda$ or if

(10.5.28) $\int_{\Gamma_c\backslash N_c} f(ng)dn = 0$ for almost all $g \in G$,

then

(10.5.29) $|\underline{r}(\phi)f(g_c na(y)k)| \ll y^{-d}\|f\|_2$ for $y \to \infty$

uniformly for $n \in N$, $k \in K$, for each $d \geqslant 2$. In particular, $\underline{r}(\phi)f$ is a bounded function at this cusp.

Proof. This is proved in the same way as proposition 10.5.5. Condition (10.5.28) replaces $F_{(c,0)}f = 0$, for $F_{(c,0)}$ has been defined for C^∞-functions only.

Remark. For $f \in L^2$ and $\phi \in C_c^\infty(G_0)$ the convolution $f * \phi(g)$ is well defined for each $g \in G$. This function $f * \phi$ represents $\underline{r}(\check{\phi})f$; it is pointwise continuous in f.

10.5.8. Corollary. If V is a $\underline{r}(G)$-irreducible subspace of $L^2(\Gamma\backslash G, \chi)$ and $f \in V_K$, the subspace of K-finite elements, then f satisfies the growth condition (4.5.2) with $d = 1$.

Remark. This finishes the proof of proposition 10.3.3.

Proof. f is an eigenfunction of $\underline{r}(\underline{\omega})f$. We may suppose that f is a weight function. Then it is weakly an eigenfunction of an elliptic operator, so it is an analytic function and $\underline{\omega}.f = \underline{r}(\underline{\omega})f$. Now apply proposition 10.2.6. On each component of G we may write $f = \underline{r}(\phi)f$ for some $\phi \in C_c^\infty(G_0)$. The corollary follows from (10.5.27).

10.6. The cuspidal part of $L^2(\Gamma\backslash G, \chi)$

Let $\lambda \in \Lambda^0$. By proposition 8.2.3 the function $\Theta_\lambda f$ is well defined for all $f \in {}^0S_\tau$; see definition 3.6.1. The sum defining $\Theta_\lambda f$ is in fact finite and $\Theta_\lambda {}^0S_\tau \subset C_c^\infty(\Gamma\backslash G,\chi)$.

10.6.1. Definition. ${}^eL^2(\Gamma\backslash G, \chi)$ is the closure in $L^2(\Gamma\backslash G, \chi)$ of the space spanned by $\bigcup_{\lambda\in\Lambda^0} \Theta_\lambda {}^0S_\tau$. ${}^0L^2(\Gamma\backslash G, \chi) = {}^eL^2(\Gamma\backslash G, \chi)^\perp$ is called the cuspidal subspace of $L^2(\Gamma\backslash G, \chi)$.

I shall often write ${}^eL^2$ and ${}^0L^2$ instead of ${}^eL^2(\Gamma\backslash G, \chi)$ and ${}^0L^2(\Gamma\backslash G, \chi)$.

Let $\lambda \in \Lambda^0$, $f \in {}^0S_\tau$ and $\phi \in L^2$, then by Fubini's theorem

(10.6.1) $\langle\phi, \Theta_\lambda f\rangle = \int_{Z_0 N\backslash G} \int_{\Gamma_\lambda\backslash N_\lambda} \phi(ng_\lambda g)dn\, P_1 f(g)d\dot{g}$

So $\phi \in {}^0L^2$ if and only if

(10.6.2) $\int_{\Gamma_\lambda\backslash N_\lambda} \phi(ng)dn = 0$ for almost all $g \in G$,

for all $\lambda \in \Lambda^0$. This implies that ${}^0L^2$ and ${}^eL^2$ are $\underline{r}(G)$-invariant subspaces of L^2.

10.6.2. Theorem. The representation $\underline{r}$ of G in ${}^0L^2(\Gamma\backslash G, \chi)$ is completely reducible; each irreducible component occurs with finite multiplicity.

The proof is the same one as given in [21], Ch. XII §4. Let $\phi \in C_c^\infty(G_0)$ be given. Trom proposition 10.5.7 follows that $|\underline{r}(\phi)f(g)| \leqslant c_\phi\|f\|_2$ for all $g \in G$ for each $f \in {}^0L^2$. So the operator $\underline{r}(\phi)$ is compact in ${}^0L^2$. This implies the complete reducibility for $\underline{r}(G_0)$, and hence for $\underline{r}(G)$.

10.6.3. Definition. Let U be a g-K-space contained in $H^{\tau}(s)$ for some $s \in \mathbb{C}$. An automorphic model $T \in A(U)$ is called a cuspidal model of U if $TU \subset {}^0L^2(\Gamma\backslash G, \chi)$. The elements of a cuspidal model are called cusp forms. $A^0(U)$ is the space of cuspidal models of U.

$A^0(U)$ is a subspace of $A^{sq}(U)$. For each $T \in A^0(U)$:

(10.6.3) $\quad {}^*FT = 0$

by (10.6.2). So the automorphic forms on h associated to $A^0(U)$ by the maps in the propositions 4.5.6-8 have vanishing Fourier coefficients of order zero, as cusp forms ought to have.

It is clear that (10.6.3) characterizes $A^0(U)$ as subspace of $A^{sq}(U)$. But we have even the following result:

10.6.4. Proposition. Let U be a g-K-space contained in some $H^{\tau}(s)$. If $T \in A(U)$ and ${}^*FT = 0$, then $T \in A^0(U)$.

Proof. The square integrability of T has to be checked. This is easily seen by the propositions 10.2.6, 10.5.5 and 10.5.6.

If $A^0(U) \neq \{0\}$, then U has a unitary structure, so it is one of the spaces discussed in 3.4.1-3. For later use I fix some notations.

Notations. If s general, $s \in i\mathbb{R}_{\geq 0}$ or $0 < s < \frac{1}{2}(1-\tau)$, then

$$\left.\begin{aligned} &(10.6.4) \quad U(s) = H^1_\tau(s) \oplus H^{-1}_\tau(s) \\ &\qquad\qquad A^0(s) = A^0(H^1_\tau(s)) \oplus A^0(H^{-1}_\tau(s)) \end{aligned}\right\} \text{if } \tau = 0,\ 1$$

$$\left.\begin{aligned} &(10.6.5) \quad U(s) = H_\tau(s) \\ &\qquad\qquad A^0(s) = A^0(H_\tau(s)) \end{aligned}\right\} \text{if } 0 < \tau < 1.$$

If $b \in Y_\tau$ or $b = \tau = 1$, then

$$(10.6.6) \quad U(\tfrac{1}{2}(b-1)) = D_b, \qquad A^0(\tfrac{1}{2}(b-1)) = A^0(D_b).$$

Remark. D^1_0 and D^{-1}_0 cannot occur in L^0, see proposition 4.5.5.

Remark. If $\tau = 0,1$, s general, then $A^0(H^1_\tau(s)) \cong A^0(H^{-1}_\tau(s))$ as Hilbert spaces. Both are isomorphic to the same space of automorphic forms on h by the map in proposition 4.5.6. See also proposition 10.4.1.

10.6.5. Proposition. There exists a discrete set $S^0 \subset i\mathbb{R}_{\geq 0} \cup (0, \frac{1}{2}(1-\tau)) \cup$ $\cup \{\frac{1}{2}(b-1) \mid b \equiv +\tau(2),\ b > 0\}$ such that

$$(10.6.7) \quad {}^0L^2(\Gamma\backslash G, \chi) = \bigoplus_{s \in S^0} \bigoplus_T TU(s),$$

where T runs through an orthonormal basis of $A^0(s)$ for each $s \in S^0$. Each $A^0(s)$ is finite dimensional.

Proof. Except for the discreteness this follows from theorem 10.6.2 and proposition 10.3.3. The discreteness can be proved by going into more details in the proof of theorem 10.6.2. I do not go into this, but refer to Roelcke's Satz 8.1 in [37].

10.6.6. Proposition. Each cusp form ψ satisfies $|\psi(g_c na(y)k)| \ll y^{-d}$ for $y \to \infty$ uniformly in $n \in N$, $k \in K$, for each cusp c and for each $d \in \mathbb{R}$.

Proof. Clear from the propositions 10.2.6, 10.5.5 and (10.6.2).

11. The analytic continuation of the Eisenstein model

11.1. Introduction

This section gives the analytic continuation of the Eisenstein model $\underline{E}(s)$. I use Selberg's method, as given in [22] and [11]. Roelcke states the result, see p. 293 of [37]; as far as I know a proof for the case of G, the universal covering group of $Sl_2(\mathbb{R}) \cup \begin{pmatrix} 1 & 0 \\ 0 & -1 \end{pmatrix} Sl_2(\mathbb{R})$, and Γ-equivariant functions has never been published. As can be seen in this section the method of [22] and [11] works in this case. The only thing I have done is taking the rank-one-case out of [22] and [11], and writing it down in the language of automorphic models. One may also compare [14]. Another approach is given in [6] and [21], Ch. XIV.

In subsection 11.2 I consider spaces of automorphic models orthogonal to the cusp forms. These models are completely determined by their Fourier coefficients of order zero. This correspondence turns out to be continuous in some sense. Subsection 11.3 deals with the spectral decomposition of ${}^0S_\tau$ and some larger $\underline{g}$-K-space of functions on $N\backslash G$. These results I need in 11.4 to give integral representations for scalar products of some elements of ${}^eL^2(\Gamma\backslash G, \chi)$. These integrals contain the Fourier coefficient $\underline{M}(s)$ of the Eisenstein model. The subsections 11.5-7 give successive steps of the analytical continuation of $\underline{M}(s)$ and $\underline{E}(s)$.

11.2. Limit formulas for automorphic models

The content of this subsection is more or less the lemmas 5.1 (p. 100) and 5.2 (p. 105) of [22]; compare also the lemma on p. 326 of [22].

Consider the Eisenstein model $\underline{E}(s)\underline{e} \in A(H^\tau(s))$ for Re $s > \frac{1}{2}$ and $\underline{e} \in E$. For each $\phi \in H^\tau(s)$ and each cusp form ψ the function

(11.2.1) $$g \to \underline{E}(s)\underline{e}\, \phi(g)\, \overline{\psi(g)}$$

is bounded on $\Gamma\backslash G$, as follows from the propositions 10.6.6, 10.5.6 and 9.2.5.

(11.2.2) $$\langle \underline{E}(s)\underline{e}\phi, \psi \rangle = \langle St(s)\phi \otimes \underline{e}, {}^*F\psi \rangle = 0,$$

see proposition 8.3.3 and (7.5.1). I have extended $\langle.,.\rangle$ to E-valued functions in the natural way. We may say that $\underline{E}(s)$ is orthogonal to the cusp forms. For any automorphic model the function corresponding to the one in (11.2.1) is integrable; so the condition "orthogonal to the cusp forms" is meaningful in general.

The Eisenstein models have another property.

11.2.1. Definition. $C^0 = \{f \in C_c^\infty(G_0) \mid f(kgk^{-1}) = f(g) \text{ for all } g \in G_0,\ k \in K_0\}$.

Take $f \in C^0$. Let ϕ be a constant section of H^τ of weight r;

(11.2.3) $$\begin{aligned} St(s)\phi(s) * f(na(y)j_\varepsilon k(\theta)) &= \\ &= y^{\frac{1}{2}+s} e^{ir\theta} \int_{G_0} St(s)\phi(s)(j_\varepsilon x) f(x^{-1}) dx \\ &= y^{\frac{1}{2}+s} e^{ir\theta} \int_N \int_A \int_{K_0} St(s)\phi(s)(n_1^{-1} a j_\varepsilon k) f(k^{-1} a^{-1} n_1^{-1}) \alpha(a)^{-1} dn_1\, da\, dk \end{aligned}$$

$$= St(s)\phi(s)(na(y)j_\varepsilon k(\theta)).$$

$$.\int_N\int_A\int_{-\infty}^{+\infty} \alpha(a)^{-\frac{1}{2}+s} e^{ir\theta_1} f(k(-\theta_1)a^{-1}n_1^{-1})dn_1 da \frac{1}{2\pi} d\theta_1$$

$$= St(s)\phi(s)(na(y)j_\varepsilon k(\theta))\alpha_f(s, \phi),$$

for some $\alpha_f(s, \phi)$, depending holomorphically on s. The sum defining $\underline{E}(s)$ converges uniformly on compact sets, so

(11.2.4) $\quad \underline{E}(s)\underline{e}\phi(s) * f = \alpha_f(s, \phi)\ \underline{E}(s)\underline{e}\phi(s)$

for all $f \in \overline{C}^0$, Re $s > \frac{1}{2}$. This is the other property to be considered.

11.2.2. Definition. Let U be a g-K-space contained in $H^\tau(s)$ for some $s \in \mathbb{C}$ or $U = E_1^\delta$. The space $A^e(U)$ consists of those $T \in A(U)$ for which

(11.2.5) $\quad \langle T\phi, \psi\rangle = 0$ for all $\phi \in U$ and all cusp forms ψ

(11.2.6) $\quad T\phi * f = \alpha_f(s, \phi)T\phi$ for all $\phi \in U$ and all $f \in C^0$.

We have seen that $\underline{E}(s)\underline{e} \in A^e(H^\tau(s))$ for all $\underline{e} \in E$, Re $s > \frac{1}{2}$.

The map

(11.2.7) $\quad T \to (\underline{a}_\tau^{(\delta)}(s, T), \underline{b}_\tau^{(\delta)}(s, T); s)$

discussed in (7.5.8) and definition 7.5.8 may be restricted to a map $A^e(H_\tau^{(\delta)}(s)) \to$ $\to N(s)$. From proposition 10.6.4 follows that it is injective.

11.2.3. Proposition. Let $\tau \in [0, 1]$, $\delta = \pm 1$ if $\tau = 0$ or 1. Let Re $s > \frac{1}{2}$. Then

(11.2.8) $\quad A^e(H_\tau^{(\delta)}(s)) = \{\underline{E}(s)\underline{e}|H_\tau^{(\delta)}(s) \mid \underline{e} \in E\}$.

Proof. By (9.4.9)

(11.2.9) $\quad \dim\{\underline{E}(s)\underline{e}|H_\tau^{(\delta)}(s) \mid \underline{e} \in E\} = \dim E$

for general s; in 9.4.5 we see that it also holds for nongeneral s. As the right hand side of (11.2.8) is contained in the left hand side we see that

(11.2.10) $\quad \dim E \leq \dim A^e | H_\tau^{(\delta)}(s) \leq \dim N(s)$

for Re $s > \frac{1}{2}$. From the Maass-Selberg relation follows

(11.2.11) $\quad \dim N(s) + \dim N(\overline{s}) \leq 2 \dim E$,

see (7.6.9). This proves the proposition.

11.2.4. Proposition. Let $s \in \mathbb{C}$ and let (s_i) be a sequence of general points with limit s. Let $T_i \in A^e(H_\tau^{(\delta)}(s_i))$, let $\underline{n}_i \in N(s_i)$ be the image of T_i under the map in (11.2.7). Suppose $\lim_{i\to\infty} \underline{n}_i = \underline{n} \in E_2(s)$. If not $\tau = 1$, $s = 0$ then for each constant section ϕ of $H_\tau^{(\delta)}$ the sequence $T_i\phi(s_i)$ converges uniformly on compact sets and

$$T\phi(s) = \lim_{i\to\infty} T_i\phi(s_i)$$

defines an element $T \in A^e(H_\tau^{(\delta)}(s))$, which satisfies

(11.2.12) $\quad {}^*FT = \underline{St}_\tau^{(\delta)}(\underline{n})$.

If $s = 0$, $\tau = 1$, let $\tilde{T}_i$ be defined by ${}^R(T_i\phi_1^\delta(s_i))^+ = {}^R(\tilde{T}_i\phi_1^\delta(-s_i))^+$, and consider constant sections ϕ as in (3.3.28). Then $T_i \oplus \Gamma(s_i)\Gamma(-s_i)^{-1}\tilde{T}_i\phi$ converges uniformly on compact sets, defining $T \in A^e(E_1^\delta)$ satisfying ${}^*FT = \widehat{\underline{St}}_1^\delta(\underline{n})$.

Remark that this proposition says that convergence of a sequence of automorphic models may be read from the convergence of the Fourier coefficients of order zero. The proof of proposition 11.2.4 will be given with help of a series of lemmas. In all these lemmas I take Siegel domains S_c as in proposition 10.5.6. By ϕ I indicate a constant section of $H_\tau^{(\delta)}$ or a section as in (3.3.28) if $\tau = 1$ and we are working in a neighbourhood of 0. If for a general $s \in \mathbb{C}$ we have $T \in H_\tau^{(\delta)}(s)$, then $\hat{T}$ denotes just T, except in the case that $\tau = 1$ and we are working near 0; then $\hat{T} = T \oplus \Gamma(s)\Gamma(-s)^{-1}\tilde{T}$ as in (7.5.12).

11.2.5. Lemma. Let $s_0 \in \mathbb{C}$, $d > \frac{1}{2} + |\mathrm{Re}\ s_0|$, $d_1 \in \mathbb{R}$. Take for $T \in A^e(H_\tau^{(\delta)}(s))$

(11.2.13) $$C_\phi(T) = \sup_{\substack{c \text{ cusp}, n\in N\\ y\geq y_c, k\in K}} y^{-d}|\hat{T}\phi(s)(g_c na(y)k)|.$$

There is a neighbourhood U of s_0 and a constant D such that for each general $s \in U$ and each $T \in A^e(H_\tau^\delta(s))$:

(11.2.14) $$C_\phi(T) < \infty$$

(11.2.15) $$|T\phi(s)(g_c na(y)k)| \leq C_\phi(T)Dy^{-d_1}$$

for all cusps c with $(c, 0) \notin \Lambda$, $n \in N$, $y \geq y_c$, $k \in K$;

(11.2.16) $$|\hat{T}\phi(s)(g_c na(y)k) - F_{(c,0)}\hat{T}\phi(s)(a(y)k)| \leq C_\phi(T)Dy^{-d_1}$$

for all cusps c with $(c, 0) \in \Lambda^0$, $n \in N$, $y \geq y_c$, $k \in K$. Furthermore, there is a positive number η such that for all general $s \in U$, $T \in A^e(H_\tau^{(\delta)}(s))$

(11.2.17) $$|\hat{T}\phi(s)(g_c na(y)k)| \leq \tfrac{1}{2}y^d C_\phi(T)$$

for all cusps c, $n \in N$, $y \geq \eta$, $k \in K$.

Proof: Choose $f \in C^0$ such that $\alpha_f(s, \phi) \neq 0$ for s near s_0. For automorphic forms ψ we have $*F(\psi * f) = (*F\psi) * f$. Applying this to an Eisenstein series we see that $\alpha_f(-s, \phi) = \alpha_f(s, \phi)$. So for $T \in A^e(H_\tau^{(\delta)}(s))$ we have $\hat{T}\phi(s) * f = \alpha_f(s, \phi)\hat{T}\phi(s)$, even if $\hat{T} \neq T$.

Now apply proposition 10.5.5 to $\hat{T}\phi(s)$ and f. We see that the Fourier coefficients of order zero determine whether $C_\phi(T) < \infty$. If $|\mathrm{Re}\ s| < d - \frac{1}{2}$ there is no problem. We can find a neighbourhood of s_0 such that $|\alpha_f(s, \phi)|^{-1}$ is bounded on this neighbourhood. With proposition 10.5.5 the assertions (11.2.14-16) are now clear. For the last assertion start with taking η_0 big enough to have $D\eta_0^{-d-d_1} < \frac{1}{4}$. For cusps with $(c, 0) \notin \Lambda$ it is sufficient to take $\eta \geq \eta_0$ to obtain (11.2.17). Now consider the case $(c, 0) \in \Lambda^0$. It is sufficient to prove that

(11.2.18) $$|F_{(c,0)}\hat{T}\phi(s)(a(y)k)| \leq \tfrac{1}{4}y^d C_\phi(T)$$

for all sufficiently large y, all $k \in K$, all general $s \in U$ and all $T \in A^e(H_\tau^{(\delta)}(s))$. By making U smaller we can arrange that U is compact. Let $\mathcal{B}_d$ be the Banach space of functions on $[\eta_0, \infty)$ bounded for the norm

(11.2.19) $$\nu_d(t) = \sup_{t\geq\eta_0} y^{-d}|t(y)|.$$

Let for $q > 0$ the subset B_q of $\mathcal{B}_d$ consist of all functions

(11.2.20) $$y \to F_{(c,0)}\hat{T}\phi(s)(a(y)k)$$

with $s \in U$ general, $T \in A^e(H_T^{(\delta)}(s))$, $k \in K$ and $C_\phi(T) = q$. Once we know that B_q is relatively compact in $\mathcal{B}_d$ we can find a finite set $E \subset B_q$ such that for each $b \in B_q$ there is a $b_0 \in E$ with $\nu_d(b-b_0) \leq \frac{1}{8}q$. As $d > |\mathrm{Re}\, s| + \frac{1}{2}$ for all $s \in U$ it is clear that we can find η_q such that

(11.2.21) $\quad y^{-d}|b_0(y)| \leq \frac{1}{8}q$

for all $b_0 \in E$, $y \geq \eta_q$. To prove (11.2.18) remark that

(11.2.22) $\quad y^{-d}|b(y)| \leq y^{-d}|b_0(y)| + \nu_d(b-b_0) \leq \frac{1}{4}q$

for each $b \in B_c$ and some $b_0 \in E$.

To prove that B_q is relatively compact in $\mathcal{B}_d$ remark that B_q is contained in the image of $A\colon U \times \mathbb{C}^2 \to \mathcal{B}_d$, defined by

(11.2.23)
$$\begin{aligned} &A(s, a, b)(y) = ay^{\frac{1}{2}+s} + by^{\frac{1}{2}-s} && \text{if } 0 \notin U \\ &A(s, a, b)(y) = ay^{\frac{1}{2}+s} + \frac{1}{2s}\, b(y^{\frac{1}{2}+s} - y^{\frac{1}{2}-s}) \text{ if } s \neq 0 && \\ &A(0, a, b)(y) = ay^{\frac{1}{2}} + by^{\frac{1}{2}}\ln y && \end{aligned} \Bigg\}\ \text{if } 0 \in U$$

This map A is continuous. The inverse image of the set $\{t \in \mathcal{B}_d | \nu_d(t) \leq q_1\}$ is closed in $U \times \mathbb{C}^2$; moreover, it also bounded in $U \times \mathbb{C}^2$. This boundedness one sees by taking $y_1 > y_2 \geq \eta_0$ and solving the following equations in a, b

(11.2.24) $\quad |a\, y_i^{\frac{1}{2}-s} + b\, y_i^{\frac{1}{2}-s}| \leq q_1 \qquad i = 1, 2;$

or a similar equation if $0 \in U$.

Now it is clear that B_q is contained in the continuous image of a compact set.

11.2.6. Lemma. Take s, s_i, T_i and ϕ as in proposition 11.2.4. Let $d > |\mathrm{Re}\, s| + \frac{1}{2}$, $d > |\mathrm{Re}\, s_i| + \frac{1}{2}$ for all i. Define $C_\phi(T_i)$ by (11.2.13). If $\sup_i C_\phi(T_i) < \infty$, then there is a subsequence $T_{i'}$ such that $T_{i'}\hat{\phi}(s_{i'})$ converges uniformly on compact sets in G.

Proof. It is sufficient to show that the $T_i\hat{\phi}(s_i)$ form a relatively compact set with respect to the supnorm on each compact set $B \subset G$. By Ascoli's theorem (see e.g. [20], p. 211) this is equivalent to

(11.2.25) $\quad$ the set $\{T_i\hat{\phi}(s_i)\}$ is equicontinuous on B

(11.2.26) $\quad$ the set $\{T_i\hat{\phi}(s_i)\}$ is uniformly bounded on B.

Condition (11.2.26) follows directly from the boundedness of the $C_\phi(T_i)$. Choose $f \in C^0$ such that $\alpha_f(s, \phi) = 1$, see proposition 10.2.6. After going over to a subsequence we may assume that $|\alpha_f(s_i, \phi)| \geq \frac{1}{2}$ for all i. Let $\underline{X} \in \underline{g}_{\mathbb{R}}$

(11.2.27) $\quad \underline{X}T_i\hat{\phi}(s_i) = \alpha_f(s_i, \phi)^{-1}\, T_i\hat{\phi}(s_i) * \underline{X}f.$

So for all $g \in B$ and all $\underline{X} \in \underline{g}_{\mathbb{R}}$

(11.2.28) $\quad |\underline{X}T_i\hat{\phi}(s_i)(g)| \leq 2 \sup_{x \in \mathrm{supp}\, f} |T_i\hat{\phi}(s_i)(gx)|.C(\underline{X}f),$

with $C(\underline{X}f)$ a constant depending on $\underline{X}f$.

The right hand side is bounded uniformly for all i and all $\underline{X}$ in a bounded neighbourhood of 0 in $\underline{g}_{\mathbb{R}}$. The equicontinuity is now clear.

11.2.7. Lemma. Take s_i, s, T_i, $\underline{n}_i$, $\underline{n}$ as in proposition 11.2.4. Take d as in lemma 11.2.6. Suppose that the $C_\phi(T_i)$ defined in (11.2.13) are bounded. Let $\psi(g) = \lim_{i'\to\infty} T_{i'}\hat{\phi}(s_{i'})$ for some subsequence converging on compact sets. Then

(11.2.29) $\quad \psi(\gamma g) = \chi(\gamma)\psi(g) \qquad$ *for* $\gamma \in \Gamma,\ g \in G$;

(11.2.30) $$\int_{\Gamma_\lambda \backslash N_\lambda} \psi(n g_\lambda g)dn = \begin{cases} ((\underline{St}_\tau^{(\delta)}(\underline{n})\phi(s)(g),\ \underline{e}_\lambda)) & \text{if not } \tau = 1,\ s = 0 \\ ((\widehat{\underline{St}}_1^{(\delta)}(\underline{n})\phi(0)(g),\ \underline{e}_\lambda)) & \text{if } \tau = 1,\ s = 0 \end{cases}$$

for $\lambda \in \Lambda^0$, $g \in G$;

(11.2.31) $\quad |\psi(g_c n a(y) k)| \ll y^d \qquad$ *for* $y \to \infty$,

uniformly in $n \in N$, $k \in K$; *if* $\underline{n} = 0$, *then* $\psi = 0$.

Proof. The first assertion is clear from the definition of ψ; the second one follows from proposition 7.5.9. Estimate (11.2.31) follows from the boundedness of the $C_\phi(T_i)$. To get the last assertion we take a cusp form ψ_1 and consider

(11.2.32) $$\int_{\Gamma\backslash G} \hat{T}_{i'}\phi(s_{i'})(g)\overline{\psi_1(g)}dg.$$

From the boundedness of the $C_\phi(T_{i'})$ follows that we may apply Lebesgue's theorem; the limit of the expression in (11.2.32) equals

(11.2.33) $$\int_{\Gamma\backslash G} \psi(g)\overline{\psi_1(g)}dg.$$

I claim that this expression is zero. If $\hat{T}_{i'} = T_{i'}$ it is clear that the expressions in (11.2.32) is zero. In the other case we need that $\tilde{T} \in A^e(H_\tau^{+\delta}(-s))$ if $T \in A^e(H_\tau^\delta(s))$; this I shall prove in the next lemma. So ψ is orthogonal to the cusp forms. If $\underline{n} = 0$, then ψ is bounded; this one sees by taking the limit of (11.2.15) and (11.2.16) and applying proposition 7.5.9. So ψ is square integrable, satisfies (10.6.2), hence $\psi \in {}^0L^2$. As it is also orthogonal to ${}^0L^2$, it vanishes.

11.2.8. *Lemma. Let* $\tau = 1$, *let* s *be general. If* $T \in A\ (H_1^\delta(s))$, *let* $\tilde{T} \in A(H_1^\delta(-s))$ *be defined by* ${}^R(T\phi_1^\delta(s))^+ = {}^R(\tilde{T}\phi_1^\delta(-s))^+$. *If* $T \in A^e(H_1^\delta(s))$, *then* $\tilde{T} \in A^e(H_1^\delta(-s))$.

Proof. The only cusp forms which might not be orthogonal to $\tilde{T}H_1^\delta(-s)$ are elements of $T_1H_1^\delta(-s)$ with $T_1 \in A^0(H_1^\delta(-s))$. From (4.5.18) follows

(11.2.34) $$\tilde{T}\phi_1^\delta(-s)(g_0 j_\varepsilon) = \varepsilon T\phi_1^\delta(s)(g_0 j_\varepsilon) \qquad g_0 \in G_0,\ \varepsilon = \pm 1.$$

The same holds for $T_1 \in A^0(H_1^\delta(-s))$. We get

(11.2.35) $$\langle \tilde{T}\phi_1^\delta(-s),\ T_1\phi_1^\delta(-s)\rangle = \sum_{\varepsilon=\pm 1} \int_{\Gamma\backslash G_0} \tilde{T}\phi_1^\delta(-s)(g_0 j_\varepsilon)\overline{T_1\phi_1^\delta(-s)(g_0 j_\varepsilon)}dg_0$$

$$= \langle T\phi_1^\delta(s),\ \tilde{T}_1\phi_1^\delta(s)\rangle = 0,$$

for $\tilde{T}_1$ is cuspidal if T_1 is.

So in weight 1 we have the desired orthogonality. Other weights are reached with help of E^+ and E^-. From (11.2.34) follows that for $f \in C^0$

(11.2.36) $$\tilde{T}\phi_1^\delta(-s) * f = \alpha_f(+s,\ \phi_1^\delta)\tilde{T}\phi_1^\delta(-s) = \alpha_f(-s,\ \phi_1^\delta)\tilde{T}\phi_1^\delta(-s).$$

By repeatedly applying E^+ and E^- to (11.2.34) one gets relations of the form

(11.2.37) $$\tilde{T}\phi_r^\delta(-s)(g_0 j\ ) = \varepsilon\ .\ \text{rational function in } s.\ T\phi_r^\delta(s)(g_0 j_\varepsilon).$$

So for each weight we may derive the relation given in (11.2.36) for weight 1.

11.2.9. *Lemma. Under the assumptions of proposition* 11.2.4 *we may find* $d \in \mathbb{R}$ *such that for the* $C_\phi(T_i)$ *defined in* (11.2.13) *we have*

(11.2.38) $$\sup_i C_\phi(T_i) < \infty.$$

Proof. By taking d larger than $\sup_j |\mathrm{Re}\ s_j| + \frac{1}{2}$ we can arrange that $C_\phi(T_i) < \infty$ for each i. Suppose, after going over to a subsequence, that $C_\phi(T_i) \nearrow \infty$. Apply lemma

11.2.6 to the sequence $C_\phi(T_i)^{-1}T_i$. We know that there are subsequences $C_\phi(T_{i'})^{-1}\hat{T}_{i'}\phi(s_{i'})$ converging on compact sets. By the last assertion of lemma 11.2.7 the limit function is zero. From the Γ-equivariance follows that the convergence is uniform on sets of the form

(11.2.39) $\quad g_c N\{a(y)\,|\,y_c \leqslant y \leqslant \eta\}K.$

So

(11.2.40) $$\sup_{\substack{c,\,n\in N\\ y_c\leqslant y\leqslant\eta,\,k\in K}} y^{-d}\,|C_\phi(T_{i'})^{-1}\hat{T}_{i'}\phi(s_{i'})(g_c n a(y)k)|$$

goes to zero for each $\eta > 0$. By assertion (11.2.17) the value of C_ϕ is determined on sets as in (11.2.39), hence

(11.2.41) $$\lim_{i'\to\infty} C_\phi(C_\phi(T_{i'})^{-1}T_{i'}) = 0.$$

But $C_\phi(C_\phi(T_{i'})^{-1}T_{i'}) = 1$ for all i'. So the assumption we started with is untenable.

Proof of proposition 11.2.4. By lemma 11.2.9 and lemma 11.2.6 there are subsequences $\hat{T}_{i'}\phi(s_{i'})$ converging to limit functions with properties described in lemma 11.2.7. From (11.2.30) follows that the difference of two such functions satisfies condition (10.6.2). Further (11.2.15-16) and proposition 7.5.9 imply that the difference is bounded. So the same reasoning as in the proof of lemma 11.2.7 shows that all limit functions are equal. Lemma 11.2.6 ensures that there cannot be subsequences not containing converging subsequences. So $\hat{T}_i\phi(s_i)$ converges uniformly on compact sets to a function denoted $T\phi(s)$. Conditions (4.5.1) and (4.5.2) in the definition of automorphic models are satisfied, see lemma 11.2.7. By the theorem of Lebesgue the orthogonality to the cusp forms is satisfied. From (11.2.30) follows that $*FT\phi(s) = \underline{St}_\tau^{(\delta)}(\underline{n})$ or $*FT\phi(s) = \underline{\hat{St}}_1^{\delta}(\underline{n})$ in the case $\tau = 1$, $s = 0$.

Clearly T is a linear operator. From the uniform convergence on compact sets it is clear that T is an intertwining operator for the K-action and satisfies condition (11.2.6) in definition 11.2.2.

The only thing left to check is the $\underline{g}$-action. Take $f \in C^0$ such that $\alpha_f(s,\phi) = 1$. For each $\underline{X} \in \underline{g}_{\mathbb{R}}$ there are constant sections $\phi_1 \dots \phi_l$ and holomorphic functions $\beta_1 \dots \beta_l$ such that

(11.2.42) $\quad \underline{X}\phi(z) = \sum_j \beta_j(z)\phi_j(z) \qquad$ for $z \in \mathbb{C}$.

Hence, on compact sets:

(11.2.43)
$$\begin{aligned}\underline{X}T\phi(s) &= \underline{X}(T\phi(s) * f)\\ &= T\phi(s) * \underline{X}f\\ &= \lim \hat{T}_i\phi(s_i) * \underline{X}f\\ &= \lim \underline{X}(\hat{T}_i\phi(s_i) * f) = \lim \alpha_f(s_i,\phi)\,\underline{X}\hat{T}_i\phi(s_i)\\ &= \lim \alpha_f(s_i,\phi)\sum_j \beta_j(s_i)\hat{T}_i\phi_j(s_i)\\ &= 1\sum_j \beta_j(s)T\phi_j(s) = T\underline{X}\phi(s).\end{aligned}$$

11.3. Spectral decomposition of functions on $N\backslash G$

This subsection discusses the spectral decomposition of the $\mathfrak{g}$-K-space ${}^0S_\tau$ defined in 3.6. We need this for the spectral decomposition of ${}^eL^2$ and the analytic continuation of the Eisenstein model. Moreover, it is the example on which subsection 13.4 has been modelled. The material in this subsection is well known; compare e.g. [8], p. 220-222.

The integral in (8.3.2) defines a scalar product for functions on $N\backslash G$, provided it converges. In 3.6 we have seen that it gives a unitary $\mathfrak{g}$-K-structure on ${}^0S_\tau$.

11.3.1. Proposition. *Let $\sigma_1 < \sigma_2$. Suppose $f \in C^\infty(N\backslash G)$ is of type τ, is K-finite on the right and satisfies*

(11.3.1) $\quad |f(a(y)k)| \ll y^{\frac{1}{2}+\sigma_1}$ *for $y \to \infty$, uniformly in $k \in K$*

(11.3.2) $\quad |f(a(y)k)| \ll y^{\frac{1}{2}+\sigma_2}$ *for $y \searrow 0$, uniformly in $k \in K$.*

Then

(11.3.3) $\quad {}^0\omega f(s) = \sum_\psi \langle f, \mathrm{St}(-\bar{s})\psi(-\bar{s})\rangle \psi(s),$

with ψ running through the standard basis of H^τ, defines a holomorphic section of H^τ on $\sigma_1 < \mathrm{Re}\, s < \sigma_2$.

(11.3.4) $\quad k.{}^0\omega f = {}^0\omega(k.f)$ *for all $k \in K$.*

If $X \in \mathfrak{g}_{\mathbb{R}}$ and Xf satisfies the estimates (11.3.1) and (11.3.2), then

(11.3.5) $\quad X.{}^0\omega f = {}^0\omega(Xf)$.

Proof. Conditions (11.3.1) and (11.3.2) ensure that $\langle f, \mathrm{St}(-\bar{s})\psi(-\bar{s})\rangle$ is well defined, see (8.3.3); the holomorphy is also clear. The K-behaviour gives no problems. The $\mathfrak{g}_{\mathbb{R}}$-behaviour is obtained by differentiation under the integral. To see that this gives no problem, remark that $k\, e^{h_1 X} \in p(z)K$ with z in a small neighbourhood of $i \in \mathfrak{h}$, for $k \in K$ and X fixed and h_1 small.

11.3.2. Corollary. *${}^0\omega$ is a $\mathfrak{g}$-K-intertwining operator from ${}^0S_\tau$ into the holomorphic global sections of H^τ.*

I denote the Mellin transform by M:

(11.3.6) $\quad M\phi(s) = \int_0^\infty \phi(y)\, y^{s-1} dy.$

Every $f \in {}^0S_\tau$ may be written

(11.3.7) $\quad f(n(x)a(y)j_\varepsilon k(\theta)) = \sum_{r \equiv \pm\tau(2)} f_r^\varepsilon(y) e^{ir\theta},$

with $f_r^\varepsilon \in C_c^\infty(0, \infty)$. Then

(11.3.8) $\quad {}^0\omega f(s) = \sum_{\delta=\pm1} \sum_{r\equiv\tau(2)} \frac{1}{2} \sum_{\varepsilon=\pm1} (\delta,\varepsilon) Mf_r^\varepsilon(-\tfrac{1}{2}-s)\, \phi_r^\delta(s) \qquad$ if $\tau = 0, 1$

(11.3.9) $\quad {}^0\omega f(s) = \sum_{\delta=\pm1} \sum_{\varepsilon=\pm1} \sum_{r\equiv\varepsilon\delta\tau(2)} \frac{1}{\sqrt{2}} Mf_r^\varepsilon(-\tfrac{1}{2}-s)\, \phi_r^\delta(s) \qquad$ if $0 < \tau < 1$.

So ${}^0\omega\,{}^0S_\tau$ consists of rapidly decreasing sections, i.e. all coefficients are estimated on vertical strips by

(11.3.10) $\quad (1 + |\mathrm{Im}\, s|)^{-a}$

for each $a \in \mathbb{R}$.

11.3.3. Proposition. Take $\sigma_1 < \sigma_2$. Let ξ be a holomorphic rapidly decreasing section of H^τ over the vertical strip $\sigma_1 \leqslant \mathrm{Re}\ s \leqslant \sigma_2$. Then the function ${}^0\Phi\xi$ on $N\backslash G$ is defined by

$$(11.3.11) \qquad {}^0\Phi\xi = \frac{1}{2\pi i} \int_{\mathrm{Re}\ s = \sigma} \mathrm{St}(s)\xi(s)ds,$$

independent of $\sigma \in [\sigma_1, \sigma_2]$. It inverts ${}^0\omega$ and it satisfies

$$(11.3.12) \qquad |{}^0\Phi\xi(a(y)k)| \ll y^{\frac{1}{2}+\sigma}$$

for all $y > 0$, $k \in K$, for each $\sigma \in [\sigma_1, \sigma_2]$.

Proof. Clear by Mellin inversion.

11.3.4. Proposition. Take $\sigma_1 < \sigma_2$. Let ξ and η be holomorphic rapidly decreasing sections of H^τ, ξ over $\sigma_1 \leqslant \mathrm{Re}\ s \leqslant \sigma_2$ and η over $-\sigma_2 \leqslant \mathrm{Re}\ s \leqslant -\sigma_1$. Then for each $\sigma \in [\sigma_1, \sigma_2]$

$$(11.3.13) \qquad \frac{1}{2\pi i} \int_{\mathrm{Re}\ s = \sigma} \{\xi, \eta\}(s)ds = \langle {}^0\Phi\xi, {}^0\Phi\eta \rangle.$$

Proof. Plancherel formula for the Mellin transform.

11.3.5. Proposition. Let $\sigma > 0$. Suppose ξ is a holomorphic rapidly decreasing section of H^τ on $|\mathrm{Re}\ s| \leqslant \sigma$. Let $t > 0$ and $\varepsilon = \pm 1$. Then

$$(11.3.14) \qquad {}_0V^\varepsilon(t)\,{}^0\Phi\xi = \frac{1}{2\pi i} \int_{\mathrm{Re}\ s = \sigma_1} 2^{1-2s}\Gamma(2s)\sin \pi(\tfrac{1}{2} + s + \tfrac{1}{2}\tau) . t^{\frac{1}{2}+s}\mathrm{St}(-s)\iota(s)X^\varepsilon(s)\xi(s)ds$$

$$(11.3.15) \qquad {}_{\frac{1}{2}}V^\varepsilon(t)\,{}^0\Phi\xi = \frac{1}{2\pi i} \int_{\mathrm{Re}\ s = \sigma_1} 2^{1-2s}\Gamma(2s)\sin \pi(\tfrac{1}{2} + s + \tfrac{1}{2}\tau) . W(s)X^\varepsilon(s)\xi(s)ds$$

for each $\sigma_1 \in (0, \sigma]$.

Proof. The integral in the definition of ${}_uV^\varepsilon(t)$, see 6.2.3, may be taken inside the integral in (11.3.11). Then apply the propositions 6.8.2 and 6.8.3.

11.4. Integral formulas for ${}^eL^2(\Gamma\backslash G, \chi)$

The subspace ${}^eL^2$ of L^2 is generated by the $\Theta_\lambda f$ with $f \in {}^0S_\tau$, $\lambda \in \Lambda^0$; see 10.6.1. For any holomorphic rapidly decreasing section ξ of H^τ on $|\mathrm{Re}\ s| \leqslant \sigma$ with $\sigma > \frac{1}{2}$, we have $\Theta_\lambda {}^0\Phi\xi \in L^2$; see propositions 11.3.3 and 8.2.3. By proposition 8.3.3 and condition (10.6.2) we get even $\Theta_\lambda {}^0\Phi\xi \in ({}^0L)^\perp = {}^eL$. We shall describe the scalar product in ${}^eL^2$ of such functions.

11.4.1. Notations. $*S_\tau = {}^0S_\tau \otimes E$, $*H^\tau = H^\tau \otimes E$; a section ξ of $*H^\tau$ is holomorphic, rapidly decreasing if its coefficient sections with respect to $\{\underline{e}_\lambda | \lambda \in \Lambda^0\}$ are holomorphic and rapidly decreasing. For sections ξ of $*H^\tau$ we define $*\Phi\xi$ by

$$(11.4.1) \qquad ((*\Phi\xi, \underline{e}_\lambda)) = {}^0\Phi((\xi, \underline{e}_\lambda)) \qquad \text{for } \lambda \in \Lambda^0.$$

For sections ξ, η of $*H^\tau$

$$(11.4.2) \qquad \{\xi, \eta\}(s) = \sum_{\lambda \in \Lambda^0} \{((\xi, \underline{e}_\lambda)), ((\eta, \underline{e}_\lambda))\}(s).$$

$\pi_1 : *H^\tau \to *H^\tau$ is given by

$$(11.4.3) \qquad ((\pi_1\xi, \underline{e}_\lambda)) = \pi_1((\xi, \underline{e}_\lambda)) \qquad \text{for } \lambda \in \Lambda^0;$$

see definition 9.4.1. The maps ι and X^ε act in $*H^\tau$ by their action on the first factor.

11.4.2. Proposition. Take $\sigma > \frac{1}{2}$. Let ξ and η be holomorphic rapidly decreasing sections of $*H^{\tau}$ on $|\mathrm{Re}\ s| \leqslant \sigma$. Then

$$(11.4.4)\qquad \langle *\Theta*\Phi\xi,\ *\Theta*\Phi\eta\rangle = \frac{1}{2\pi i}\int_{\mathrm{Re}\ s = \sigma_1} \{\pi_1\xi,\ \eta\}(s)ds + \frac{1}{2\pi i}\int_{\mathrm{Re}\ s = \sigma_2} \{\iota\underline{M}\ \pi_1\xi,\ \eta\}(s)ds,$$

for each $\sigma_1 \in [-\sigma, \sigma]$, $\sigma_2 \in [-\sigma, -\frac{1}{2}]$.

Remark. $\iota\underline{M}\ \pi_1\xi$ is a section of $*H^{\tau}$, defined by

$$(11.4.5)\qquad \iota\underline{M}\ \pi_1\xi(s) = \iota(-s)\underline{M}(-s)\ \pi_1\xi(-s).$$

From proposition 8.5.3 and definition 9.4.2 follows that $\iota\underline{M}\ \pi_1\xi$ is holomorphic rapidly decreasing on $-\sigma \leqslant \mathrm{Re}\ s < -\frac{1}{2}$.

Proof. Application of propositions 8.3.3 and 8.5.1 gives

$$(11.4.6)\qquad \langle *\Theta*\Phi\xi,\ *\Theta*\Phi\eta\rangle = \sum_{\lambda,\mu\in\Lambda^0} {}_0\langle F_{\mu}{}^0{}_{\lambda}\,{}^0\Phi((\xi,\ \underline{e}_{\lambda})),\ {}^0\Phi((\eta,\ \underline{e}_{\mu}))\rangle =$$
$$\sum_{\lambda,\mu\in\Lambda^0} {}_0\langle \delta_{\lambda\mu} P_1\,{}^0\Phi((\xi,\ \underline{e}_{\lambda})),\ {}^0\Phi((\eta,\ \underline{e}_{\mu}))\rangle +$$
$$\sum_{\lambda,\mu\in\Lambda^0} {}_0\langle \sum_{c\in C_{\lambda\mu}} S(\lambda,\ \mu;\ c)\,{}_0V^1(c^{-2})P_1\,{}^0\Phi((\xi,\ \underline{e}_{\lambda})),\ {}^0\Phi((\eta,\ \underline{e}_{\mu}))\rangle.$$

The first term equals

$$(11.4.7)\qquad \frac{1}{2\pi i}\int_{\mathrm{Re}\ s = \sigma_1} \{\pi_1\xi,\ \eta\}(s)ds;$$

use proposition 11.3.5 and $P_1{}^0\Phi = {}^0\Phi\pi_1$.

By propositions 11.3.5 and 8.5.3:

$$(11.4.8)\qquad \sum_{c\in C_{\lambda\mu}} S(\lambda,\ \mu;\ c)\,{}_0V^1(c^{-2})P_1\,{}^0\Phi((\xi,\ \underline{e}_{\lambda})) =$$
$$\sum_{c\in C_{\lambda\mu}}^{\lambda\mu} S(\lambda,\ \mu;\ c)\ \frac{1}{2\pi i}\int_{\mathrm{Re}\ s = -\sigma_2} 2^{1-2s}\Gamma(2s)\sin\pi(\tfrac{1}{2} + s + \tfrac{1}{2}\tau).$$
$$.\ c^{-1-2s}St(-s)\iota(s)X^1(s)\pi_1((\xi(s),\ \underline{e}_{\lambda}))ds =$$
$$\frac{1}{2\pi i}\int_{\mathrm{Re}\ s = \sigma_2} St(s)((\underline{M}(-s)\underline{e}_{\lambda},\ \underline{e}_{\mu}))\iota(-s)((\pi_1\xi(-s),\ \underline{e}_{\lambda}))ds.$$

The second term in the right hand side of (11.4.6) equals

$$(11.4.9)\qquad \sum_{\lambda,\mu\in\Lambda^0} \frac{1}{2\pi i}\int_{\mathrm{Re}\ s = \sigma_2} ((\underline{M}(-s)\underline{e}_{\lambda},\ \underline{e}_{\mu}))\{((\iota\ \pi_1\xi,\ \underline{e}_{\lambda})),\ ((\eta,\ \underline{e}_{\mu}))\}(s)ds$$
$$= \frac{1}{2\pi i}\int_{\mathrm{Re}\ s = \sigma_2} \{\iota\underline{M}\ \pi_1\xi,\eta\}(s)ds.$$

11.4.3. Remark. The form

$$(11.4.10)\qquad (\xi,\ \eta) \mapsto \langle *\Theta*\Phi\xi,\ *\Theta*\Phi\eta\rangle$$

gives a pre-hilbert space structure on the rapidly decreasing holomorphic sections of $*H^{\tau}$ on $|\mathrm{Re}\ s| \leqslant \sigma$. After dividing out a null space and completion we get a hilbert space H, isomorphic to ${}^{e}L^2$.

11.4.4. Remark. Holomorphic sections η of $*H^\tau$ on $|\mathrm{Re}\ s| \leq \sigma$ estimated by $(1+|\mathrm{Im}\ s|)^{-1}$ still give rise to elements of H for which (11.4.4) stays valid. To see this approximate η by $\eta_v(s) = e^{vs^2}\eta(s)$, $v > 0$.

11.5. Holomorphic continuation of $\underline{M}$ and $\underline{E}$ to $\mathcal{D}$

$\mathcal{D}$ is the region $\{s \in \mathbb{C} | \mathrm{Re}\ s > 0,\ s \notin (0,\frac{1}{2}]\}$.
$\underline{M}$ and $\underline{E}$ are holomorphic for $\mathrm{Re}\ s > \frac{1}{2}$. The continuation to $\mathcal{D}$ is proved by the method given in [22], p. 321-325.

11.5.1. Lemma. $\underline{M}$ has a holomorphic continuation to the region $\mathcal{D}$.

Proof. Compare [22], p. 321-322 and p. 127-128.
Multiplication with s^2 of sections of $*H^\tau$ gives an unbounded selfadjoint operator A in H; see remark 11.4.3. Its resolvent $R(z^2,A)$ is given by multiplication with the function $s \to (z^2-s^2)^{-1}$. The resolvent $R(z^2,A)$ may be extended to a holomorphic function on $\mathcal{D}$, for A is selfadjoint.
Fix $\lambda,\mu \in \Lambda^0$ and a constant section ψ of H^τ as indicated in (3.5.18), (3.5.19) such that $\pi_1\psi = \psi$. Take

$$\text{(11.5.1)}\qquad \begin{aligned}\xi(s) &= e^{s^2}\,\psi(s)\otimes \underline{e}_{\lambda}\\ \eta(s) &= e^{s^2}\,\psi(s)\otimes \underline{e}_{\mu}.\end{aligned}$$

Denote the corresponding elements of H by $\tilde{\xi}$ and $\tilde{\eta}$. For $\frac{1}{2} < \sigma < \mathrm{Re}\ z$

$$\text{(11.5.2)}\qquad \langle R(z^2,A)\tilde{\xi},\tilde{\eta}\rangle_H = \frac{1}{2\pi i}\int_{\mathrm{Re}\ z=\sigma_1} (z^2-s^2)^{-1}\,\delta_{\lambda\mu}\,e^{2s^2}\,ds$$
$$+ \frac{1}{2\pi i}\int_{\mathrm{Re}\ z=\sigma_2} (z^2-s^2)^{-1}\,\gamma_\psi(-s)\,e^{2s^2}\,((\underline{M}(-s)\underline{e}_\lambda,\underline{e}_\mu))ds$$

with $|\sigma_1| \leq \sigma$, $-\sigma \leq \sigma_2 < -\frac{1}{2}$

$$= \frac{1}{2z}(e^{2z^2} + \gamma_\psi(z)\,e^{2z^2}((\underline{M}(z)\underline{e}_\lambda,\underline{e}_\mu)))$$

$+$ the same integrals with $\sigma_1 > \mathrm{Re}\ z$, $\sigma_2 < -\mathrm{Re}\ z$.

Remark that $\gamma_\psi(z)\ \underline{M}(z)$ is holomorphic in z with $\mathrm{Re}\ z > \frac{1}{2}$, for it is part of the Fourier coefficient of order zero of $\underline{E}(z)$. The left hand side of (11.5.2) is holomorphic in $z \in \mathcal{D}$, the integrals are also holomorphic in $z \in \mathcal{D}$. So $z \mapsto \gamma_\psi(z)\ \underline{M}(z)$ is holomorphic on $\mathcal{D}$. Possible zeros of $\gamma_\psi(z)$ with $0 < \mathrm{Re}\ z \leq \frac{1}{2}$ occur in $(0,\frac{1}{2}]$. This proves the lemma.

11.5.2. Proposition. Let ψ be a constant section of $*H^\tau$. For each $g \in G$ the function $s \to \underline{E}(s)\ \psi(s)\ (g)$ has a holomorphic extension to $\mathcal{D}$; the complex differentiability with respect to s is uniform for g in compact subsets of G.
The operator $\underline{E}(s) : \psi(s) \to \underline{E}(s)\psi(s)$ so obtained for $s \in \mathcal{D}$, $\mathrm{Re}\ s \leq \frac{1}{2}$ is an automorphic model of $*H^\tau(s)$. For each $\underline{e} \in E$: $\underline{E}(s)\underline{e} \in A^e(H^\tau(s))$.

Remark. All Fourier coefficients of $\underline{E}(s)$ are holomorphic on $\mathcal{D}$, as follows from the uniform differentiability. So not only $\underline{M}$, but also the $\underline{f}_\alpha$ have holomorphic extensions to $\mathcal{D}$.

The proof of proposition 11.5.2 is prepared in a series of lemmas. These lemmas concern the continuation of $s \to \underline{E}(s)\psi(s)\otimes\underline{e}_\lambda(g)$. We may assume that ψ is a standard basis element of H^τ, with $\pi_1\psi = \psi$; otherwise $\underline{E}(s)\psi(s)\otimes\underline{e}_\lambda = 0$. Of course $\lambda \in \Lambda^0$. Define for $s \in \mathbb{C}$ the functions $f_1(s)$ and $f_2(s)$ on G by

(11.5.3)
$$f_1(s)(n\,a(y)k) = \begin{cases} St(s)\psi(s)(a(y)k) & \text{if } y \leqslant 1 \\ 0 & \text{if } y > 1 \end{cases}$$

$$f_2(s)(na(y)k) = \begin{cases} 0 & \text{if } y \leqslant 1 \\ St(s)\psi(s)(a(y)k) & \text{if } y > 1 \end{cases}$$

for $n \in N$, $k \in K$. The function $f_1(s)$ corresponds to the function F' on p. 323 of [22]. For Re $s > \frac{1}{2}$

(11.5.4) $\quad \underline{E}(s)\psi(s) \otimes \underline{e}_\lambda = \Theta_\lambda f_1(s) + \Theta_\lambda f_2(s)$.

11.5.3. Lemma. $\Theta_\lambda f_2(s)$ is well defined for all $s \in \mathbb{C}$. It satisfies

(11.5.5) $\quad |\Theta_\lambda f_2(s)(g_c\; n\; a(y)k)| \ll \max(1, y^{\frac{1}{2}+\text{Re } s})$ for $y \to \infty$

uniformly in $n \in N$, $k \in K$ for each cusp c.

(11.5.6) $\quad \langle \Theta_\lambda f_2(s), \psi_1 \rangle = 0$ for all cups forms ψ_1.

Proof. Use propositions 8.2.3 and 8.3.3.

11.5.4. Lemma. $s \to \Theta_\lambda f_2(s)(g)$ is holomorphic on $\mathbb{C}$; the complex differentiability is uniform for g in compact sets in G. For each $\phi \in C_c^\infty(G)$ the function $\Theta_\lambda f_2(s) * \phi$ satisfies the estimate (11.5.5).

Proof. From the geometry of fundamental domains for $\bar{\Gamma}\backslash h$, see e.g. [30], follows that $\Theta_\lambda f_2(s)(g)$ is given by a finite sum; the number of terms, for g in a compact set, is finite. This proves the first assertion. The estimate for $\Theta_\lambda f_2(s)*\phi$ may be proved directly or by the methods in 10.5. In the direct proof one shows that $\Theta_\lambda f_2(s)*\phi = \Theta_\lambda(f_2(s)*\phi)$ and that $f_2(s)*\phi(n\;a(y)k)$ is zero for small y and is $\mathcal{O}(y^{\frac{1}{2}+\text{Re } s})$ for y large.

11.5.5. Lemma. Let $s \in \mathcal{D}$ and $-\text{Re } s < \sigma_1 < \sigma_2$. Define the section η_s of ${}^*H^\tau$ by

(11.5.7) $\quad \eta_s(z) = (s-z)^{-1}\psi(z)\otimes\underline{e}_\lambda$.

Then for rapidly decreasing holomorphic sections ξ of H^τ on $\sigma_1 \leqslant \text{Re } s \leqslant \sigma_2$

(11.5.8) $$\langle {}^\circ\phi\xi, f_1(s)\rangle = \frac{1}{2\pi i}\int_{\text{Re } z=\sigma_0} \{\xi, ((\eta_s, \underline{e}_\lambda))\}(z)\,dz$$

for $\sigma_0 \in [\sigma_1, \sigma_2]$.

Proof.

(11.5.9) $$\langle {}^\circ\Phi\xi, f_1(s)\rangle = \frac{1}{2\pi i}\int_{\mathrm{Re}\, z=\sigma_0}\int_{Z_0\backslash K}\int_0^1 \mathrm{St}(z)\xi(z)(k)\,\overline{\mathrm{St}(s)\psi(s)(k)}\,. $$
$$\cdot\; y^{\frac12+z+\frac12+\bar s-2}\,dy\,dk\,dz$$
$$= \frac{1}{2\pi i}\int_{\mathrm{Re}\, z=\sigma_0}(z+\bar s)^{-1}\{\xi,\psi\}(z)\,dz =$$
$$= \frac{1}{2\pi i}\int_{\mathrm{Re}\, z=\sigma_0}\{\xi,((\eta_s,\underline{e}_\lambda))\}(z)\,dz.$$

11.5.6. Lemma. Let $\mathrm{Re}\, s > \frac12$; take $\frac12 < \sigma < \mathrm{Re}\, s$. For each rapidly decreasing holomorphic section ξ of ${}^*H^\tau$ on $|\mathrm{Re}\, z| \leqslant \sigma$:

(11.5.10) $$\langle {}^*\Theta{}^*\Phi\xi,\ \Theta_\lambda f_1(s)\rangle = \frac{1}{2\pi i}\int_{\mathrm{Re}\, z=-\sigma}\{(1+\iota\underline{M})\pi_1\xi,\eta_s\}(z)\,dz$$

Proof. In the same way as in the proof of proposition 11.4.2 we get

(11.5.11) $$\langle {}^*\Theta{}^*\Phi\xi,\ \Theta_\lambda f_1(s)\rangle = \langle {}^\circ\Phi(((1+\iota\underline{M})\pi_1\xi,\underline{e}_\lambda)),\ f_1(s)\rangle.$$

Apply the previous lemma.

11.5.7. Lemma. $h_s = \Theta_\lambda f_1(s)$ defines a holomorphic map

(11.5.12) $$\{s\in\mathbb{C}\,|\,\mathrm{Re}\, s>\tfrac12\}\to {}^eL^2 \;:\; s\mapsto h_s.$$

Proof. $\Theta_\lambda f_1(s)$ represents an element of ${}^eL^2$; this follows from propositions 8.2.3, 10.5.6, 8.3.3 and condition (10.6.2). By looking at (11.5.3) we see that if s_0 is given, with $\mathrm{Re}\, s_0 > \frac12$, there is a neighbourhood U of s_0 such that

(11.5.13) $$\left|\frac{d}{ds} f_1(s)(n\,a(y)k)\right| \ll \begin{cases} y^{1+\varepsilon}\mathrm{St}(s_0)\psi(s_0)(k) & \text{if } y\leqslant 1\\ 0 & \text{if } y>1\end{cases}$$

uniformly for $s\in U$, $n\in N$, $k\in K$. Now we may apply propositions 8.2.3 and 10.5.6 to see that

(11.5.14) $$\lim_{s\to s_0}\|(s-s_0)^{-1}(\Theta_\lambda f_1(s)-\Theta_\lambda f_1(s_0))-\Theta_\lambda\tfrac{d}{ds}f_1(s_0)\| = 0.$$

11.5.8. Lemma. For $\mathrm{Re}\, s>\frac12$, $\mathrm{Re}\, u>\frac12$, $s\neq\bar u$

(11.5.15) $$\langle h_s,h_u\rangle = -(s+\bar u)^{-1}$$
$$+\ (\bar u-s)^{-1}\gamma_\psi(s)((\underline{M}(s)\underline{e}_\lambda,\underline{e}_\lambda))+(s-\bar u)^{-1}\gamma_\psi(\bar u)((\underline{M}(\bar u)\underline{e}_\lambda,\underline{e}_\lambda)).$$

Proof. Take $\sigma>\frac12$, $\sigma<\mathrm{Re}\, s$, $\sigma<\mathrm{Re}\, u$. By lemma 11.5.6 h_s is represented by η_s. Using remark 11.4.4 we see that

(11.5.16) $$\langle h_s,h_u\rangle = \frac{1}{2\pi i}\int_{\mathrm{Re}\, s=-\sigma}(((s-z)^{-1}\underline{e}_\lambda+\gamma_\psi(-z)\underline{M}(-z)(s+z)^{-1}\underline{e}_\lambda,(u+\bar z)^{-1}\underline{e}_\lambda))\,dz$$
$$= \frac{1}{2\pi i}\int_{\mathrm{Re}\, s=-\sigma}(s-z)^{-1}(\bar u+z)^{-1}\,dz$$
$$+\ \frac{1}{2\pi i}\int_{\mathrm{Re}\, s=-\sigma}(s+z)^{-1}(\bar u+z)^{-1}\gamma_\psi(-z)((\underline{M}(-z)\underline{e}_\lambda,\underline{e}_\lambda))\,dz$$

Moving the line of integration to the left we obtain the lemma.

11.5.9. Lemma. The map $s\to h_s$ has a holomorphic continuation to $\mathcal{D}$.

Proof. We see in lemma 11.5.8 that there is a holomorphic function ω on $\mathcal{D}\times\mathcal{D}\backslash\{(s,s)\,|\,s\in\mathcal{D}\}$ such that $\langle h_s,h_u\rangle = \omega(s,\bar u)$ for $\mathrm{Re}\, s$, $\mathrm{Re}\, u>\frac12$, $\bar u\neq s$.

Furthermore, for Re $s > \frac{1}{2}$

(11.5.17) $$\left\|\frac{d^n}{ds^n} h_s\right\|^2 = \frac{d^{2n}}{ds^n d\bar{s}^n} \omega(s,\bar{s}).$$

Take s_0 with Re $s_0 > \frac{1}{2}$, $s_0 \notin \mathbb{R}$. Considering the power series at s_0 of h_s and at $(s_0,\bar{s}_0)$ of $\omega(s,\bar{s})$ we see that the first one certainly converges for those s with $(s,\bar{s})$ in the interior of the circle of convergence of the latter one. Successive application of this argument ensures that h_s is holomorphic whereever $\omega(s,\bar{s})$ is. This proves the lemma.

11.5.10. Lemma. <u>For each $\phi \in C_c^\infty(G_0)$ and $g \in G$ the function $s \to \underline{r}(\phi)h_s(g)$ is complex differentiable on $\mathcal{D}$, uniformly for g in compact sets.</u>

<u>Proof</u>. By proposition 10.5.7.

(11.5.18) $$(s_1-s)^{-1}(\underline{r}(\phi)h_{s_1}(g)-\underline{r}(\phi)h_s(g)) - \underline{r}(\phi)\frac{d}{ds}h_s(g)$$

tends to zero for $s_1 \to s$, uniformly for g in compact sets.

<u>Proof</u> of proposition 11.5.2. Denote for each $\phi \in C_c^\infty(G)$

(11.5.19) $$\phi_1(g) = \phi(g^{-1}).$$

Let $\phi \in C^0$. For Re $s > \frac{1}{2}$

(11.5.20) $$(\underline{E}(s)\psi(s)\otimes\underline{e}_\lambda)*\phi = \alpha_\phi(s,\psi)\underline{E}(s)\psi(s)\otimes\underline{e}_\lambda,$$

on the other hand, by (11.5.4) and lemma 11.5.7

(11.5.21) $$(\underline{E}(s)\psi(s)\otimes\underline{e}_\lambda)*\phi = \Theta_\lambda f_2(s)*\phi + \underline{r}(\phi_1)h_s.$$

So the meromorphic pointwise extension of $E(s)\psi(s)\otimes\underline{e}_\lambda$ is given by

(11.5.22) $$\underline{E}(s)\psi(s)\otimes\underline{e}_\lambda(g) = \alpha_\phi(s,\psi)^{-1}(\Theta_\lambda f_2(s)*\phi(g) + \underline{r}(\phi_1)h_s(g)).$$

The extension does not depend on $\phi \in C^0$, as the left hand side is independent of ϕ for Re $s > \frac{1}{2}$. For any $s_0 \in \mathcal{D}$ we can choose ϕ such that $\alpha_\phi(s,\psi) \neq 0$ in a neighbourhood of s_0 (see proposition 10.2.6). So the extension is in fact holomorphic. From the lemmas 11.5.4 and 11.5.10 follows that the complex differentiability is uniform for g in compact sets.

Let $\chi \in C^0$. Then

(11.5.23) $$\begin{aligned}\underline{E}(s)\psi(s)\otimes\underline{e}_\lambda*\chi &= \alpha_\phi(s,\psi)^{-1}(\Theta_\lambda f_2(s)*(\phi*\chi) + \underline{r}((\phi*\chi)_1)h_s) = \\ &= \alpha_\phi(s,\psi)^{-1}\,\alpha_{\phi*\chi}(s,\psi)\underline{E}(s)\psi(s)\otimes\underline{e}_\lambda = \alpha_\chi(s,\psi)\underline{E}(s)\psi(s)\otimes\underline{e}_\lambda.\end{aligned}$$

$\underline{E}(s)\psi(s)\otimes\underline{e}_\lambda$ is orthogonal to the cusp forms, for $\Theta_\lambda f_2(s)$ and h_s are, and $\underline{r}(C^0)$ maps cusp forms into cusp forms. So (11.2.5) and (11.2.6) are satisfied. To see that $\underline{E}(s)\psi(s)\otimes\underline{e}_\lambda$ satisfies the growth condition (4.5.2) use lemma 11.5.4 and proposition 10.5.7. The Γ-equivariance in (4.5.1) is satisfied by $\underline{E}(s)\psi(s)\otimes\underline{e}_\lambda$ for Re $s > \frac{1}{2}$ and is retained on continuation; the same holds for the K-behaviour.

To finish the proof the behaviour under differentiation has to be checked. Let $\underline{X} \in \underline{g}$, then

(11.5.24) $$\underline{X}\psi(s) = \sum_j \beta_j(s)\psi_j(s)$$

for some holomorphic β_j and constant sections ψ_j.

The function $\underline{E}(s)\psi(s)\otimes\underline{e}_\lambda$ is differentiable, for

(11.5.25) $\quad \underline{X}\ \underline{E}(s)\psi(s)\otimes\underline{e}_\lambda(g) =$

$$\alpha_\phi(s,\psi)^{-1}\ \underline{E}(s)\psi(s)\otimes\underline{e}_\lambda * \underline{X}\ \phi(g).$$

Furthermore, the right hand side of (11.5.25) is meromorphic for $s \in \mathcal{D}$, as $\underline{X}\phi$ has compact support. For Re $s > \frac{1}{2}$ we may write out the left hand side of (11.5.25) using (11.5.24). The resulting equality stays valid after continuation.

11.6. Meromorphic continuation of $\underline{M}$ and $\underline{E}$ to $\mathcal{D}_*$.

$\mathcal{D}_*$ is the region $\mathbb{C}\setminus[-\frac{1}{2},\frac{1}{2}]$. Compare for this subsection [22], p. 325-326, 331.

11.6.1. Proposition. $\underline{M}$ has a meromorphic continuation to $\mathcal{D}_*$; it is holomorphic on $i\mathbb{R}\setminus\{0\}$. It satisfies

(11.6.1) $\quad \underline{M}(-s)\underline{M}(s) = I.$

The proof is prepared by some lemmas.

I fix a basis section ψ as in the proof of lemma 11.5.1. For Re $u > \frac{1}{2}$ and $\underline{e} \in E$ the function $f_3(u,\underline{e})$ on G with values in E is defined by

(11.6.2)
$$f_3(u,\underline{e})(n\ a(y)k) = \begin{cases} St(u)\psi(u)(n\ a(y)k)\underline{e} & \text{if } y \leq 1 \\ -\gamma_\psi(u)St(-u)\psi(-u)(n\ a(y)k)\underline{M}(u)\underline{e} & \text{if } y > 1 \end{cases}$$

Compare the function F" on p. 323 of [22]. Remark that

(11.6.3) $\quad f_3(u,\underline{e}) = f_1(u)\underline{e} - f_2(-u)\gamma_\psi(u)\underline{M}(u)\underline{e}.$

11.6.2. Lemma. Let Re $u > \frac{1}{2}$, $\underline{e} \in E, \frac{1}{2} < \sigma <$ Re u. For any rapidly decreasing holomorphic section ξ of $^*H^\tau$

(11.6.4) $\quad \langle {}^*\Theta{}^*\phi\xi, {}^*\Theta f_3(u,\underline{e})\rangle = \frac{1}{2\pi i}\int_{\mathrm{Re}\ s=-\sigma} \{(1+\iota M)\pi_1\xi,\ \eta_{u,\underline{e}}\}(s)ds$

with

(11.6.5) $\quad \eta_{u,\underline{e}}(s) = (u-s)^{-1}\ \psi(s)\otimes\underline{e} - (u+s)^{-1}\ \psi(s)\otimes\gamma_\psi(u)\underline{M}(u)\underline{e}.$

Proof. As in the case of lemma 11.5.6.

11.6.3. Lemma. Let $\underline{e} \in E$. For Re $u > \frac{1}{2}$ the function $^*\Theta f_3(u,\underline{e})$ represents an element $l_{u,\underline{e}} \in {}^eL^2$. The map $u \to l_{u,\underline{e}}$ has a holomorphic extension to $\mathcal{D}$.

Proof. We handle $^*\Theta f_1(u)\underline{e}$ and $^*\Theta f_2(-u)\gamma_\psi(u)\underline{M}(u)\underline{e}$ separately. For the first function use lemma 11.5.9. The other case may be handled as in the proof of lemma 11.5.4; as Re$(\frac{1}{2}-u) < \frac{1}{2}$ for $u \in \mathcal{D}$ we see that the differentiation is well defined in the L^2-sense.

11.6.4. Lemma. For $\underline{e}$, $\underline{e}_1 \in E$, $u,z \in \mathcal{D}$, $u \neq \bar{z}$

(11.6.6)
$$\langle l_{u,\underline{e}}, l_{z,\underline{e}_1}\rangle = (u+\bar{z})^{-1}[((\underline{e},\underline{e}_1)) - ((\gamma_\psi(u)\underline{M}(u)\underline{e},\gamma_\psi(z)\underline{M}(z)\underline{e}_1))] +$$
$$+ (u-\bar{z})^{-1}[((\underline{e},\gamma_\psi(z)\underline{M}(z)\underline{e}_1)) - ((\gamma_\psi(u)\underline{M}(u)\underline{e},\underline{e}_1))].$$

Proof. By lemma 11.6.3 it is sufficient to prove it for Re u, Re $z > \frac{1}{2}$. The computation is done as in the proof of lemma 11.5.8. Compare p. 134-135 of [22].

11.6.5. Lemma. For $z = x+iy \in \mathcal{D}$, $y \neq 0$:

(11.6.7) $$\|\gamma_\psi(z)\underline{M}(z)\| \leq 1 + 2x|y|^{-1}.$$

Proof. Apply lemma 11.6.4 to $\langle h_{z,e}, h_{z,e}\rangle$ for $\underline{e} \in E$ with $\|\underline{e}\| = 1$. We see that

(11.6.8) $$(2x)^{-1}(1-\|\gamma_\psi(z)\underline{M}(z)\underline{e}\|^2) + y^{-1}\,\mathrm{Im}((\underline{e},\gamma_\psi(z)\underline{M}(z)\underline{e})) \geq 0.$$

So

(11.6.9) $$\|\gamma_\psi(z)\underline{M}(z)\underline{e}\|^2 \leq 1 + 2x|y|^{-1}\|\gamma_\psi(z)\underline{M}(z)\underline{e}\|.$$

This implies (11.6.7).

11.6.6. Lemma. $1_{z,\underline{e}}$ is represented by a bounded function for all $z \in \mathcal{D}$, $\underline{e} \in E$. For each $y_0 \neq 0$ there is a neighbourhood U of iy_0 such that this bound is uniform for $z \in U \cap \mathcal{D}$.

Proof. For Re $z > \frac{1}{2}$

(11.6.10) $$*\Theta f_3(z,\underline{e})(g) = \underline{E}(z)\psi(z)\otimes\underline{e}(g) - *\Theta f_4(z,\underline{e})(g)$$

with

(11.6.11) $$f_4(z,\underline{e})(n\;a(y)k) = \begin{cases} 0 & \text{if } y \leq 1 \\ *F\;\underline{E}(z)\psi(z)\otimes\underline{e} & \text{if } y > 1. \end{cases}$$

$*\Theta f_4(z,\underline{e})(g)$ is given by a finite sum for all $z \in \mathcal{D}$; hence (11.6.10) may be used to extend $*\Theta f_3(z,\underline{e})$. By taking the scalar product with elements of $C_c^\infty(\Gamma\backslash G,\chi)$ we see that $*\Theta f_3(z,\underline{e})$ represents $1_{z,\underline{e}}$ for all $z \in \mathcal{D}$.

We know that for $\lambda = (c,0) \in \Lambda^0$ and $\phi \in C^0$

(11.6.12) $$\begin{aligned} &|\alpha_\phi(z,\psi)|\,|*\Theta f_3(z,\underline{e})(g_c\;n\;a(y)k)| \leq \\ &\leq |\underline{E}(z)\psi(z)\otimes\underline{e}*\phi(g_c n\;a(y)k) - F_{(c,0)}\underline{E}(z)\psi(z)\otimes\underline{e}*\phi(n\;a(y)k)| \\ &+ |((*F\underline{E}(z)\psi(z)\otimes\underline{e}(n\;a(y)k),\underline{e}_\lambda)) - *\Theta f_4(z,\underline{e})(g_c n\;a(y)k)|\,|\alpha_\phi(z,\psi)| \end{aligned}$$

which is seen to be bounded, by propositions 10.5.5, 9.4.3 and 8.2.3.

In the case $(c,0) \notin \Lambda$ we have

(11.6.13) $$\begin{aligned} &|\alpha_\phi(z,\psi)|\,|*\Theta f_3(z,\underline{e})(g_c n\;a(y)k)| \leq |\underline{E}(z)\psi(z)\otimes\underline{e}*\phi(g_c n\;a(y)k)| + \\ &+ |*\Theta f_4(z,\underline{e})(g_c n\;a(y)k)|\,|\alpha_\phi(z,\psi)|, \end{aligned}$$

which is also bounded.

Now let z_j tend to iy_0 and suppose

(11.6.14) $$\sup_{g\in G} |*\Theta f_3(z_j,\underline{e})(g)| \nearrow \infty.$$

By proposition 8.2.3 the second terms in the right hand sides of (11.6.12) and (11.6.13) are uniformly bounded for all j. Let us take $\phi \in C^0$ such that $\alpha_\phi(iy_0,\psi) = 1$. Then lemma 11.2.5 implies that the $C_\psi(\underline{E}(z_j)\underline{e})$ defined in (11.2.13) are unbounded. Select a subsequence for which the $C_\psi(\underline{E}(z_j)\underline{e})$ are increasing. By lemma 11.6.5 the $\gamma_\psi(z_j)\underline{M}(z_j)$ are bounded, so we may go over once again to a subsequence for which the $\underline{M}(z_j)$ converge. So this subsequence satisfies the conditions of proposition 11.2.4, and the $C(\underline{E}(z_j)\underline{e})$ have to be bounded by lemma 11.2.8. So assumption (11.6.14) leads to a contradiction.

Proof of proposition 11.6.1. Take $y_0 \neq 0$. Let U be a neighbourhood of iy_0 as in lemma 11.6.6. Take $\underline{e}, \underline{e}_1 \in E$. From lemma 11.6.6 follows that $\langle 1_{z,\underline{e}}, 1_{z,\underline{e}_1} \rangle$ is bounded for $z \in U \cap \mathcal{D}$. So from the lemmas 11.6.4 and 11.6.5 follows that

(11.6.15) $$(z+z^{-1})[((\underline{e},\underline{e}_1)) - ((\gamma_\psi(z)\underline{M}(z)\underline{e}, \gamma_\psi(z)\underline{M}(z)\underline{e}_1))]$$

is bounded for $z \in U \cap \mathcal{D}$. So

(11.6.16) $$\lim_{\substack{z \to iy_0 \\ z \in \mathcal{D}}} ((\gamma_\psi(z)\underline{M}(z)\underline{e}, \gamma_\psi(z)\underline{M}(z)\underline{e}_1)) = ((\underline{e},\underline{e}_1)).$$

As $\lim_{z \to iy_0} \gamma_\psi(\bar{z})\gamma_\psi(z) = 1$ this implies

(11.6.17) $$\lim_{\substack{z \to iy_0 \\ z \in \mathcal{D}}} \underline{M}(\bar{z})\underline{M}(z) = I.$$

The reasoning on p. 139 of [22] now leads to the meromorphic continuation of $\underline{M}$ to $\mathcal{D}_*$ and to the functional equation.

11.6.7. Proposition. Let ψ be a constant section of ${}^*H^\tau$. For each $g \in G$ the function $s \to \underline{E}(s)\psi(s)(g)$ has a meromorphic continuation to $\mathcal{D}_*$; it is holomorphic everywhere where $\underline{M}$ is. The complex differentiation is uniform for g in compact subsets of G. The operator $\underline{E}(s) : \psi(s) \to \underline{E}(s)\psi(s)$ so obtained gives an automorphic model in $A^e({}^*H^\tau(s))$. The extended Eisenstein model $\underline{E}(s)$ satisfies the functional equation

(11.6.18) $$\underline{E}(-s) = \underline{E}(s)\iota(-s)\underline{M}(-s).$$

Proof. For $y_0 \neq 0$ one defines

(11.6.19) $$\underline{E}(iy_0)\psi(iy_0)\otimes\underline{e} = \lim_{\substack{s \to iy_0 \\ s \in \mathcal{D}}} \underline{E}(s)\psi(s)\otimes\underline{e},$$

using proposition 11.2.4. The limit is the same for all sequences in $\mathcal{D}$ tending to iy_0, for the difference of two limits is cuspidal and orthogonal to ${}^0L^2$. For $s \in \mathcal{D}$ one defines

(11.6.20) $$\underline{E}(-s) = E(s)\iota(-s)\underline{M}(-s).$$

On $-\mathcal{D}$ the assertions of the proposition are clear. On $i\mathbb{R}$ we have to check whether

(11.6.21) $$\lim_{\substack{s \to iy_0 \\ s \in \mathcal{D}}} \underline{E}(-s) = \underline{E}(-iy_0).$$

This follows from a consideration of the Fourier coefficients of order zero and the fact that the limit gives a model in $A^e({}^*H^\tau(-iy_0))$. The differentiability follows from proposition 11.2.4.

11.7. Meromorphic continuation of $\underline{M}$ and $\underline{E}$ to $\mathbb{C}$.

Compare this subsection with p. 99-105 of [11].

11.7.1. Lemma. For every $s \in \mathbb{C}$: $\dim N(s) + \dim N(\bar{s}) \leqslant 2 \dim E$.

For every $s \in \mathbb{R}$: $\dim N(s) \leqslant \dim E$.

Proof: directly from the Maass-Selberg relation (7.6.9).

11.7.2. Lemma. $s \in \mathbb{C}$, U is a neighbourhood of s contained in U^g or U_b; $\Omega_U : E^2 \times U \to E_2$ is the corresponding restriction of Ω^g or Ω_b. The scalar product on E gives one on E^2, which is transported to $E_2(s^*)$ for every $s^* \in U$ by means of Ω_U. Let $s_i \in U$ tend to s; let for every i $\underline{\underline{b}}_i : E \to E_2(s_i)$ be unitary. Then there is a subsequence such that $\lim_{i' \to \infty} \underline{\underline{b}}_{i'} : E \to E_2(s)$ is unitary.

Proof. $\{\underline{\underline{b}}_i \underline{e}_\lambda \mid \lambda \in \Lambda^o\}$ is an orthonormal system in $E_2(s_i)$; $\{\underline{\underline{b}}_i \underline{e}_\lambda \mid i \in \mathbb{N}\}$ is a bounded set. By successively taking subsequences we get a sequence of orthonormal systems in $E_2(s_i)$ converging to an orthonormal system in $E_2(s)$.

11.7.3. Lemma. For every $s_0 \in \mathbb{C}$:

$$\dim N(s_0) = \dim E \tag{11.7.1}$$

and $N(s_0)$ is the intersection of $E_2(s_0)$ with the closure in E_2 of $\bigcup_{s \neq s_0} N(s)$.

Proof. Consider a general $s_0 \in \mathcal{D}_*$ at which $\underline{M}$ is holomorphic; then $\underline{M}(\bar{s}_0) = M(s_0)^*$ is well defined as well. The map $\underline{e} \mapsto (\underline{e}, \underline{M}(s_0)\underline{e}; s_0)$ gives an injection $E \to N(s_0)$, so $\dim N(s_0) \geqslant \dim E$; similarly $\dim N(\bar{s}_0) \geqslant \dim E$, so $\dim N(s_0) = \dim E$ by lemma 11.7.1. If $\underline{M}$ has a pole at $s_0 \in \mathcal{D}_*$, then $\underline{M}$ is regular at $-s_0$. So we may use the isomorphism

$$N(s_0) \to N(-s_0) : (\underline{a}, \underline{b}; s_0) \to (\underline{b}, \underline{a}; -s_0), \tag{11.7.2}$$

corresponding to

$$\begin{aligned} &A(H_0^\delta(s_0)) \to A(H_0^\delta(-s_0)) : T \to T\, \iota_0^\delta(-s_0) && \text{if } \tau = 0 \\ &A(H_1^\delta(s_0)) \to A(H_1^\delta(-s_0)) : T \to \Gamma(s_0)\Gamma(-s_0)^{-1}\tilde{T} && \text{if } \tau = 1, \text{ see (7.5.11)} \\ &A(H_\tau(s_0)) \to A(H_\tau(-s_0)) : T \to T\, \iota_\tau^1(-s_0) && \text{if } 0 < \tau < 1. \end{aligned} \tag{11.7.3}$$

Before I prove (11.7.1) for other s_0 I consider the second statement. From proposition 11.2.4 and definition 7.5.10 follows that $N(s_0) \supset \mathrm{Cl}(\bigcup_{s \neq s_0} N(s)) \cap E_2(s_0)$. We may find a sequence s_i of general points converging to s_0 such that $\underline{M}$ is holomorphic at each s_j. The systems $\{(\underline{e}_\lambda, \underline{M}(s_i)\underline{e}_\lambda; s_i) \mid \lambda \in \Lambda^o\}$ may be orthonormalized; so we get unitary maps $\underline{\underline{b}}_i : E \to N(s_i) \subset E_2(s_i)$. By lemma 11.7.2 a subsequence converges to a unitary map $\underline{\underline{b}} : E \to E_2(s_0)$. But as $N(s_0) \supset \mathrm{Cl}(\bigcup_{s \neq s_0} N(s)) \cap E_2(s_0)$ we have $\underline{\underline{b}} : E \to N(s_0)$. So $N(s_0)$ contains a subspace with the same dimension as E. By lemma 11.7.1 if $s_0 \in \mathbb{R}$ and by (11.7.1) if $s_0 \notin \mathbb{R}$ we get $\underline{\underline{b}}(E) = N(s_0)$ and $\dim N(s_0) = \dim E$.

11.7.4. Lemma. $s \in \mathbb{R}$, U is a neighbourhood of s contained in U^g or U_b; Ω_U as in lemma 11.7.2. $N(s^*)$ for $s^* \in U$ gets a scalar product from $E_2(s^*)$. For $\mu \in \mathbb{C}$ and $s^* \in U$

(11.7.4) $\quad p_\mu(s^*) : E_2(s^*) \to E$

is defined by

(11.7.5) $\quad p_\mu(s^*)\Omega_U(\underline{a},\underline{b};s^*) = \underline{a} + \mu\underline{b}.$

Then there is a $\mu \in \mathbb{C}$ and a neighbourhood U_1 of s contained in U such that $p_\mu(s^*) : N(s^*) \to E$ is bijective with uniformly bounded inverse for $s^* \in U_1$.

This is lemma 97 of [11].

Proof. To prove that $p_\mu(s) : N(s) \to E$ is bijective, it is sufficient to prove injectivity, see lemma 11.7.3.

Define $\pi_1, \pi_2 : E_2(s) \to E$ by

(11.7.6) $\quad \pi_1\Omega_U(\underline{a},\underline{b};s) = \underline{a}, \ \pi_2\Omega_U(\underline{a},\underline{b};s) = \underline{b}$

Let $W = \ker(\pi_1|N(s))$, $W' = W^\perp \cap N(s)$. The Maass-Selberg relation (7.6.9) implies that $\pi_1 W' \perp \pi_2 W$. Furthermore, $\pi_2 : W \to E$ and $\pi_1 : W' \to E$ are injective. So

(11.7.7) $\quad \dim \pi_1 W' + \dim \pi_2 W = \dim W' + \dim W = \dim N(s) = \dim E;$

hence $\pi_1 W' \oplus \pi_2 W = E$. Further $p_\mu(s)|W = \mu\pi_2|W$, so if $\mu \neq 0$, then $p_\mu(s)$ is injective on W. Let $q : E \to \pi_1 W'$ be the orthogonal projection. We need to prove injectivity of $qp_\mu : W' \to \pi_1 W'$. But on W'

(11.7.8) $\quad qp_\mu = \pi_1 + \mu q\pi_2.$

As π_1 is injective on W' it is clear that qp_μ is injective for all but a finite number of $\mu \in \mathbb{C}$.

So we have proved that $p_\mu(s) : N(s) \to E$ is bijective for a suitably chosen $\mu \in \mathbb{C}$.

If the assertion of the lemma were not true, then there would be a sequence of points s_i converging to s and $\underline{n}_i \in N(s_i)$ such that

(11.7.9) $\quad \lim_{i\to\infty} \|\underline{n}_i\| = \infty$

(11.7.10) $\quad \|p_\mu(s_i)\underline{n}_i\| \leq 1$ for all i.

Now take $\underline{m}_i = \|\underline{n}_i\|^{-1}\underline{n}_i$. Then the $\underline{m}_i$ are bounded; after going over to a subsequence we may assume that the $\underline{m}_i$ converge to some $\underline{m} \in N(s)$, with $\|\underline{m}\| = 1$. But

(11.7.11) $\quad \lim p_\mu(s_i)\underline{m}_i = 0$

and

(11.7.12) $\quad \lim p_\mu(s_i)\underline{m}_i = p_\mu(s)\underline{m}$;

this is in contradiction with the bijectivity of $p_\mu(s) : N(s) \to E$.

11.7.5. Lemma. Let s and U be as in lemma 11.7.4. Let $U_1 = \{s_0 \in U \cap U^g | s_0 \notin [-\frac{1}{2},\frac{1}{2}]$, $\underline{M}$ is holomorphic at $s_0\}$. Define for $s_0 \in U_1$

(11.7.13) $\quad \phi(s_0) : E \to N(s_0) : \underline{e} \mapsto (\underline{e},\underline{M}(s_0)\underline{e};s_0).$

This is a bijection and $s_0 \mapsto \phi(s_0)$ is holomorphic on U_1.

Proof: clear.

U_1 is open in U, its closure is a neighbourhood of s. We may assume that this closure is equal to U.

11.7.6. Lemma. Let s, U, U_1 and ϕ be as in lemma 11.7.5. Define for $s_0 \in U_1$ and μ as in lemma 11.7.4:

(11.7.14) $\psi(s_0) = p_\mu \phi(s_0) : E \to E.$

This is a bijection and $s_0 \mapsto \psi(s_0)$ is holomorphic on U_1. The inverse $\psi(s_0)^{-1}$ also depends holomorphically on s_0.

Proof. By the lemmas 11.7.4 and 11.7.5 the bijectivity is clear. The holomorphy of $\psi(s_0)$ follows from the definitions. As det $\psi(s_0)$ is nonzero and holomorphic on U_1 the lemma follows.

11.7.7. Lemma. With the notations of the previous lemmas define for $s_0 \in U_1$

(11.7.15) $C(s_0) = \phi(s_0)\psi(s_0)^{-1} : E \to N(s_0).$

Then $C(s_0)$ may be continued holomorphically to $U \cap [-\frac{1}{2},\frac{1}{2}]$.

Proof. $C(s_0)$ is holomorphic on U_1. Let $s^* \in U \cap [-\frac{1}{2},\frac{1}{2}]$ be given. Suppose $s_i \in U_1$ tend to s^*. Take $\underline{e} \in E$ and $\underline{h}_i = C(s_i)\underline{e}$; then $p_\mu(s_i)\underline{h}_i = \underline{e}$ and $\|\underline{h}_i\| = \|(p_\mu(s_i)|N(s_i))^{-1}\underline{e}\|$ is bounded. Any limit point $\underline{h}$ of the $\underline{h}_i$ satisfies $p_\mu(s^*)\underline{h} = \underline{e}$ and is an element of $N(s^*)$. This implies that $\underline{h}$ is unique and $\underline{h} = \lim_{i\to\infty} \underline{h}_i$. So

(11.7.16) $$C(s^*)\underline{e} = \lim_{\substack{s\in U_1\\ s\to s^*}} C(s)\underline{e}$$

defines a map inverse to $p_\mu(s^*)|N(s^*)$. This extension of C to s^* is holomorphic at s^* provided we know that C is holomorphic on the intersection of $\mathbb{C}\setminus\mathbb{R}$ with a neighbourhood of s^*. This is the case if $s^* \in (0,\frac{1}{2}]$. If $s^* \in [-\frac{1}{2},0]$ we have to worry about a possible sequence $z_j \in U\setminus U_1$ tending to s^*. Then $\phi(s)$, $\psi(s)$ and $\psi(s)^{-1}$ are meromorphic at each z_j. So $\|C(s)\underline{e}_\lambda\|$ goes to infinity for $s \to z_j$ at least for one $\lambda \in \Lambda^0$. After going over to a subsequence, we find $s_j \in U_1$ with $\lim |s_j - z_j| = 0$ such that $\|C(s_j)\underline{e}\| \nearrow \infty$ for some $\underline{e} \in E$. But $s_j \to s^*$ and we have seen above that $\|C(s_j)\underline{e}\|$ stays bounded. So even if $s^* \in [-\frac{1}{2},0]$ there is not a sequence of points in $\mathbb{C}\setminus\mathbb{R}$ tending to s^* in which $C(s)$ is singular.

11.7.8. Lemma. Let s, U, U_1 and ϕ be as in lemma 11.7.5. $s_0 \mapsto \phi(s_0)$ has a meromorphic continuation to U.

Proof. Define for general $s^* \in \mathbb{C}$

(11.7.17) $q(s^*) : E_2(s^*) \to E : (\underline{a},\underline{b};s^*) \to \underline{a}.$

From definition 7.5.6 follows that $s^* \to q(s^*)$ is meromorphic on $\mathbb{C}$. Further $q(s_0)\phi(s_0) = I_E$ for all $s_0 \in U_1$. This implies that $\psi(s_0)^{-1} = q(s_0)C(s_0)$ is meromorphic for $s_0 \in U$, and so is $\phi(s_0) = C(s_0)\psi(s_0)$.

11.7.9. Proposition. The function $s \mapsto \underline{M}(s)$ has a meromorphic continuation to $\mathbb{C}$. Its poles in the region Re $s \geq 0$ are situated in $(0,\frac{1}{2}(1-\tau))$ and have first order.

Proof. From lemma 11.7.8, (11.7.13) and definition 7.5.6 follows that $\underline{M}$ is meromorphic on $\mathbb{C}$. From lemma 11.5.1 follows that its poles with Re $s \geq 0$ are situated in $[0,\frac{1}{2}]$.

Let us first consider $s_0 = 0$. If ψ is a constant section of H^τ then $\gamma_\psi(s)^{-1}$ is bounded for $s \to 0$. By lemma 11.6.5 we see that $\|M(t(1+i))\|$ is bounded for $t \searrow 0$, so $\underline{M}$ is holomorphic at $s_0 = 0$. A similar argument works for $s_0 \in (0,\frac{1}{2}]$, s_0 general: In this case

(11.7.18) $\quad \|\underline{M}(s_0+it)\| \ll |t|^{-1}$,

so $\underline{M}$ has at most a pole of order one at s_0. If $\tau \neq 1$, the only non-general point in $(0,\frac{1}{2}]$ is $\frac{1}{2}(1-\tau)$. We may choose ψ of weight τ to give γ_ψ a first order pole at $\frac{1}{2}(1-\tau)$. So $\underline{M}$ has to be bounded near $\frac{1}{2}(1-\tau)$. Now let $s_0 \in (0,\frac{1}{2}]$ be general and suppose that $\underline{M}$ has a first order pole at s_0. Take $\underline{e} \in E$ such that
$\lim_{s\to s_0} (s-s_0)\underline{M}(s)\underline{e} = \underline{m} \neq 0.$

Consider a sequence $s_i \notin \mathbb{R}$ tending to s_0. By proposition 11.2.4

(11.7.19) $\quad T = \lim_{i\to\infty} (s_i-s_0)\underline{E}(s_i)\underline{e}$

is an element of $A^e(H^\tau(s_0))$, with Fourier coefficient of order zero given by $(0,\underline{m};s_0) \in E_2$. From (8.3.11) follows that the elements of $TH^\tau(s_0)$ are in $L^2(\Gamma\backslash G,\chi)$. So $H^\tau(s_0)$ has to have an invariant subspace with unitary structure; hence $0 < s_0 < \frac{1}{2}(1-\tau)$, see proposition 3.4.5.

11.7.10. Definition. Let U be a g-K-space contained in some $H^\tau(s)$. Then $A^{esq}(U) = A^e(U) \cap A^{sq}(U)$.

Let $s_0 \in (0, \frac{1}{2}(1-\tau)]$; so $\tau \neq 1$.

(11.7.20)
$$\underline{M}_{s_0} = \begin{cases} \operatorname{res}_{s\to s_0} \underline{M}(s) & \text{if } s_0 \neq \frac{1}{2}(1-\tau) \\ \underline{M}(s_0) & \text{if } s_0 = \frac{1}{2}(1-\tau) \end{cases}$$
$$E(s_0) = (\ker \underline{M}_{s_0})^\perp$$

So $E(s_0) \neq 0$ for only a finite number of $s_0 \in (0,\frac{1}{2}(1-\tau)]$.

11.7.11. Proposition. Let $s_0 \in (0,\frac{1}{2}(1-\tau)]$ and let $\underline{e} \in E(s_0)$

(11.7.21)
$$\underline{E}_{s_0}\underline{e} = \lim_{\substack{s\to s_0\\ s\notin\mathbb{R}}} (s-s_0)\underline{E}(s)\underline{e}$$

defines an automorphic model $\underline{E}_{s_0}\underline{e} \in A^{esq}(H^\tau(s_0))$. All elements of $A^{esq}(H^\tau(s_0))$ are of this form.

Proof. If s_i is a sequence with limit s_0 we see that $\lim_{i\to\infty} (s_i-s_0)\underline{E}(s_i)\underline{e}$ exists by proposition 11.2.4.

Two sequences give the same limit, for the difference of the limits would be cuspidal and orthogonal to the cusp forms at the same time. The square integrability of $\underline{E}_{s_0}\underline{e}$ follows from proposition 10.5.5. If $\underline{e}_1 \in \ker \underline{M}_{s_0}$ then $\lim_{s\to s_0} \underline{E}(s)\underline{e}_1$ gives an element of $A^e(H^\tau(s_0))$ which is not square integrable, unless $\underline{e}_1 = 0$. By lemma 11.7.3 we obtain in this way the whole of $A^e(H^\tau(s_0))$; so any element of $A^{esq}(H^\tau(s_0))$ is of the form (11.7.21).

11.7.12. Corollary. Let $\alpha \in \Lambda^1$, $0 < s_0 < \frac{1}{2}(1-\tau)$, $\underline{e} \in E(s_0)$. Then

(11.7.22) $\alpha(H_0^\delta(s_0),\ \underline{E}_{s_0}\underline{e}|H_0^\delta(s_0)) = (\delta,\varepsilon(\alpha))((\underline{e},\ \operatorname{res}_{s\to s_0} \underline{f}_\alpha(s)))$ if $\tau = 0$

(11.7.23) $\alpha(H_\tau(s_0), \underline{E}_{s_0}\underline{e}) = c_{\varepsilon(\alpha)}(-s_0)e^{-\frac{1}{2}\pi i\tau}((\underline{e},\ \operatorname{res}_{s\to s_0} \underline{f}_\alpha(s)))$ if $0 < \tau < 1$

Proof. Use propositions 9.5.2 and 7.4.2.

11.7.13. Corollary. Let $\tau = 0$, $s_0 = \frac{1}{2}$, $\underline{e} \in E(\frac{1}{2})$. Then

(11.7.24) $\underline{E}_{\frac{1}{2}}\ \underline{e}|(D_2+\tilde{D}_2) = 0$

(11.7.25) $F_\alpha\ \underline{E}_{\frac{1}{2}}\ \underline{e} = 0$ for all $\alpha \in \Lambda^1$.

Proof. The only way to intertwine $H_0^\delta(\frac{1}{2})$ in a nontrivial way with a space with unitary $\underline{g}$-K-structure, is to use the isomorphism $H_0^\delta(\frac{1}{2})$ mod $D_2^\delta \cong D_0^\delta$. The last assertion follows from proposition 7.4.2.

11.7.14. Corollary. Let $0 < \tau < 1$, $s_0 = \frac{1}{2}(1-\tau)$, $\underline{e} \in E(s_0)$. Then

(11.7.26) $\underline{E}_{s_0}\ \underline{e}|\tilde{D}_{2-\tau} = 0$;

if $\alpha \in \Lambda^1$ then

(11.7.27)
$$\alpha(H_\tau(s_0), \underline{E}_{s_0}\underline{e}|H_\tau(s_0)) = \begin{cases} e^{-\frac{1}{2}\pi i\tau}((\underline{e}, \underline{f}_\alpha(s_0))) & \text{if } \varepsilon(\alpha) = 1 \\ 0 & \text{if } \varepsilon(\alpha) = -1 \end{cases}$$

Proof. $\underline{E}_{s_0}\underline{e}$ is only nonzero on the first component of $H^\tau(s_0)$. As it is square integrable, it has to be zero on $\tilde{D}_{2-\tau} \subset H_\tau(s_0)$. The vanishing in (11.7.27) for $\varepsilon(\alpha) = -1$ follows from proposition 7.4.3. If we write out the expression in proposition 9.5.2 we see that the Whittaker model already provides a first order pole, so $\underline{f}_\alpha$ has to be regular.

11.7.15. Proposition. Let ψ be a constant section of $*H^\tau$. For each $g \in G$ the function $s \to \underline{E}(s)\psi(s)(g)$ has a meromorphic continuation to $\mathbb{C}$. It is holomorphic outside the set

(11.7.28) $\{s \in \mathbb{C}|\underline{M}$ has a pole at $s\} \cup \{\frac{1}{2}(b-1)|b \in Y_\tau,\ b = -\tau(2)$ or $b = \tau = 1\}$.

The differentiation with respect to s is uniform for g in compact sets. For s outside the set in (11.7.28)

(11.7.29) $\underline{E}(s) : \psi(s) \to \underline{E}(s)\psi(s)$ defines an element of $A^e(*H^\tau(s_0))$.

The only poles of $\underline{E}(s)$ with Re $s \geq 0$ occur in $(0,\frac{1}{2}(1-\tau)]$ if $\tau \neq 1$ and $\{0\}$ if $\tau = 1$. The poles in $(0,\frac{1}{2}(1-\tau)]$ have first order. $\underline{E}$ has the functional equation

(11.7.30) $\underline{E}(-s) = \underline{E}(s)\iota(-s)\underline{M}(-s)$.

Proof. The continuation of E follows from proposition 11.2.4, as was the case in the proof of proposition 11.6.7. Inspection of definition 7.5.6. shows that in $\frac{1}{2}(b-1)$ with $b \in Y_\tau$, $b = -\tau(2)$ or $b = \tau = 1$, there may turn up singularities of the Fourier coefficient of order zero which are not due to $\underline{M}$.

11.7.16. Proposition. Let $\alpha \in \Lambda^1$. The Fourier coefficient $\underline{f}_\alpha$ of $\underline{E}$ has a meromorphic continuation to $\mathbb{C}$. Its poles in the region Re $s \geq 0$ are situated in those points of

$(0,\frac{1}{2}(1-\tau))$ <u>for which $\underline{M}$ has a pole, and have first order.</u> $\underline{f}_\alpha$ <u>satisfies</u>

(11.7.31) $$\underline{f}_\alpha(-s) = c_{\varepsilon(\alpha)}(-s)\underline{M}(-s)\underline{f}_\alpha(s).$$

<u>Proof</u>. Proposition 9.5.2 gives

(11.7.32) $$F_\alpha \underline{E}(s)\underline{e} = ((\underline{e},\underline{f}_\alpha(\bar{s})))W(s)X^{\varepsilon(\alpha)}(s)\pi_1$$

Apply this to $\phi = \phi_\tau^\delta$ if $\tau = 0,1$, $\phi = \phi_\tau^1$ if $0 < \tau < 1$, and consider it in $a(y)j_\varepsilon$ with $y > 0$, $\varepsilon = 1$ if $\tau = 0,1$, $\varepsilon = \varepsilon(\alpha)$ if $0 < \tau < 1$. Inspection of the resulting formula gives the statements on the meromorphy and the poles of $\underline{f}_\alpha$. To get the functional equation use (5.5.3) and

(11.7.33) $$\iota(s)X^\varepsilon(s)\pi_1 = c_\varepsilon(-s)X^\varepsilon(-s)\pi_1\,\iota(s).$$

12. Spectral decomposition of ${}^{e}L^2(\Gamma\backslash G,\chi)$.

12.1. Introduction.

Now we have got the analytical continuation of the Eisenstein model, we may derive the spectral decomposition of ${}^{e}L^2$. The points $s_0 \in (0,\frac{1}{2}(1-\tau)]$ at which $\underline{E}(s)$ has a pole give non-cuspidal invariant subspaces of ${}^{e}L^2$ generated by the $\underline{E}_{s_0}\underline{e}$ with $\underline{e} \in E(s_0)$. The orthogonal complement in ${}^{e}L^2$ of these subspaces is given by a direct integral.

For functions of the form $*\Theta*\Phi\xi$ with ξ a holomorphic rapidly decreasing section of $*H^\tau$ I give a decomposition of the scalar product in 12.2. From this decomposition it becomes clear how ${}^{e}L^2$ is built up. It has some closed invariant subspaces, the orthogonal complement ${}^{c}L^2$ is a continuous integral of representation of the principal series. This space ${}^{c}L^2$ I discuss in 12.3. We need a scalar product formula for more general functions than considered in 12.2; this extension is given in 12.5. The Eisenstein transform occurring in this formula is introduced in 12.4. The results of this section are reformations in our notations of know results; compare [37], Satz 11.2 and Satz 12.3.

12.2. Integral formulas

Let $\sigma > \frac{1}{2}$. From proposition 11.4.2 we know that for holomorphic rapidly decreasing sections ξ and η of $*H^\tau$ on $|\mathrm{Re}\ s| \leqslant \sigma$

$$\langle *\Theta*\Phi\xi, *\Theta*\Phi\eta\rangle = \frac{1}{2\pi i}\int_{\mathrm{Re}\ s=-\sigma}\{(1+\underline{M})\pi_1\xi,\eta\}(s)ds. \tag{12.2.1}$$

On the line $\mathrm{Re}\ s = 0$ the operator $\underline{M}(s)$ is unitary, hence bounded; lemma 11.6.5 gives boundedness on the strip $0 < \mathrm{Re}\ s \leqslant \sigma$. This implies that $\{(1+\iota\underline{M})\pi_1\xi,\eta\}(s)$ is rapidly decreasing for $|\mathrm{Im}\ s| \to \infty$ on the strip $-\sigma \leqslant \mathrm{Re}\ s \leqslant 0$ outside a neighbourhood of the poles. We can change the path of integration to the line $\mathrm{Re}\ s = 0$. Doing this there occur residues at the points in the following set:

12.2.1. Definition. $S^e = \{s_0 \in (0,\frac{1}{2}(1-\tau)]\ |\ E(s_0) \neq \{0\}\}$.

On the line $\mathrm{Re}\ s = 0$ I rewrite

$$\{(1+\iota\underline{M})\pi_1\xi,\eta\}(s) = \tfrac{1}{2}\{(1+\iota\underline{M})^2\pi_1\xi,\pi_1\eta\}(s) \quad \text{(see proposition 9.4.4)} \tag{12.2.2}$$

$$= \tfrac{1}{2}\{(1+\iota\underline{M})\pi_1\xi,\pi_1\eta\}(s) + \tfrac{1}{2}\{(1+\iota\underline{M})\pi_1\xi,\iota\underline{M}\pi_1\eta\}(-s) \quad \text{(see (3.5.17))}.$$

So we get

$$\frac{1}{2\pi i}\int_{\mathrm{Re}\ s=0}\tfrac{1}{2}\{(1+\iota\underline{M})\pi_1\xi,\pi_1\eta\}(s)ds + \frac{1}{2\pi i}\int_{\mathrm{Re}\ s=0}\tfrac{1}{2}\{(1+\iota\underline{M})\pi_1\xi,\iota\underline{M}\pi_1\eta\}(s)ds$$

$$= \frac{1}{4\pi i}\int_{\mathrm{Re}\ s=0}\{(1+\iota\underline{M})\pi_1\xi,\ (1+\iota\underline{M})\pi_1\eta\}(s)ds.$$

The integrand is invariant under $s \to -s$, so we may integrate over $(0,i\infty) = \{it \mid t > 0\}$. We get the following result:

12.2.2. Lemma. Let $\sigma > \frac{1}{2}$. For holomorphic rapidly decreasing sections ξ and η of $*H^\tau$ on $|\mathrm{Re}\, s| \leq \sigma$:

$$\langle *\Theta*\Phi\xi, *\Theta*\Phi\eta \rangle = \frac{1}{2\pi i} \int_0^{i\infty} \{(1+\iota\underline{M})\pi_1\xi(s), (1+\iota\underline{M})\pi_1\eta(s)\}_s ds + \tag{12.2.3}$$

$$+ \sum_{s_0 \in S^e} \{\operatorname*{res}_{s\to s_0} \iota(s)\underline{M}(s).\pi_1\xi(s_0), \pi_1\eta(s_0)\}_{-s_0}.$$

The $*\Theta*\Phi\xi$ form a dense set in ${}^eL^2$. Let $s_0 \in S_0$, $\underline{e} \in E(s_0)$ and $\phi(s_0) \in \pi_1 H^\tau(s_0)$. Then $\underline{E}_{s_0}\, \underline{e}\, \phi(s_0) \in {}^eL^2$, see proposition 11.7.11. It is not difficult to approximate $\underline{E}_{s_0}\, \underline{e}\, \phi(s_0)$ by elements of the form $*\Theta*\Phi\xi$:

12.2.3. Lemma. Let $s_0 \in S_0$, $\underline{e} \in E(s_0)$, ϕ a constant section of $\pi_1 H^\tau$. Define the section ξ_v by

$$\xi_v(s) = e^{-v(s^2-s_0^2)^2}\, \phi(s)\otimes\underline{e}. \tag{12.2.4}$$

For $v > 0$ this is a holomorphic rapidly decreasing section of $*H^\tau$ on any strip $|\mathrm{Re}\, s| \leq \sigma$ and

$$\lim_{v\to\infty} *\Theta*\Phi\xi_v = \underline{E}_{s_0}\, \underline{e}\, \phi(s_0) \text{ in } {}^eL^2(\Gamma\backslash G,\chi). \tag{12.2.5}$$

Proof. It is clear that for any holomorphic rapidly decreasing section ξ of $*H^\tau$

$$\lim_{v\to\infty} \langle *\Theta*\Phi\xi, *\Theta*\Phi\xi_v \rangle = \{\operatorname*{res}_{s\to s_0} \iota(s)\underline{M}(s).\pi_1\xi(s_0), (s_0)\otimes\underline{e}\}_{-s_0}. \tag{12.2.6}$$

Suppose in addition that $*\Phi\xi \in *S_\tau$, see 11.4.1. The $*\Theta*\Phi\xi$ of this form already are dense in ${}^eL^2$. Now $*\Theta*\Phi\xi \in C_c^\infty(\Gamma\backslash G,\chi)$, so we get

$$\langle *\Theta*\Phi\xi, \underline{E}_{s_0}\, \underline{e}\, \phi(s_0) \rangle = \lim_{s\to s_0} (\bar{s}-s_0)\langle P_1^*\Phi\xi, *F\underline{E}(s)\phi(s)\otimes\underline{e} \rangle \quad \text{(see propositions 8.3.3 and 11.7.11)} \tag{12.2.7}$$

$$= \lim_{s\to s_0} (\bar{s}-s_0)\langle P_1^*\Phi\xi, \mathrm{St}(-s)\iota(s)\underline{M}(s)\phi(s)\otimes\underline{e} \rangle$$

$$= \lim_{s\to s_0} (\bar{s}-s_0)\{\pi_1\xi(\bar{s}), \iota(s)\underline{M}(s)\phi(s)\otimes\underline{e}\}_{\bar{s}} \quad \text{(see propositions 11.3.1 and 11.3.3)}$$

$$= \lim_{s\to s_0} (\bar{s}-s_0)\{\iota(\bar{s})\underline{M}(\bar{s})\pi_1\xi(\bar{s}), \phi(s)\otimes\underline{e}\}_{-\bar{s}_0}$$

$$= \{\operatorname*{res}_{s\to s_0} \iota(s)\underline{M}(s).\pi_1\xi(s_0), \phi(s_0)\otimes\underline{e}\}_{-s_0}$$

Comparison of (12.2.6) and (12.2.7) proves the lemma.

12.2.4. Definition. Let $s_0 \in S^e$. Define

$${}^eL^2(s_0) = \mathrm{Cl}\{\underline{E}_{s_0}\, \underline{e}\, \phi(s_0) \mid \underline{e} \in E(s_0), \phi(s_0) \in \pi_1 H^\tau(s_0)\} \tag{12.2.8}$$

This is a noncuspidal invariant subspace of L^2.

12.2.5. Lemma. Let $s_0 \in S^e$, let ξ and η be as in lemma 12.2.2 and let $(*\Theta*\Phi\xi)_{s_0}$ and $(*\Theta*\Phi\eta)_{s_0}$ be the projections of $*\Theta*\Phi\xi$ and $*\Theta*\Phi\eta$ in ${}^eL^2(s_0)$. Then

$$\langle (*\Theta*\Phi\xi)_{s_0}, (*\Theta*\Phi\eta)_{s_0} \rangle = \{\operatorname*{res}_{s\to s_0} \iota(s)\underline{M}(s).\pi_1\xi(s_0), \pi_1\eta(s_0)\}_{-s_0}. \tag{12.2.9}$$

Proof. Let us first construct an orthonormal basis of ${}^{e}L^2(s_0)$. For ϕ,ϕ_1 basis sections of H^τ and $\underline{e},\underline{e}_1 \in E(s_0)$

$$(12.2.10)\qquad \langle \underline{E}_{s_0}\,\underline{e}\;\phi(s_0),\ \underline{E}_{s_0}\,\underline{e}_1\;\phi_1(s_0)\rangle = \{\iota_{s_0}\underline{M}_{s_0}\phi(s_0)\otimes\underline{e},\phi_1(s_0)\otimes\underline{e}_1\}_{-s_0} =$$

$$= \{\iota_{s_0}\phi(s_0),\phi_1(s_0)\}_{-s_0}((\underline{M}_{s_0}\underline{e},\underline{e}_1))$$

with

$$(12.2.11)\qquad \iota_{s_0} = \begin{cases} \iota(s_0) & \text{if } s_0 < \tfrac12(1-\tau) \\ \operatorname*{res}_{s\to s_0} \iota(s) & \text{if } s_0 = \tfrac12(1-\tau) \end{cases}$$

For the first step in (12.2.10) use lemma 12.2.3. If $\phi \neq \phi_1$ then $\{\iota_{s_0}\phi(s_0),\phi_1(s_0)\}_{-s_0} = 0$. So we may let ϕ run through those elements of the standard basis of $\pi_1 H^\tau$ for which $\iota_{s_0}\phi(s_0) \neq 0$. From an extension of proposition 9.4.4 follows that $\underline{M}_{s_0}$ is hermitian and maps $E(s_0)$ into itself. So for a given ϕ we let $\underline{e}$ run though an orthogonal basis of $E(s_0)$, consisting of eigenvectors of $\underline{M}_{s_0}$ and satisfying

$$(12.2.12)\qquad ((\underline{M}_{s_0}\underline{e},\underline{e})) = \{\iota_{s_0}\phi(s_0),\phi(s_0)\}^{-1}_{-s_0}.$$

Remark that these numbers are real. Now write

$$(12.2.13)\qquad \pi_1\xi(s) = \sum_{\psi}(\pi_1\xi)_\psi(s)\psi(s),$$

with ψ running through the standard basis of $\pi_1 H^\tau$. Then with help of lemma 12.2.3:

$$(12.2.14)\qquad \langle *\Theta*\Phi\xi,\underline{E}_{s_0}\,\underline{e}\;\phi(s_0)\rangle = \{\iota_{s_0}\phi(s_0),\phi(s_0)\}_{-s_0}((\underline{M}_{s_0}(\pi_1\xi)_\phi(s_0),\underline{e})) =$$

$$= ((\underline{M}_{s_0}\underline{e},\underline{e}))^{-1}((\underline{M}_{s_0}(\pi_1\xi)_\phi(s_0),\underline{e})).$$

So we get

$$(12.2.15)\qquad \langle (*\Theta*\Phi\xi)_{s_0},(*\Theta*\Phi\eta)_{s_0}\rangle = \sum_{\phi,\underline{e}}((\underline{M}_{s_0}\underline{e},\underline{e}))^{-2}(((\pi_1\xi)_\phi(s_0),\underline{M}_{s_0}\underline{e}))$$

$$\cdot\ \overline{(((\pi_1\eta)_\phi(s_0),\underline{M}_{s_0}\underline{e}))}.$$

On the other hand

$$(12.2.16)\qquad \{\operatorname*{res}_{s\to s_0}\iota(s)\underline{M}(s).\pi_1\xi(s_0),\pi_1\eta(s_0)\}_{-s_0} =$$

$$= \sum_{\phi}\{\iota_{s_0}\underline{M}_{s_0}(\pi_1\xi)_\phi(s_0)\phi(s_0),(\pi_1\eta)_\phi(s_0)\phi(s_0)\}_{-s_0}$$

for the other components are killed by ι_{s_0}

$$= \sum_{\phi}\{\iota_{s_0}\phi(s_0),\phi(s_0)\}_{-s_0}((\underline{M}_{s_0}(\pi_1\xi)_\phi(s_0),\overline{(\pi_1\eta)_\phi(s_0)}))$$

$$= \sum_{\phi,\underline{e}}((M_{s_0}\underline{e},\underline{e}))^{-1}((\underline{M}_{s_0}(\pi_1\xi)_\phi(s_0),\underline{e}))\overline{(((\pi_1\eta)_\phi(s_0)\underline{e}))((\underline{e},\underline{e}))^{-1}}$$

$$= \sum_{\phi,\underline{e}}((\underline{M}_{s_0}\underline{e},\underline{e}))^{-2}((\underline{M}_{s_0}(\pi_1\xi)_\phi(s_0),\underline{e}))((\underline{M}_{s_0}(\pi_1\eta)_\phi(s_0),\underline{e})).$$

Comparision of (12.2.15) and (12.2.16) completes the proof.

Remark. For $s_0 \in S^e$, $\underline{e} \in E(s_0)$, $\underline{e} \neq 0$ and $\iota_{s_0}\phi(s_0) \neq 0$ we have

(12.2.17) $\qquad \{\iota_{s_0}\underline{M}_{s_0}\phi(s_0)\otimes\underline{e},\ \phi(s_0)\otimes\underline{e}\}_{-s_0} > 0,$

for it is equal to $\|\underline{E}_{s_0}\ \underline{e}\ \phi(s_0)\|^2$. After extension of the scalar products in 3.4 to E-valued $\underline{g}$-K-spaces we get

(12.2.18) $\qquad \{\iota_{s_0}\underline{M}_{s_0}\phi(s_0)\otimes\underline{e},\phi(s_0)\otimes\underline{e}\}_{-s_0} =$

$$\begin{cases} [\underline{M}_{s_0}\phi(s_0)\otimes\underline{e},\phi(s_0)\otimes\underline{e}]_{s_0} & \text{if } s_0 < \tfrac{1}{2}(1-\tau) \\ -\langle\iota_2^*\underline{M}_{s_0}\phi(s_0)\otimes\underline{e},\iota_2^*\phi(s_0)\otimes\underline{e}\rangle_0 & \text{if } \tau = 0,\ s_0 = \tfrac{1}{2} \\ -\pi^{-1}\sin\pi\tau\langle\iota_{2-\tau}^*\underline{M}_{s_0}\phi(s_0)\otimes\underline{e},\iota_{2-\tau}^*\phi(s_0)\otimes\underline{e}\rangle_\tau & \text{if } 0<\tau<1,\ s_0 = \tfrac{1}{2}(1-\tau) \end{cases}$$

So $\underline{M}_{s_0}$ is positive definite on $E(s_0)$ if $s_0 < \frac{1}{2}(1-\tau)$ and negative definite if $s_0 = \frac{1}{2}(1-\tau)$.

12.3. The subspace ${}^cL^2(\Gamma\backslash G,\chi)$.

12.3.1. Definition. The space ${}^cL^2 = {}^cL^2(\Gamma\backslash G,\chi)$ is defined as the orthogonal complement in ${}^eL^2(\Gamma\backslash G,\chi)$ of $\bigoplus_{s_0\in S^e} {}^eL^2(s_0)$.

From the lemmas 12.2.2 and 12.2.5 it is clear that for ξ and η as in lemma 12.2.2 for the ${}^cL^2$-components holds

(12.3.1) $\qquad \langle(*\Theta*\Phi\xi)_c,(*\Theta*\Phi\eta)_c\rangle = \frac{1}{2\pi i}\int_0^{i\infty}\langle(1+\iota\underline{M})\pi_1\xi(s),(1+\iota\underline{M})\pi_1\eta(s)\rangle_s ds.$

Here I have extended definition (3.4.3) in the natural way.
$(1+\iota\underline{M})\pi_1\xi$ is a section of $\pi_1^*H^\tau$ on $(0,i\infty)$. For general sections of $\pi_1^*H^\tau$ on $(0,i\infty)$ we may write down the integral in the right hand side of (12.3.1). It converges at least for sections of the form

(12.3.2) $\qquad \phi(s) = \sum_\psi \phi_\psi(s)\psi(s),$

where ψ runs through a finite subset of the standard basis of π_1H^τ and ϕ_ψ is a square integrable function on $(0,i\infty)$ with values in E. The integral

(12.3.3) $\qquad \frac{1}{2\pi i}\int_0^{i\infty}\langle\phi(s),\phi_1(s)\rangle_s ds$

gives a scalar product. After completion we get a Hilbert space, which I call ${}^cH$. It is not difficult to see that the $(1+\underline{M})\pi_1\xi$ with ξ a holomorphic rapidly decreasing section of $*H^\tau$ on some strip, are dense in ${}^cH$. For if $h \in C_c^\infty(0,i\infty)$, ψ a basis section of π_1H^τ and $\underline{e} \in E$, then ϕ defined by

(12.3.4) $\qquad \phi(s) = h(s)\psi(s)\otimes\underline{e}$

is approximated by $(1+\iota\underline{M})\xi_v$
with

(12.3.5) $\qquad \xi_v(s) = \pi^{-\frac{1}{2}}\int_0^\infty v^{\frac{1}{2}}e^{v(s-iu)^2}h(iu)du.\psi(s)\otimes\underline{e},$

v positive and large.

So the map

(12.3.6) $\quad (*\Theta*\Phi\xi)_c \to (1+\iota\underline{M})\pi_1\xi$

extends to a unitary isomorphism ${}^cL^2 \to {}^c\mathcal{H}$.

12.4. The Eisenstein transform.

12.4.1. Definition. Take $\sigma > \frac{1}{2}$. Let the continuous function f on G satisfy

(12.4.1) $\quad f(\gamma g) = \chi(\gamma)f(g)$ for $\gamma \in \Gamma$, $g \in G$

(12.4.2) $\quad$ f is K-finite on the right

(12.4.3) $\quad |f(g_c\, na(y)k)| \ll y^{\frac{1}{2}-\sigma}$ for $y \to \infty$

uniformly in $n \in N$, $k \in K$ for each cusp c; then the Eisenstein transform ηf is defined by

(12.4.4) $\quad \eta f(s) = \sum_{\psi,\lambda\in\Lambda^0} \langle f, \underline{E}(-s)\psi(-\bar{s})\otimes\underline{e}_\lambda\rangle \psi(s)\otimes\underline{e}_\lambda$,

ψ running through the standard basis of H^τ, for $|\mathrm{Re}\, s| \leq \sigma$, and $\underline{E}$ holomorphic at $-\bar{s}$. Clearly ηf is a section of $*H^\tau$. If $|f| \in C_c(\Gamma\backslash G)$, then ηf is holomorphic in the same region as $\underline{E}$ and has only first order poles at $s_0 \in S^e$.

12.4.2. Lemma. For f as in definition 12.4.1

(12.4.5) $\quad \eta f = \iota\underline{M}\eta f$.

Proof. Directly from (11.6.18).

12.4.3. Notation. For $g \in *S_\tau = {}^0S_\tau\otimes E$ I define $*\omega g$ by

(12.4.6) $\quad ((*\omega g, \underline{e})) = {}^0\omega((g,\underline{e}))$ for $\underline{e} \in E$.

12.4.4. Lemma. For $g \in *S_\tau$

(12.4.7) $\quad \eta*\Theta g = (1+\iota\underline{M})\pi_1 *\omega g$.

Proof. It is sufficient to consider g with $*\omega g(s) = h(s)\psi(s)$ for some basis section ψ.

(12.4.8)
$$\eta*\Theta g(s) = \sum_{\lambda\in\Lambda^0}\langle *\Theta g, \underline{E}(-\bar{s})\psi(-\bar{s})\otimes\underline{e}_\lambda\rangle\psi(s)\otimes\underline{e}_\lambda =$$
$$= \sum_{\lambda\in\Lambda^0}\langle P_1 g, \mathrm{St}(-\bar{s})\psi(-\bar{s})\otimes\underline{e}_\lambda\rangle\psi(s)\otimes\underline{e}_\lambda +$$
$$+ \sum_{\lambda\in\Lambda^0}\langle P_1 g, \mathrm{St}(\bar{s})\psi(\bar{s})\otimes\underline{M}(-\bar{s})\underline{e}_\lambda\rangle\gamma_\psi(-s)\psi(s)\otimes\underline{e}_\lambda$$
$$= \pi_1 *\omega g(s) + \iota(-s)\underline{M}(-s)\pi_1 *\omega g(-s).$$

12.4.5. Lemma. Let $s_0 \in S^e$. For $g \in *S_\tau$,

(12.4.9) $\quad \operatorname*{res}_{s\to s_0} \eta*\Theta g(-s) = \iota_{s_0}\underline{M}_{s_0}\pi_1 *\omega g(s_0)$

ι_{s_0} has been defined in (12.2.11).

Proof. Direct from the previous lemma.

12.4.6. Lemma. For $g, g_1 \in *S_\tau$ the scalar product of the ${}^cL^2$-components of $*\Theta g$ and

$*\Theta g_1$ is given by

(12.4.10) $$\langle (*\Theta g)_c, (*\Theta g_1)_c \rangle = \frac{1}{2\pi i} \int_0^{i\infty} \langle \eta{*}\Theta g(s), \eta{*}\Theta g_1(s)\rangle_s \, ds$$

Proof. See (12.3.1) and lemma 12.4.4.

I would like to describe also the scalar product of the ${}^eL^2(s_0)$-components with help of the Eisenstein transform. Take $g, g_1 \in *S_\tau$ and $s_0 \in S^e$. By lemma 12.2.4

(12.4.11) $$\langle (*\Theta g)_{s_0}, (*\Theta g_1)_{s_0} \rangle = \{\iota_{s_0} \underline{M}_{-s_0} \pi_1 {*}\omega g(s_0), \pi_1 {*}\omega g_1(s_0)\}_{-s_0} =$$
$$= \{\underline{M}_{-s_0} \pi_1 {*}\omega g(s_0), \iota_{s_0} \pi_1 {*}\omega g_1(s_0)\}_{s_0}.$$

To use lemma 12.4.5 I need an inverse of $\underline{M}_{-s_0}$.

12.4.7. Definition. Let $s_0 \in S^e$. The inverse of $\underline{M}_{-s_0} : E(s_0) \to E(s_0)$ is denoted $\underline{N}_{-s_0}$.

12.4.8. Remark. $\underline{N}_{-s_0}$ is hermitian. It is positive definite if $s_0 < \frac{1}{2}(1-\tau)$ and negative definite if $s_0 = \frac{1}{2}(1-\tau)$. See the end of subsection 12.2.

Now let us first consider the case $s_0 = \frac{1}{2}(1-\tau)$. We have

(12.4.12) $$\begin{cases} \iota_{s_0} = \operatorname{res}_{s\to\frac{1}{2}} \iota(s) = \iota^*_2 & \text{if } \tau = 0 \\ \iota_{s_0}\pi_1 = \operatorname{res}_{s\to\frac{1}{2}(1-\tau)} \iota(s)\pi_1 = -\pi^{-1} \sin \pi\tau \; \iota^*_{2-\tau}\pi_1 & \text{if } 0 < \tau < 1 \end{cases}$$

Using (3.4.6), (3.4.8) and (3.4.13) we see that the expression in (12.4.11) equals

(12.4.13) $$-\langle \underline{N}_{-s_0} \operatorname{res}_{s\to s_0} \eta{*}\Theta g(-s), \operatorname{res}_{s\to s_0} \eta{*}\Theta g_1(-s)\rangle_0 \qquad \text{if } \tau = 0$$
$$-\frac{\pi}{\sin \pi\tau} \langle \underline{N}_{-s_0} \operatorname{res}_{s\to s_0} \eta{*}\Theta g(-s), \operatorname{res}_{s\to s_0} \eta{*}\Theta g_1(-s)\rangle_\tau \qquad \text{if } 0 < \tau < 1$$

In the case $s_0 < \frac{1}{2}(1-\tau)$ the expression in (12.4.11) equals

(12.4.14) $$\{\underline{N}_{-s_0} \underline{M}_{-s_0} \pi_1 {*}\omega g(s_0), \iota_{s_0} \underline{M}_{-s_0} \pi_1 {*}\omega g_1(s_0)\}_{s_0} =$$
$$= \{\iota_{s_0} \underline{N}_{-s_0} \underline{M}_{-s_0} \pi_1 {*}\omega g(s_0), \iota(-s_0)\iota_{s_0} \underline{M}_{-s_0} \pi_1 {*}\omega g_1(s_0)\}_{-s_0} =$$
$$= [\underline{N}_{-s_0} \operatorname{res}_{s\to s_0} \eta{*}\Theta g(-s), \operatorname{res}_{s\to s_0} \eta{*}\Theta g_1(-s)]_{-s_0}$$

12.4.9. Lemma. Let $s_0 \in S^e$, g, $g_1 \in *S_\tau$. The scalar product of the projections of $\Theta{*}g$ and $*\Theta g_1$ on ${}^eL^2(s_0)$ is given by (12.4.14) if $s_0 < \frac{1}{2}(1-\tau)$ and by (12.4.13) if $s_0 = \frac{1}{2}(1-\tau)$.

12.5. Spectral decomposition.

In the lemmas 12.4.6 and 12.4.9 we have seen how the scalar product of elements in $*\Theta{*}S_\tau$ may be expressed by the Eisenstein transform. But this transform has been defined for a wider class of functions. The scalar product formula turns out to be valid for some of these functions.

12.5.1. Proposition. Let f and g satisfy the conditions in definition 12.4.1, suppose that ηf and ηg are continuous on $(0,i\infty)$ and meromorphic at the points $-s_0$ for $s_0 \in S^e$ with first order poles. Let f_e and g_e be the projections of f and g in ${}^eL^2(\Gamma\backslash G,\chi)$. Then

(12.5.1) $$\langle f_e,g_e\rangle = \frac{1}{2\pi i}\int_0^{i\infty} \langle \eta f(s),\eta g(s)\rangle_s\, ds +$$

$$+ \sum_{\substack{s_0\in S^e\\ s_0<\frac12(1-\tau)}} [\underline{N}_{s_0} \operatorname*{res}_{s\to s_0} \eta f(-s), \operatorname*{res}_{s\to s_0} \eta g(-s)]_{-s_0} +$$

$$+ \begin{cases} -\langle \underline{N}_{\frac12} \operatorname*{res}_{s\to\frac12} \eta f(-s), \operatorname*{res}_{s\to\frac12} \eta g(-s)\rangle_0 & \text{if } \tau = 0 \text{ and } \frac12 \in S^e \\ -\dfrac{\pi}{\sin \pi\tau}\langle \underline{N}_{\frac12(1-\tau)} \operatorname*{res}_{s\to\frac12(1-\tau)} \eta f(-s), \operatorname*{res}_{s\to\frac12(1-\tau)} \eta g(-s)\rangle_\tau & \text{if } 0<\tau<1 \text{ and } \frac12(1-\tau) \in S^e \end{cases}$$

Remark that condition (12.4.3) implies that $f,g \in L^2(\Gamma\backslash G,\chi)$.

Proof. We write f_c and g_c for the ${}^cL^2$-components of f and g and f_{s_0}, g_{s_0} for the ${}^eL^2(s_0)$-components. We shall consider $\langle f_{s_0},g_{s_0}\rangle$ and $\langle f_c,g_c\rangle$ separately.

Fix $s_0 \in S^e$.

12.5.2. Lemma. Take f as in proposition 12.5.1. Then

(12.5.2) $$\operatorname*{res}_{s\to s_0} \eta f(-s) \in \iota_{s_0}\pi_1 H^\tau(s_0)\otimes E(s_0)$$

and for $\underline{e} \in E(s_0)$ and ϕ a constant section of $\pi_1 H^\tau$

(12.5.3) $$\langle f,\underline{E}_{s_0}\underline{e}\,\phi(s_0)\rangle = \{\operatorname*{res}_{s\to s_0} \eta f(-s),\phi(s_0)\otimes\underline{e}\}_{-s_0}.$$

Proof. For any basis section ψ of H^τ, and $\underline{e} \in E$, $(s-s_0)\underline{E}(+s)\psi(s)\otimes\underline{e}$ satisfies for s near s_0 a growth condition enabling us to take the limit in

(12.5.4) $$\lim_{s\to s_0} (\bar{s}-s_0)\langle f,\underline{E}(s)\psi(s)\otimes\underline{e}\rangle$$

inside the integral; to see this consult lemma 11.2.5. So we get

(12.5.5) $$\begin{aligned}\operatorname*{res}_{s\to s_0} \eta f(-s) &= \lim_{s\to s_0} (s-s_0) \sum_{\psi,\lambda}\langle f,\underline{E}(\bar{s})\psi(\bar{s})\otimes\underline{e}_\lambda\rangle\psi(-s)\otimes\underline{e}_\lambda \\ &= \sum_{\psi,\lambda}\langle f, \operatorname*{res}_{s\to s_0} \underline{E}(\bar{s})\psi(\bar{s})\otimes\underline{e}_\lambda\rangle\psi(-s_0)\otimes\underline{e}_\lambda\end{aligned}$$

From this follows that only ψ with $\iota_{s_0}\psi(s_0) \neq 0$ occur and, furthermore, that

(12.5.6) $$\operatorname*{res}_{s\to s_0} \eta f(-s) = \sum_{\phi,\underline{e}} \langle f,\underline{E}_{s_0}\underline{e}\,\phi(s_0)\rangle\phi(-s_0)\otimes\underline{e},$$

where ϕ and $\underline{e}$ are chosen as in the proof of lemma 12.2.5. To see this, remark that $\{\underline{e}_\lambda \mid \lambda \in \Lambda^0\}$ in (12.4.4) may be replaced by any other orthonormal basis of E. Both assertions in the lemma are clear from (12.5.6).

Continuation of the proof of proposition 12.5.1. By lemma 12.5.2 we may construct a rapidly decreasing holomorphic section ξ_f of $\pi_1{}^*H^\tau$ such that

(12.5.7) $$\operatorname*{res}_{s\to s_0} \eta f(-s) = \iota_{s_0}\underline{M}_{s_0}\pi_1\xi_f(s_0).$$

From the lemmas 12.5.2 and 12.2.3 follows that $*\Theta*\Phi\xi_f$ and f have the same scalar products with all $E_{s_0}\ \underline{e}\ \phi(s_0)$. Doing the same for g we see from lemma 12.2.5 that

(12.5.8) $$\langle f_{s_0}, g_{s_0}\rangle = \langle (*\Theta*\Phi\ \xi_f)s_0, (*\Theta*\Phi\xi_g)s_0\rangle =$$

$$= \{\iota_{s_0}\underline{M}_{s_0}\ \pi_1\xi_f(s_0), \pi_1\xi_g(s_0)\}_{-s_0}.$$

The same computations as led to lemma 12.4.9 show that $\langle f_{s_0}, g_{s_0}\rangle$ has the form indicated in (12.5.1).

To prove that $\langle f_c, g_c\rangle = \frac{1}{2\pi i}\int_0^{i\infty} \langle \eta f(s), \eta g(s)\rangle_s\, ds$ we have to show that ηf represents f_c in ${}^cH$ under the isomorphisms ${}^cL^2 \cong {}^cH$ discussed in 12.3.

12.5.3. Lemma. Let ϕ be a continuous section of $\pi_1 *H^\tau$ on $(0, i\infty)$ with compact support. The element ${}^c\phi \in {}^cL^2$ corresponding to ϕ under the isomorphism ${}^cL^2 \cong {}^cH$ is represented by the function

(12.5.9) $$x \to \frac{1}{2\pi i}\int_0^{i\infty} \underline{E}(s)\phi(s)(x)ds.$$

Proof. As $*\Theta*S_\tau$ is dense in ${}^eL^2$ it is sufficient to consider $\langle {}^c\phi, *\Theta g\rangle$ for $g \in *S_\tau$.

(12.5.10) $$\langle {}^c\phi, *\Theta g\rangle = \frac{1}{2\pi i}\int_0^{i\infty} \langle \phi(s), \eta*\Theta g(s)\rangle_s\, ds =$$

$$= \sum_{\psi,\lambda} \frac{1}{2\pi i}\int_0^{i\infty} \overline{\langle *\Theta g, \underline{E}(-\bar{s})\psi(-\bar{s})\otimes\underline{e}_\lambda\rangle}\{\phi(s), \psi(s)\otimes\underline{e}_\lambda\}_s\, ds$$

$$= \sum_{\psi,\lambda}\int_{\Gamma\backslash G} \overline{*\Theta g(x)}\ \frac{1}{2\pi i}\int_0^{i\infty} \underline{E}(s)\psi(s)\otimes\underline{e}_\lambda(x)\ \{\phi(s), \psi(s)\otimes\underline{e}_\lambda\}_s\, dsdx$$

$$= \int_{\Gamma\backslash G} \overline{*\Theta g(x)}\ \frac{1}{2\pi i}\int_0^{i\infty} \underline{E}(s)\phi(s)(x)dsdx.$$

As $*\Theta g$ and ϕ have compact support the convergence is absolute, so interchanging the order of integration gives no problem.

Continuation of the proof of proposition 12.5.1. Let ϕ be as in lemma 12.5.3.

(12.5.11) $$\langle f, {}^c\phi\rangle = \int_{\Gamma\backslash G} f(x)\ \frac{1}{2\pi i}\int_0^{i\infty} \overline{\underline{E}(s)\phi(s)(x)}\ dsdg.$$

In order to interchange the order of integration here we may apply lemma 11.2.5. In neighbourhoods of points of $(0, i\infty)$ we have estimates for $\underline{E}(s)\phi(s)(x)$ ensuring absolute convergence. As supp ϕ is compact this is sufficient. So we get

(12.5.12) $$\langle f, {}^c\phi\rangle = \frac{1}{2\pi i}\int_0^{i\infty}\int_{\Gamma\backslash G} f(g)\ \overline{\underline{E}(s)\phi(s)(g)}\ dgdt$$

$$= \frac{1}{2\pi i}\int_0^{i\infty} \sum_{\psi,\lambda}\int_{\Gamma\backslash G} f(g)\ \overline{\underline{E}(-\bar{s})\psi(-\bar{s})\otimes\underline{e}_\lambda(g)}\ \overline{\{\phi(s), \psi(s)\otimes\underline{e}_\lambda\}_s}\ dgds$$

$$= \frac{1}{2\pi i}\int_0^{i\infty} \{\eta f(s), \phi(s)\}_s\, ds.$$

So indeed f_c is represented in ${}^cH$ by ηf.

12.5.4. Remark. If f and g are only bounded, and satisfy the other conditions in definition 12.4.1, then everying goes through except the contribution $\langle f_{\frac{1}{2}}, g_{\frac{1}{2}}\rangle$ if $s_0 = \frac{1}{2} \in S^e$.

13. The space H_σ^τ

13.1. Introduction

In subsection 11.3 the spectral decomposition of the $\underline{g}$-K-space ${}^{0}S_\tau$ has been given. The principal ingredient was the Mellin transform. The spectral decomposition of ${}^{\frac{1}{2}}S_\tau$ in 13.4 will follow along the same lines. The Mellin transform is replaced by a Whittaker transform, studied in 13.3. In 13.2 I have gathered other results on Whittaker functions.

The space ${}^{\frac{1}{2}}S_\tau$ is not large enough to serve as space of test functions in the proof of the sum formula. In 13.5 I discuss a larger space H_σ^τ. This may be compared with the situation in sections 11 and 12. There we needed general rapidly decreasing holomorphic sections of ${}^{*}H^\tau$; we could not manage with only those in $*\omega\ {}^{*}S_\tau$.

Except for (13.2.12) the facts in 13.2 are well-known, see [25] or other books on special functions. For $\kappa = 0$ formula (13.2.12) amounts to Basset's formula, see [43], p.172. The Whittaker transform studied in 13.3 amounts to the Lebedev transform in the case $\kappa = 0$, see [25], p.398. The general case is new as far as I know. The idea of handling it is similar to Kuznetsov's treatment of a Bessel transform in [15], thm. 5, §2.5.

13.2. Whittaker functions

In section 5 I have used several results on Whittaker functions and I shall need some more facts. In this subsection I state these results. Mosttimes I refer to [40], but the results can be found in most books on special functions, e.g. [25]. In this section κ is a real number and s a complex one.

The Whittaker differential equation is

(13.2.1) $$f''(y) + (-\tfrac{1}{4}+\kappa y^{-1}+(\tfrac{1}{4}-s^2)y^{-2})f(y) = 0;$$

it is the same equation for s and -s. It always has a solution decaying rapidly for $y \to \infty$ and another one increasing rapidly.

A solution is given by

(13.2.2) $$M_{\kappa,s}(y) = y^{\frac{1}{2}+s}\, e^{-\frac{1}{2}y} \sum_{n=0}^{\infty} \frac{(\frac{1}{2}+s-\kappa)_n}{n!\,(1+2s)_n}\, y^n$$

with

(13.2.3) $$(a)_n = \prod_{j=0}^{n-1} (a+j), \quad (a)_0 = 1;$$

see [40], p.9. $M_{\kappa,s}$ is defined for $s \notin -\frac{1}{2}\mathbb{N}$.

(13.2.4) $$|M_{\kappa,s}(y)| \ll y^{\frac{1}{2}+\mathrm{Re}\, s} \quad \text{for } y \downarrow 0.$$

In general its growth for $y \to \infty$ is exponential.
Another solution is

(13.2.5) $$W_{\kappa,s}(y) = \Gamma(2s)\Gamma(\tfrac12+s-\kappa)^{-1}M_{\kappa,-s}(y) + \Gamma(-2s)\Gamma(\tfrac12-s-\kappa)^{-1}M_{\kappa,s}(y),$$

see [40], p.10. It can be extended to an even holomorphic function of $s \in \mathbb{C}$.
On p.61 of [40] an asymptotic expansion is given from which follows that

(13.2.6) $$|W_{\kappa,s}(y)| \ll y^{\kappa}\, e^{-\frac12 y} \qquad \text{for } y \to \infty .$$

In general $M_{\kappa,s}$ and $W_{\kappa,s}$ are linearly independent. From (13.2.2) follows that for $a \in \mathbb{N}$:

(13.2.7) $$\operatorname*{res}_{s\to-\frac12 a} M_{\kappa,s}(y) = -\tfrac12((a-1)!a!)^{-1}(\kappa+\tfrac12-\tfrac12 a)_a\, M_{\kappa,\frac12 a}(y).$$

Let $b \equiv 2\kappa(2)$, $b \leqslant 2\kappa$, if $2\kappa \in \mathbb{Z}$ then $b \geqslant 1$, then

(13.2.8) $$W_{\kappa,\frac12(b-1)}(y) = (-1)^{\kappa-\frac12 b}\,\Gamma(b)^{-1}\,\Gamma(\tfrac12 b+\kappa)\,M_{\kappa,\frac12(b-1)}(y).$$

For $\kappa = \frac12 b$, $b \geqslant 1$ if $b \in \mathbb{Z}$:

(13.2.9) $$W_{\frac12 b,\frac12(b-1)}(y) = M_{\frac12 b,\frac12(b-1)}(y) = y^{\frac12 b}\, e^{-\frac12 y}.$$

For κ and s general

(13.2.10) $$\begin{aligned} yM'_{\kappa,s}(y) &= (\tfrac12+s+\kappa)M_{\kappa+1,s}(y) + (\tfrac12 y-\kappa)M_{\kappa,s}(y) \\ &= (\tfrac12+s-\kappa)M_{\kappa-1,s}(y) + (\kappa-\tfrac12 y)M_{\kappa,s}(y). \end{aligned}$$

(13.2.11) $$\begin{aligned} yW'_{\kappa,s}(y) &= -W_{\kappa+1,s}(y) + (\tfrac12 y-\kappa)W_{\kappa,s}(y) \\ &= (\tfrac12+s-\kappa)(\tfrac12-s-\kappa)W_{\kappa-1,s}(y) + (\kappa-\tfrac12 y)W_{\kappa,s}(y); \end{aligned}$$

see [40], p.24,25.
Corollary 6.7.5 gives for Re $s > 0$, $\frac12 + s + \kappa \notin -\mathbb{N}$ the following integral representation

(13.2.12) $$W_{\kappa,s}(y) =$$
$$\pi^{-1}2^{s-1}\Gamma(\tfrac12+s+\kappa)y^{\frac12-s}\int_{-\infty}^{+\infty} e^{-\frac12 ixy}(1+x^2)^{-\frac12-s}e^{2\kappa i\arctan x}dx.$$

This implies

(13.2.13) $$|W_{\kappa,s}(y)| \ll y^{\frac12-\sigma}\,|\Gamma(\tfrac12+s-\kappa)| \qquad \text{for } y \to \infty$$

uniformly for Re $s = \sigma$, with $\sigma > 0$.

13.3 Whittaker transforms

13.3.1. Definition For $f \in C_c^\infty(0,\infty)$ the Whittaker transforms $w_\kappa f$ and $m_\kappa f$ are defined by

(13.3.1) $$w_\kappa f(s) = \int_0^\infty f(y)\, W_{\kappa,s}(y)\, y^{-2}dy$$

(13.3.2) $$m_\kappa f(s) = \int_0^\infty f(y) \quad M_{\kappa,s}(y)\, y^{-2} dy .$$

Remark that $w_\kappa f$ is holomorphic on $\mathbb{C}$ and that $m_\kappa f$ may have poles in $-\frac{1}{2}\mathbb{N}$. Both transforms are related by the equation corresponding to (13.2.5).

Put

(13.3.3) $$A_n(s) = (-1)^n 2^{-n} (n!)^{-1} \sum_{l=0}^{n} (-1)^l \binom{n}{l} 2^l \frac{(\frac{1}{2}+s-\kappa)_l}{(1+2s)_l},$$

then $m_\kappa f$ may be expressed by

(13.3.4) $$m_\kappa f(s) = \sum_{n=0}^{\infty} A_n(s)\, \mathcal{M}f(-\tfrac{1}{2}+s+n);$$

here $\mathcal{M}$ is the Mellin transform given in (11.3.6).

13.3.2. Lemma. On the region $|\arg s| < \frac{5}{6}\pi$

(13.3.5) $$|A_n(s)| \ll (n!)^{-1} C^n$$

for some $C > 0$.

Proof. Let $j \geq 0$, $j \in \mathbb{Z}$, then $|2s+1+j| \geq \frac{1}{2}(j+1)$, so

(13.3.6) $$\left|\frac{\frac{1}{2}+s-\kappa+j}{1+2s+j}\right| \leq \tfrac{1}{2}\left(1+ \frac{|-2\kappa+j|}{|1+2s+j|}\right) \leq \tfrac{1}{2}\left(1+ \frac{j+2|\kappa|}{\frac{1}{2}(j+1)}\right) \leq \tfrac{1}{2} + 2|\kappa| .$$

So

(13.3.7) $$\left|\frac{(\frac{1}{2}+s-\kappa)_l}{(1+2s)_l}\right| \leq (\tfrac{1}{2}+2|\kappa|)^l,$$

hence the lemma follows with $C = 1+2|\kappa|$.

13.3.3. Lemma. Uniformly for $|\arg s| < \frac{5}{6}\pi$

(13.3.8) $$|m_\kappa f(s)| \ll C^{|\mathrm{Re}\, s - \frac{1}{2}|} (1+|\mathrm{Im}\, s|)^{-a}$$

for all $a \in \mathbb{R}$, for some $C > 1$.

On strips of the form $|\mathrm{Re}\, s| \leq \sigma$

(13.3.9) $$|w_\kappa f(s)| \ll e^{-\frac{1}{2}\pi|\mathrm{Im}\, s|} (1+|\mathrm{Im}\, s|)^{-a}$$

for each $a \in \mathbb{R}$.

Proof. There is a constant C_f such that

(13.3.10) $$|\mathcal{M}f(z)| \ll C_f^{|\mathrm{Re}\, z|} (1+|\mathrm{Im}\, z|)^{-a}$$

for each $a \in \mathbb{R}$. Use this and lemma 13.3.2 to estimate the series in (13.3.4). This gives (13.3.8).

To get (13.3.9) use (13.2.5).

13.3.4. Lemma. Let $n \geq 1$. For $|s| \to \infty$

(13.3.11) $$|A_n(s)| \ll |s|^{-1}.$$

Proof. Straightforward.

13.3.5. Proposition. For $f, g \in C_c^\infty(0,\infty)$

(13.3.12) $$\int_0^\infty f(y)\overline{g(y)}\, y^{-2}dy =$$

$$\frac{1}{2\pi i}\int_0^{i\infty} w_\kappa f(s)\,\overline{w_\kappa g(s)}\,|\Gamma(2s)^{-1}\Gamma(\tfrac12+s-\kappa)|^2 ds$$

$$+ \sum_{\substack{b\equiv 2\kappa(2)\\ 1<b\leqslant 2\kappa}} (b-1)(\kappa-\tfrac12 b)!^{-1}\,\Gamma(\tfrac12 b+\kappa)^{-1} w_\kappa f(\tfrac12(b-1))\overline{w_\kappa g(\tfrac12(b-1))}$$

The map w_κ extends to a unitary isomorphism from $L^2((0,\infty),y^{-2}dy)$ onto

$$L^2((0,i\infty),|\Gamma(2s)^{-1}\Gamma(\tfrac12+s-\kappa)|^2ds) \oplus$$
$$L^2(\{\tfrac12(b-1)\,|\,b\equiv 2\kappa(2), 1<b\leqslant 2\kappa\},\ (b-1)(\kappa-\tfrac12 b)!^{-1}\Gamma(\tfrac12 b+\kappa)^{-1}).$$

Proof. Equality (13.3.12) follows from the lemmas 13.3.6, 13.3.7 and 13.3.8. Clearly w_κ may be extended to a unitary map, also denoted by w_κ. To get surjectivity of w_κ I copy the proof of the surjectivity of the Eisenstein transform on p. 349 of [21]. Consider on $C_c^\infty(0,\infty)$ the operator $-y^2\frac{\partial^2}{\partial_y^2} + (\tfrac14 y^2-\kappa y)$. It has in $L^2((0,\infty), y^{-2}dy)$ a self-adjoint extension A ; to see this use Ch.XIV §6 and the corollary on p.372 of [21]. Under w_κ the operator A corresponds to multiplication by $\tfrac14-s^2$. Now take $s_0 \in (0,i\infty) \cup \{\tfrac12(b-1)\,|\,b\equiv 2\kappa(2),\ 1<b\leqslant 2\kappa\}$ and some cut-off function $\phi\in C_c^\infty(0,\infty)$; $\phi\geqslant 0$. Let $f\in C_c(0,\infty)$ be given by $f(y) = \phi(y)\,\overline{W_{\kappa,s_0}(y)}$. Then $w_\kappa f(s_0) > 0$. Now define the bounded function h on $\mathbb{R}$ by $h(\tfrac14-s^2) = w_\kappa f(s)^{-1}$ for s near s_0 and $h(\tfrac14-s^2)=0$ otherwise. Then $w_\kappa(h(A)f)(s) = h(\tfrac14-s^2)w_\kappa f(s)$. So we get the characteristic function of a small neighbourhood of s_0 in the image of w_κ. Hence w_κ has dense image.

13.3.6. Lemma. Let $f,g\in C_c^\infty(0,\infty)$. Then

(13.3.13) $$\frac{1}{2\pi i}\int_0^{i\infty} w_\kappa f(s)\,\overline{w_\kappa g(s)}\,|\Gamma(2s)^{-1}\Gamma(\tfrac12+s-\kappa)|^2 ds =$$
$$= \frac{1}{2\pi i}\int_{\mathrm{Re}\, s=0} m_\kappa f(s) m_\kappa \bar g(-s)ds$$
$$+ \frac{1}{2\pi i}\int_{\mathrm{Re}\, s=0} m_\kappa f(s) m_\kappa \bar g(s)\,\frac{\Gamma(-2s)\Gamma(\tfrac12+s-\kappa)}{\Gamma(2s)\Gamma(\tfrac12-s-\kappa)}\,ds$$

and all these integrals converge absolutely.

Proof. The convergence follows from lemma 13.3.3. Using (13.2.5) two times I get

(13.3.14) $$\frac{1}{2\pi i}\int_0^{i\infty} w_\kappa f(s)\,\overline{w_\kappa g(s)}\,|\Gamma(2s)^{-1}\Gamma(\tfrac12+s-\kappa)|^2 ds =$$
$$\frac{1}{2\pi i}\int_{\mathrm{Re}\, s=0} m_\kappa f(s) w_\kappa \bar g(-s)\Gamma(2s)^{-1}\Gamma(\tfrac12+s-\kappa)ds =$$
$$\frac{1}{2\pi i}\int_{\mathrm{Re}\, s=0} m_\kappa f(s) m_\kappa \bar g(-s)ds$$
$$+ \frac{1}{2\pi i}\int_{\mathrm{Re}\, s=0} m_\kappa f(s) m_\kappa \bar g(s)\Gamma(-2s)\Gamma(2s)^{-1}\Gamma(\tfrac12+s-\kappa)\Gamma(\tfrac12-s-\kappa)^{-1}ds.$$

13.3.7. Lemma. For $f,g \in C_c^\infty(0,\infty)$

$$\frac{1}{2\pi i}\int_{\mathrm{Re}\,s=0} m_\kappa f(s) m_\kappa \bar g(-s)\,ds = \int_0^\infty f(y)\,\overline{g(y)}\,y^{-2}dy - \tag{13.3.15}$$

$$-\sum_{a=2}^{\infty} \tfrac{1}{2}\frac{(\kappa-\frac{1}{2}a+1)_{a-1}}{(a-1)!\,(a-2)!}\; m_\kappa f(\tfrac{1}{2}(a-1)) m_\kappa \bar g(\tfrac{1}{2}(a-1)).$$

Proof. By (13.3.4) and lemma 13.3.2:

$$\frac{1}{2\pi i}\int_{\mathrm{Re}\,s=0} m_\kappa f(s) m_\kappa \bar g(-s)\,ds = \tag{13.3.16}$$

$$\sum_{n,m\geq 0} \frac{1}{2\pi i}\int_{\mathrm{Re}\,s=0} A_n(s)A_m(-s)\mathcal{M}f(-\tfrac{1}{2}+s+n)\;\mathcal{M}\bar g(-\tfrac{1}{2}-s+m)\,ds.$$

The term with $n = m = 0$ gives

$$\int_0^\infty f(y)\,\overline{g(y)}\,y^{-2}dy \tag{13.3.17}$$

by the inversion formula for the Mellin transform. After some computations we shall see that the other terms give the other term in the right hand side of (13.3.15).

Let $T > 0$, $T \notin \frac{1}{2}\mathbb{Z}$. We have for $n \geq 1$ or $m \geq 1$

$$\frac{1}{2\pi i}\int_{\mathrm{Re}\,s=0} A_n(s)A_m(-s)\;\mathcal{M}f(-\tfrac{1}{2}+s+n)\;\mathcal{M}\bar g(-\tfrac{1}{2}-s+m)\,ds = K_T^{n,m}\{f,\bar g\} + o(1) \tag{13.3.18}$$

for $T \to \infty$

where $K_T^{n,m}\{.,.\}$ is the distribution on $(0,\infty)^2$ given by the function

$$K_T^{n,m}(y,t) = \frac{1}{2\pi i}\int_{-iT,\ \mathrm{Re}\,s=0}^{iT} A_n(s)\;A_m(-s)\;y^{s+n-\frac{1}{2}}t^{-s+m-\frac{1}{2}}ds \tag{13.3.19}$$

with respect to $y^{-1}dy\,t^{-1}dt$.

Let $y \leq t$. The integral over $[-iT,iT]$ is moved :

$$K_T^{n,m}(y,t) = \frac{1}{2\pi i}\int_{|s|=T,\ \mathrm{Re}\,s\geq 0} A_n(s)A_m(-s)y^{s+n-\frac{1}{2}}t^{-s+m-\frac{1}{2}}ds - \tag{13.3.20}$$

$$-\sum_{1\leq a< 2T} A_n(\tfrac{1}{2}a)\;\operatorname*{res}_{s\to\frac{1}{2}a} A_m(-s)\;y^{\frac{1}{2}a+n-\frac{1}{2}}\;t^{-\frac{1}{2}a+m-\frac{1}{2}}.$$

Let us first estimate the integral. As $n \geq 1$ or $m \geq 1$ the integrand may be estimated using lemma 13.3.4 :

$$|A_n(s)A_m(-s)y^{s+n-\frac{1}{2}}t^{-s+m-\frac{1}{2}}| \ll T^{-1}\;y^{n-\frac{1}{2}}t^{m-\frac{1}{2}}(y/t)^{\mathrm{Re}\,s}.$$

We get the following estimate for the integral :

$$T^{-1}y^{n-\frac{1}{2}}t^{m-\frac{1}{2}}\min(T,\;(\log(t/y))^{-1}). \tag{13.3.21}$$

The residues contribute

$$-\tfrac{1}{2}\sum_{a=1}^{\min(m,[2T])} (\kappa+\tfrac{1}{2}-\tfrac{1}{2}a)_a\;a!^{-1}(a-1)!^{-1}\;A_n(\tfrac{1}{2}a)A_{m-a}(\tfrac{1}{2}a)y^{\frac{1}{2}a+n-\frac{1}{2}}t^{-\frac{1}{2}a+m-\frac{1}{2}} \tag{13.3.22}$$

For $y \geq t$ we move the integral to the left; we get

$$(13.3.23)\qquad K_T^{n,m}(y,t) = -\tfrac12 \sum_{a=1}^{\min(n,[2T])} (\tfrac12+\kappa-\tfrac12 a)_a\, a!^{-1}(a-1)!^{-1} A_{n-a}(\tfrac12 a)A_m(\tfrac12 a)\,.\, y^{-\frac12 a+n-\frac12}t^{\frac12 a+m-\frac12}$$
$$+\, O(T^{-1}y^{n-\frac12}t^{m-\frac12}\min(T,\log(y/t)^{-1})).$$

For given f and g both O-terms contribute to $K_T^{n,m}\{f,\bar g\}$ a term estimated by $O(T^{-1}\log T)$. So up to a term $o(1)$ for $T\to\infty$ the expression in the left hand side of (13.3.18) is given by a distribution given by

$$(13.3.24)\qquad -\tfrac12 y^{n-\frac12}t^{m-\frac12}\cdot\begin{cases}\displaystyle\sum_{a=1}^{m}(\kappa+\tfrac12-\tfrac12 a)_a\, a!^{-1}(a-1)!^{-1}A_n(\tfrac12 a)A_{m-a}(\tfrac12 a)(y/t)^{\frac12 a} & \text{if } y\leq t\\ \displaystyle\sum_{a=1}^{n}(\kappa+\tfrac12-\tfrac12 a)_a\, a!^{-1}(a-1)!^{-1}A_{n-a}(\tfrac12 a)A_m(\tfrac12 a)(t/y)^{\frac12 a} & \text{if } y\geq t.\end{cases}$$

As this distribution does not depend on T we have equality. To prove the lemma we have to sum it over $n,m\geq 0$, $(n,m)\neq(0,0)$. It turns out that we have absolute convergence, so we may rearrange the terms. In both regions $y\leq t$ and $y\geq t$ we get the same sum

$$(13.3.25)\qquad -\tfrac12\sum_{a=1}^{\infty}(a-1)!^{-1}a!^{-1}(\kappa+\tfrac12-\tfrac12 a)_a \sum_{l=0}^{\infty}A_l(\tfrac12 a)y^{\frac12 a+l-\frac12}\sum_{m=0}^{\infty}A_m(\tfrac12 a)y^{\frac12 a+m-\frac12}$$

Integrated against $f(y)\overline{g(t)}$ this gives the second term in the right hand side of the lemma.

13.3.8. Lemma. For $f,g\in C_c^\infty(0,\infty)$

$$(13.3.26)\qquad \frac{1}{2\pi i}\int_{\mathrm{Re}\, s=0} m_\kappa f(s))m_\kappa\bar g(s)\Gamma(-2s)\Gamma(2s)^{-1}\Gamma(\tfrac12+s-\kappa)\Gamma(\tfrac12-s-\kappa)^{-1}ds$$

$$=\begin{cases}\displaystyle\sum_{\substack{a\equiv 2\kappa+1\,(2)\\ a>1}}\tfrac12(a-1)!^{-1}(a-2)!^{-1}(\kappa+1-\tfrac12 a)_{a-1}m_\kappa f(\tfrac12(a-1))m_\kappa\bar g(\tfrac12(a-1))\\ \displaystyle +\sum_{\substack{1<a<2|\kappa|\\ a\equiv 2\kappa\,(2)}}-\tfrac12\operatorname{sgn}\kappa(a-1)!^{-1}(a-2)!^{-1}(\kappa+1-\tfrac12 a)_{a-1}m_\kappa f(\tfrac12(a-1))m_\kappa\bar g(\tfrac12(a-1))\\ \qquad\qquad\qquad\qquad \text{\underline{if} } \kappa\in\tfrac12\mathbb{Z}\\ \displaystyle\sum_{a=2}^{\infty}\tfrac12(a-1)!^{-1}(a-2)!^{-1}(\kappa+1-\tfrac12 a)_{a-1}m_\kappa f(\tfrac12(a-1))m_\kappa\bar g(\tfrac12(a-1))\\ \displaystyle -\sum_{\substack{1<b\leq 2\kappa\\ b\equiv 2\kappa\,(2)}}\Gamma(b)^{-1}\Gamma(b-1)^{-1}(\kappa-\tfrac12 b)!^{-1}\Gamma(\tfrac12 b+\kappa)m_\kappa f(\tfrac12(b-1))m_\kappa\bar g(\tfrac12(b-1))\\ \qquad\qquad\qquad\qquad \text{\underline{if} } \kappa\notin\tfrac12\mathbb{Z}\,.\end{cases}$$

Proof. This proof is easier than the previous one. By lemma 13.3.3 the integral tends to zero if we move it off to the right. So we are left with the contributions of the poles in the region $\mathrm{Re}\ s>0$. These poles are due to the gamma factors. In the case $\kappa\in\frac12\mathbb{Z}$ poles are possible in $s=\frac12(a-1)$ with $a>1$, $a\in\mathbb{Z}$. If $a\equiv 2\kappa+1(2)$ then the pole is due to $\Gamma(-2s)$. So we obtain the first term in the right hand side of (13.3.26). If $a \equiv 2\kappa(2)$, then the factor $\Gamma(\frac{1}{2}-s-\kappa)^{-1}$ cancels the pole of $\Gamma(-2s)$ for $a > |2\kappa|$. If $a \equiv 2\kappa(2)$ and $1 < a \leqslant 2\kappa$ then

(13.3.27) $$\operatorname*{res}_{s\to\frac{1}{2}(a-1)} \Gamma(-2s)\Gamma(2s)^{-1}\Gamma(\tfrac{1}{2}+s-\kappa)\Gamma(\tfrac{1}{2}-s-\kappa)^{-1}$$
$$= -\tfrac{1}{2}(-1)^{a-1}(a-1)!^{-1}\Gamma(a-1)^{-1}(-1)^{\frac{1}{2}a-\kappa}(\kappa-\tfrac{1}{2}a)!^{-1}(-1).(-1)^{1-\frac{1}{2}a-\kappa}(\tfrac{1}{2}a+\kappa-1)!$$
$$= \tfrac{1}{2}(a-1)!^{-1}(a-2)!^{-1}(1+\kappa-\tfrac{1}{2}a)_{a-1}.$$

If $a \equiv 2\kappa(2)$, $1 < a \leqslant -2\kappa$ then the residue in $\frac{1}{2}(a-1)$ equals

(13.3.28) $$-\tfrac{1}{2}(-1)^{a-1}(a-1)!^{-1}\Gamma(a-1)^{-1}\Gamma(\tfrac{1}{2}a-\kappa)\Gamma(1-\tfrac{1}{2}a-\kappa)^{-1}$$
$$= -\tfrac{1}{2}(a-1)!^{-1}(a-2)!^{-1}(\kappa-\tfrac{1}{2}a+1)_{a-1}.$$

In the case $\kappa \notin \frac{1}{2}\mathbb{Z}$ the factor $\Gamma(-2s)$ gives poles at $s = \frac{1}{2}(a-1), a \geqslant 2, a \in \mathbb{Z}$. The same computation as in the other case gives the corresponding term in (13.3.26). The factor $\Gamma(\frac{1}{2}+s-\kappa)$ gives poles at $s = \frac{1}{2}(b-1)$ with $1 < b \leqslant 2\kappa$, $b \equiv 2\kappa(2)$. The residue of the gamma factors at $\frac{1}{2}(b-1)$ equals

(13.3.29) $$\Gamma(-b+1)\Gamma(b-1)^{-1}(-1)^{\frac{1}{2}b-\kappa}(\kappa-\tfrac{1}{2}b)!^{-1}\Gamma(1-\tfrac{1}{2}b-\kappa)^{-1}$$
$$= \Gamma(b)^{-1}(\sin \pi b)^{-1}\,\Gamma(b-1)^{-1}(-1)^{\frac{1}{2}b-\kappa}(\kappa-\tfrac{1}{2}b)!^{-1}\Gamma(\tfrac{1}{2}b+\kappa)\sin \pi(\tfrac{1}{2}b+\kappa)$$
$$= \Gamma(b)^{-1}\Gamma(b-1)^{-1}\Gamma(\tfrac{1}{2}b+\kappa)(\kappa-\tfrac{1}{2}b)!^{-1}.$$

By lemma 13.3.3 the Whittaker transforms of elements of $C_c^\infty(0,\infty)$ are contained in the following spaces of functions:

13.3.9. Definition. Let $\sigma > 0$ be given, such that $2\sigma+1 \not\equiv \pm 2\kappa(2)$. $F_{\kappa,\sigma}$ is the space of functions ϕ on

(13.3.30) $$\{s \in \mathbb{C} \mid |\mathrm{Re}\, s| \leqslant \sigma\} \cup \{\tfrac{1}{2}(b-1) \mid 1 < b \leqslant 2\kappa,\ b \equiv 2\kappa(2)\}$$

satisfying

(13.3.31) ϕ is holomorphic on the region $|\mathrm{Re}\, s| \leqslant \sigma$

(13.3.32) $\phi(s) = \phi(-s)$ for $|\mathrm{Re}\, s| \leqslant \sigma$

(13.3.33) $|\phi(s)| \ll e^{-\frac{1}{2}\pi|\mathrm{Im}\, s|}(1+|\mathrm{Im}\, s|)^{-a}$ for all $a \in \mathbb{R}$.

The spaces $F_{\kappa,\sigma}$ are dense in the Hilbert space

$$L^2((0,i\infty),|\Gamma(2s)^{-1}\Gamma(\tfrac{1}{2}+s-\kappa)|^2 ds) \oplus L^2(\{\tfrac{1}{2}(b-1) \mid 1 < b \leqslant 2\kappa, b \equiv 2\kappa(2)\},(b-1)(\kappa-\tfrac{1}{2}b)!^{-1}.\Gamma(\tfrac{1}{2}b+\kappa)^{-1}).$$

On these spaces the inverse of the Whittaker transform is given by an absolutely converging integral :

13.3.10. Proposition. Let $F_{\kappa,\sigma}$ be as in definition 13.3.9. For $\phi \in F_{\kappa,\sigma}$ put

(13.3.34) $$f_\phi(y) = \frac{1}{2\pi i}\int_0^{i\infty} \phi(s)W_{\kappa,s}(y)|\Gamma(2s)^{-1}\Gamma(\tfrac{1}{2}+s-\kappa)|^2 ds +$$
$$+ \sum_{\substack{b\equiv 2\kappa(2)\\ 1<b\leqslant 2\kappa}} (b-1)(\kappa-\tfrac{1}{2}b)!^{-1}\Gamma(\tfrac{1}{2}b+\kappa)^{-1}\ \phi(\tfrac{1}{2}(b-1))W_{\kappa,\frac{1}{2}(b-1)}(y)\ ;$$

this integral converges absolutely. If $\phi = w_\kappa f$ for some $f \in C_c^\infty(0,\infty)$, then $f_\phi = f$. For general $\phi \in F_{\kappa,\sigma}$ the function f_ϕ represents an element of $L^2((0,\infty),y^{-2}dy)$.

Proof. The convergence of (13.3.34) is clear from (13.3.33), (13.2.5) and (13.2.2). The other assertions are consequences of (13.3.1) and proposition 13.3.5.

13.3.11. Proposition. Let $0 \leqslant \sigma_1 \leqslant \sigma$, σ as in definition 13.3.9, $2\sigma_1+1 \not\equiv \pm 2\kappa(2)$. For all $\phi \in F_{\kappa,\sigma}$

(13.3.35)
$$f_\phi(y) = \frac{1}{2\pi i} \int_{\mathrm{Re}\, s=\sigma_1} \phi(s)\Gamma(2s)^{-1}\Gamma(\tfrac12+s-\kappa)M_{\kappa,s}(y)ds + \sum_{\substack{2\sigma_1+1<b\leqslant 2\kappa \\ b\equiv 2\kappa(2)}} \phi(\tfrac12(b-1))(-1)^{\kappa-\frac12 b}\Gamma(b-1)^{-1}(\kappa-\tfrac12 b)!^{-1}M_{\kappa,\frac12(b-1)}(y)$$

(13.3.36)
$$= \frac{1}{4\pi i}\int_{\mathrm{Re}\, s=\sigma_1} \phi(s)\Gamma(2s)^{-1}\Gamma(-2s)^{-1}\Gamma(\tfrac12+s-\kappa)\Gamma(\tfrac12-s-\kappa)W_{\kappa,s}(y)ds + \sum_{\substack{2\sigma_1+1<b\leqslant 2\kappa \\ 2\kappa\equiv b(2)}} \phi(\tfrac12(b-1))(-1)^{\kappa-\frac12 b}\Gamma(b-1)^{-1}(\kappa-\tfrac12 b)!^{-1}M_{\kappa,\frac12(b-1)}(y)$$
$$\left. \begin{aligned} &+ \tfrac12 \sum_{\substack{1<b\leqslant\min(2\kappa,2\sigma_1+1) \\ b\equiv 2\kappa(2)}} \phi(\tfrac12(b-1))(-1)^{\kappa-\frac12 b}\Gamma(b-1)^{-1}(\kappa-\tfrac12 b)!^{-1}M_{\kappa,\frac12(b-1)}(y) \\ &+ \tfrac12 \sum_{\substack{1<b<2\sigma_1+1 \\ b\equiv -2\kappa(2) \\ b\geqslant 2-2\kappa}} \phi(\tfrac12(b-1))(-1)^{\kappa+\frac12 b-1}\Gamma(1-b)^{-1}(\kappa+\tfrac12 b-1)!^{-1}M_{\kappa,\frac12(1-b)}(y) \end{aligned} \right\}$$

The last two terms occur in the case $2\kappa \notin \mathbb{Z}$ only.

Proof. Use (13.2.5) to transform the integral in (13.3.34) into

(13.3.37)
$$\frac{1}{2\pi i}\int_{\mathrm{Re}\, s=0} \phi(s)\, M_{\kappa,s}(y)\, \Gamma(+2s)^{-1}\Gamma(\tfrac12+s-\kappa)ds;$$

moving the integral to the line $\mathrm{Re}\, s = \sigma_1$ we pick up some residues which cancel some terms in the sum in (13.3.34); use (13.2.8) to go from $W_{\kappa,\frac12(b-1)}$ to $M_{\kappa,\frac12(b-1)}$. The integral in (13.3.34) also equals

(13.3.38)
$$\frac{1}{4\pi i}\int_{\mathrm{Re}\, s=0} \phi(s)\, W_{\kappa,s}(y)\Gamma(2s)^{-1}\Gamma(-2s)^{-1}\Gamma(\tfrac12+s-\kappa)\Gamma(\tfrac12-s-\kappa)ds;$$

this time it is a bit more work to figure out the residues in the strip $0 < \mathrm{Re}\, s < \sigma_1$. If $2\kappa \in \mathbb{Z}$ the only poles occur in $s = \frac12(b-1)$ with $b \equiv 2\kappa(2)$, $1 < b \leqslant 2\kappa$; the residues turn out to cancel terms in the sum in (13.3.34). The case $2\kappa \notin \mathbb{Z}$ is more complicated. The poles due to the factor $\Gamma(\frac12+s-\kappa)$ are situated in $s = \frac12(b-1)$, $1 < b \leqslant 2\kappa$, $b \equiv 2\kappa(2)$. The residue cancels only one half of the corresponding term in (13.3.34). The factor $\Gamma(\frac12-s-\kappa)$ gives poles in $s = \frac12(b-1)$, $b \equiv -2\kappa(2)$, $b \geqslant 2-2\kappa$. This leads to the last term in (13.3.36).

13.3.12. Corollary. For every $\phi \in F_{\kappa,\sigma}$

(13.3.39) $\quad |f_\phi(y)| \ll \begin{cases} y^{\frac{1}{2}+\sigma} & \text{for } y \downarrow 0 \\ y^{\frac{1}{2}-\sigma} & \text{for } y \to \infty. \end{cases}$

Proof. For $y \downarrow 0$ I use the M-representation of f_ϕ with $\sigma_1 = \sigma$ in (13.3.35). As the estimate (13.2.4) is valid uniformly on the line Re $s=\sigma$, the first estimate follows. For $y \to \infty$ the W-representation with $\sigma_1 = \sigma$ in (13.3.36) is needed. The terms $M_{\kappa,\pm\frac{1}{2}(b-1)}(y)$ decay exponentially, see (13.2.8), so they certainly satisfy the estimate. To estimate the integral, use (13.2.13).

13.3.13. Proposition. Let $\phi \in F_{\kappa,\sigma}$, with σ as in 13.3.9, let $|\text{Re } s| < \sigma$ or $s = \frac{1}{2}(b-1)$, $1 < b \leq 2\kappa$, $b \equiv 2\kappa(2)$. Then the integral in

(13.3.40) $\quad w_\kappa f_\phi(s) = \int_0^\infty f_\phi(y) W_{\kappa,s}(y) y^{-2} dy$

converges absolutely. Furthermore,

(13.3.41) $\quad w_\kappa f_\phi(s) = \phi(s).$

Proof. The behaviour for $y \to \infty$ of $f_\phi(y)$ and $W_{\kappa,s}(y)$ certainly gives absolute convergence. If $|\text{Re } s| < \sigma$, then we may suppose Re $s \geq 0$. From (13.2.5) and (13.2.4) follows that

(13.3.42) $\quad |W_{\kappa,s}(y)| \ll y^{\frac{1}{2}-\text{Re } s} \quad$ for $y \downarrow 0$.

Together with corollary 13.3.12 the convergence is clear. For $s = \frac{1}{2}(b-1)$ use (13.2.4) and (13.2.8). Now we have absolute convergence, so proposition 13.3.4 implies that $w_\kappa f_\phi$ defined in (13.3.40) represents the unitary extension of w_κ on $C_c^\infty(0,\infty)$. This means that (13.3.41) is valid for $s \in (0,i\infty)$ or $s = \frac{1}{2}(b-1)$ with $1 < b \leq 2\kappa$, $b \equiv 2\kappa(2)$. To get (13.3.41) for all s with $|\text{Re } s| < \sigma$ we prove that $w_\kappa f_\phi$ is holomorphic on this region. As $w_\kappa f_\phi$ is clearly continuous, see the estimate above, it is harmless to exclude a finite set of points. Split up the integral in (13.3.40) : $\int_0^\infty = \int_0^{y_0} + \int_{y_0}^\infty$. For $\int_0^{y_0}$ use (13.2.5) and (13.2.2) to get holomorphy almost everywhere ($M_{\kappa,\pm s}$ has singularities, and so have the gamma factors). For $\int_{y_0}^\infty$ use the integral representation:

(13.3.43) $\quad W_{\kappa,s}(y) = \frac{-1}{2\pi i} \Gamma(\frac{1}{2}-s+\kappa) e^{-\frac{1}{2}y} y^\kappa \int_C (-t)^{s-\kappa-\frac{1}{2}} (1+ty^{-1})^{s+\kappa-\frac{1}{2}} e^{-t} dt$

with C a contour coming from ∞ with positive imaginary part, circling 0 and returning to ∞ with negative imaginary part; see [40], p.52.

13.4. Spectral decomposition of ${}^{\frac{1}{2}}S_\tau$

The space ${}^{\frac{1}{2}}S_\tau$ has been defined in 3.6.1, a scalar product has been given in (8.3.2).

13.4.1. Proposition. Let $s \in \mathbb{C}$, suppose W is holomorphic at $-\bar{s}$; define for $f \in {}^{\frac{1}{2}}S_\tau$

(13.4.1) $\quad \omega f(s) = \sum_\psi \langle f, W(-\bar{s})\psi(-\bar{s})\rangle \psi(s),$

ψ runs through the standard basis of H^τ. Then ωf is a meromorphic section of H^τ over $\mathbb{C}$; ω intertwines the g-K-actions on ${}^{\frac{1}{2}}S_\tau$ and H^τ.

Proof. See the proof of proposition 11.3.1. As $|f|$ has compact support in $Z_0N\backslash G$, the differentiation under the integral sign is easier. The meromorphy follows from the meromorphy of W and the fact that $|f|$ has compact support in $Z_0N\backslash G$.

13.4.2. Proposition. Let $b \in Y_\tau$ or $b = \tau = 1$. For $f \in {}^{\frac{1}{2}}S_\tau$ define

$$\pi_b f = \sum_{\eta=\pm 1} \sum_{\substack{\eta r \equiv b(2) \\ \eta r \geq b}} \frac{\Gamma(\frac{1}{2}(\eta r+b))}{\Gamma(b)(\frac{1}{2}(\eta r-b))!} \langle f, D^b\phi_r[b]\rangle \quad a_b\phi_r[b] \tag{13.4.2}$$

with $a_b = \begin{cases} |b-1|^{\frac{1}{2}} & \text{if } b \neq 1 \\ 1 & \text{if } b = 1 \end{cases}$

Then $\pi_b : {}^{\frac{1}{2}}S_\tau \to D_b$ is an intertwining operator for the g-K-structure.

Remark. If $b > 1$ then D^b is a g-K-operator from D_b into the completion of ${}^{\frac{1}{2}}S_\tau$. It multiplies the norm by $(\Gamma(b)\Gamma(b-1))^{\frac{1}{2}}$. If we project ${}^{\frac{1}{2}}S_\tau$ onto the image and go to D_b by the inverse of $(\Gamma(b)\Gamma(b-1))^{-\frac{1}{2}}D^b$ we get just π_b.

Proof of proposition 13.4.2 : just check that one can pull the g-K-action through π_b.

13.4.3. Lemma. Let $f \in {}^{\frac{1}{2}}S_\tau$ be written as

$$f(n(x)a(y)j_\varepsilon k(\theta)) = \sum_{r=\pm\tau(2)} e^{\frac{1}{2}ix} f_r^\varepsilon(y)\, e^{ir\theta}. \tag{13.4.3}$$

Then

$$\omega f(s) = \tfrac{1}{2} \sum_{\delta,\varepsilon = \pm 1} \sum_{r\equiv\tau(2)} (\delta,\varepsilon)\Gamma(\tfrac{1}{2}+s-\tfrac{1}{2}\varepsilon r) w_{\frac{1}{2}\varepsilon r} f_r^\varepsilon(s)\phi_r^\delta(s) \quad \text{if } \tau = 0,1 \tag{13.4.4}$$

$$\omega f(s) = \sum_{\substack{\eta=\pm 1 \\ \varepsilon=\pm 1}} \sum_{r\equiv\varepsilon\eta\tau(2)} \frac{1}{\sqrt{2}} \Gamma(\tfrac{1}{2}+s-\tfrac{1}{2}\varepsilon r) w_{\frac{1}{2}\varepsilon r} f_r^\varepsilon(s)\phi_r^\eta(s) \quad \text{if } 0 < \tau < 1. \tag{13.4.5}$$

Proof. I work out the computation for $0 < \tau < 1$; the other one is easier. For $\delta = \pm 1$

$$P_\delta f(n(x)a(y)j_\varepsilon k(\theta)) = \sum_{\substack{q \\ \varepsilon q=\delta\tau(2)}} e^{\frac{1}{2}ix} f_q^\varepsilon(y)\, e^{iq\theta} \tag{13.4.6}$$

and $P_\delta W(s) = W(s)\pi_\delta$. For $\eta = \pm 1$, $r \equiv \alpha\tau(2), \alpha = \pm 1$:

$$\langle f, W(-\bar{s})\phi_r^\eta(-\bar{s})\rangle = \int_{Z_0N\backslash G} P_\eta f(g)\, W(-\bar{s})\phi_r^\eta(-\bar{s})(g)d\dot{g} = \tag{13.4.7}$$

$$\frac{1}{2\pi}\int_0^\infty \sum_{\varepsilon=\pm 1} \int_0^\pi \sum_{\substack{q \\ \varepsilon q\equiv\eta\tau(2)}} f_q^\varepsilon(y)e^{iq\theta}\sqrt{2}\,\Gamma(\tfrac{1}{2}+s-\tfrac{1}{2}\varepsilon r)\,.$$

$$.W_{\frac{1}{2}\varepsilon r,-s}(y)\, e^{-ir\theta}\, \delta_{\varepsilon\alpha,\eta} d\theta\, y^{-2}\, dy$$

see (2.4.8) and definition 5.4.2

$$= \frac{1}{\sqrt{2}} \int_0^\infty f_r^{\alpha\eta}(y)\, \Gamma(\tfrac12+s-\tfrac12\alpha\eta r) W_{\frac12\alpha\eta r,s}(y) y^{-2} dy$$

$$= \frac{1}{\sqrt{2}}\, \Gamma(\tfrac12+s-\tfrac12\alpha\eta r)\, w_{\frac12\alpha\eta r}\, f_r^{\alpha\eta}(s).$$

(13.4.8) $$\omega f(s) = \sum_{\substack{\eta=\pm1\\ \alpha=\pm1}} \sum_{r\equiv\alpha\tau(2)} \frac{1}{\sqrt{2}}\, \Gamma(\tfrac12+s-\tfrac12\alpha\eta r) w_{\frac12\alpha\eta r} f_r^{\alpha\eta}(s)\phi_r^{\eta}(s)$$

$$= \sum_{\eta,\varepsilon=\pm1} \sum_{r\equiv\varepsilon\eta\tau(2)} \frac{1}{\sqrt{2}}\, \Gamma(\tfrac12+s-\tfrac12\varepsilon r) w_{\frac12\varepsilon r}\, f_r^{\varepsilon}(s)\, \phi_r^{\eta}(s).$$

13.4.4. Lemma. Let $f \in {}^{\frac12}S_\tau$ be written as in (13.4.3). Let $b \in Y_\tau$, or $b = \tau = 1$. Then

(13.4.9) $$\pi_b f = \sum_{\eta=\pm1} \sum_{\substack{r\\ \eta r\geqslant b\\ \eta r\equiv b(2)}} \frac{1}{\sqrt{2}}\, a_b (-1)^{\frac12(\eta r-b)} (\tfrac12(\eta r-b))!^{-1} w_{\frac12\eta r} f_r^{\eta}(\tfrac12(b-1))\phi_r[b].$$

Proof. In the same way as the previous lemma.

13.4.5. Lemma. Let $f \in {}^{\frac12}S_\tau$. Then $\omega f(s)$ is holomorphic at general points s; at non-general points it has poles of at most first order. For $b \in Y_\tau$ or $b = \tau = 1$

(13.4.10) $$\operatorname*{res}_{s\to\frac12(b-1)} \omega f(s) = \begin{cases} |b-1|^{-\frac12}\pi_b f & \text{if } b \neq 1\\ \pi_b f & \text{if } b = 1\,. \end{cases}$$

On vertical strips $|\mathrm{Re}\, s| \leqslant \sigma$, outside a neighbourhood of the poles, $\omega f(s)$ is estimated by

(13.4.11) $$e^{-\pi|\mathrm{Im}\, s|}(1+|\mathrm{Im}\, s|)^{-a}$$

for each $a \in \mathbb{R}$.

(13.4.12) $$\iota\omega f = \omega f.$$

Proof. The statement on the meromorphy is clear upon inspection of (13.4.4) and (13.4.5); to get (13.4.10) one carries out a computation. For (13.4.11) see (13.3.9). Finally, remark that we may replace the standard basis in (13.4.1) by the basis mentioned in (3.5.18), (3.5.19). So we obtain

(13.4.13) $$\begin{aligned} \iota\omega f(s) &= \iota(-s)\,\omega f(-s)\\ &= \sum_\psi \langle f, W(\bar s)\psi(\bar s)\rangle\ \gamma_\psi(-s)\psi(s)\\ &= \sum_\psi \langle f, W(\bar s)\gamma_\psi(-\bar s)\psi(\bar s)\rangle\ \psi(s)\\ &= \sum_\psi \langle f, W(\bar s)\iota(-\bar s)\psi(-\bar s)\rangle\ \psi(s)\\ &= \sum_\psi \langle f, W(-\bar s)\psi(-\bar s)\rangle\ \psi(s) \qquad \text{See (5.5.3)}\\ &= \omega f(s). \end{aligned}$$

13.4.6. Proposition. For $f,g \in {}^{\frac12}S_\tau$

(13.4.14) $\langle f,g\rangle = \frac{1}{2\pi i}\int_0^{i\infty} \{\omega f(s),\omega g(s)\}_s\, |\Gamma(2s)|^{-2}\, ds + \sum_{\substack{b>1\\ b\in Y}} \langle \pi_b f,\pi_b g\rangle_b$.

Proof. Write f and g as in (13.4.3), compute $\langle f,g\rangle$ in terms of the f_r^ε and g_r^ε, apply proposition 13.3.5, rearrange the terms to get the right hand side of (13.4.14), using (3.4.5), the embedding of D_b in $H^\tau(\frac{1}{2}(b-1)$ discussed in 3.5 and definition 3.5.3.

13.5. The space H_σ^τ.

Each $f \in {}^{\frac{1}{2}}S_\tau$ determines a section ωf of H^τ and elements $\pi_b f$ of D_b with properties given in lemma 13.4.5. This is the point of departure for the following definition.

13.5.1. Definition. Let $\sigma > 0$, $2\sigma+1 \not\equiv \pm\tau\,(2)$. H_σ^τ is a subspace of the direct sum of the space of meromorphic sections of H^τ on the region $|\mathrm{Re}\ s| \leq \sigma$ and $\bigoplus_{\substack{b>1-2\sigma\\ b\in Y_\tau \text{ or } b=\tau=1}} D_b$. For an element ξ of this direct sum I denote the corresponding section of H^τ also by ξ and the component in D_b by ξ_b.
ξ is an element of H_σ^τ if the following conditions hold :

(13.5.1) The section ξ of H^τ is holomorphic in general point s with $|\mathrm{Re}\ s| \leq \sigma$; at non-general points it has at most a pole of order one.

(13.5.2) If $b \in Y_\tau$, $|\frac{1}{2}(b-1)| < \sigma$, then $\operatorname{res}_{s\to\frac{1}{2}(b-1)} \xi(s) = |b-1|^{-\frac{1}{2}}\xi_b$

If $b = \tau = 1$, then $\operatorname{res}_{s\to 0} \xi(s) = \xi_1$.

(13.5.3) On the region $|\mathrm{Re}\ s| \leq \sigma$, outside a neighbourhood of the poles, $\xi(s)$ is estimated by $e^{-\pi|\mathrm{Im}\ s|}(1+|\mathrm{Im}\ s|)^{-a}$ for each $a \in \mathbb{R}$.

(13.5.4) $\iota\xi = \xi$

Clearly H_σ^τ is a g-K-space. Remark that $\xi_b = 0$ for all except a finite number of $b \in Y_\tau$. Lemma 13.4.5 enables us to define

13.5.2. Definition. $\Xi_\sigma : {}^{\frac{1}{2}}S_\tau \to H_\sigma^\tau$ is defined by

(13.5.5) $\xi = \Xi_\sigma f$ if and only if $\begin{cases} \xi(s) = \omega f(s) & \text{for } |\mathrm{Re}\ s| \leq \sigma \\ \xi_b = \pi_b f & \text{for } b \in Y_\tau \text{ or } b = \tau = 1. \end{cases}$

Proposition 13.4.6 enables us to make Ξ_σ unitary by defining :

13.5.3. Definition. For $\xi,\theta \in H_\sigma^\tau$

(13.5.6) $\langle \xi,\theta\rangle = \frac{1}{2\pi i}\int_0^{i\infty} \{\xi(s),\theta(s)\}_s |\Gamma(2s)|^{-2} ds + \sum_{b\in Y_\tau, b>1} \langle \xi_b,\theta_b\rangle_b$.

In the remainder of this subsection I study a map Φ inverting Ξ_σ.

13.5.4. Lemma. Let $\xi \in H_\sigma^\tau$. For each $\varepsilon = \pm 1$ and $r \equiv \pm\tau\,(2)$ there is a unique function $\xi_r^\varepsilon \in F_{\frac{1}{2}\varepsilon r,\sigma}$ such that $\xi(s)$ is given by (13.4.4) or (13.4.5) and ξ_b by (13.4.9), in both cases with $w_{\frac{1}{2}\varepsilon r}$ f_r^ε replaced ξ_r^ε.

Proof. Let $|\mathrm{Re}\, s| \leqslant \sigma$. The functions $\xi_r^\varepsilon(s)$ are uniquely determined by (13.4.5) in the case $0 < \tau < 1$. In the case $\tau = 0,1$ the functions $\Gamma(\frac{1}{2}+s-\frac{1}{2}r)\xi_r^1(s) \pm \Gamma(\frac{1}{2}+s+\frac{1}{2}r)$. $\xi_r^{-1}(s)$ are determined by (13.4.4); this implies the uniqueness. For $b \in Y_\tau$, $\frac{1}{2}(b-1) > \sigma$ (13.4.9) determines $\xi_r^\varepsilon(\frac{1}{2}(b-1))$. The holomorphy in general points is directly clear from (13.5.1) in the case $0 < \tau < 1$; in the case $\tau = 0,1$ we have to use that Γ-functions are nowhere zero. Now we consider $b \in Y_\tau$ with $|\frac{1}{2}(b-1)| < \sigma$ or $b = \tau = 1$. In the case $0 < \tau < 1$, $b \equiv \alpha\tau(2)$ we have by (13.5.2) holomorphy of $\Gamma(\frac{1}{2}+s-\frac{1}{2}\varepsilon r)\xi_r^\varepsilon(s)$ at $s = \frac{1}{2}(b-1)$ if $\varepsilon r \equiv -\alpha\tau(2)$ or if $\varepsilon r \equiv \alpha\tau(2)$ and $\varepsilon r < b$. If $\varepsilon r \equiv \alpha\tau(2)$ and $\varepsilon r \geqslant b$ a first order pole is admitted, but this is already provided by the Γ-factor. Clearly ξ_b is given by the analogue of (13.4.9). In the case $\tau = 0$ or 1 we know that $\Gamma(\frac{1}{2}+s-\frac{1}{2}r)\xi_r^1(s) \pm \Gamma(\frac{1}{2}+s+\frac{1}{2}r)\xi_r^{-1}(s)$ are holomorphic at $s = \frac{1}{2}(b-1)$ if $|r| < b$. This gives holomorphy of $\xi_r^{\pm 1}(s)$. Now let $|r| \geqslant b$; then

(13.5.7) $$\lim_{s\to\frac{1}{2}(b-1)} (s-\tfrac{1}{2}(b-1))\tfrac{1}{2} \sum_{\delta,\varepsilon=\pm 1} (\delta,\varepsilon)\Gamma(\tfrac{1}{2}+s-\tfrac{1}{2}\varepsilon r)\xi_r^\varepsilon(s)\phi_r^\delta(s)$$

is a multiple of $\frac{1}{\sqrt{2}}(\phi_r^1(\frac{1}{2}(b-1)) + \mathrm{sgn}\, r\ \phi_r^{-1}(\frac{1}{2}(b-1)))$. So for $\delta = \pm 1$

(13.5.8) $$\lim_{s\to\frac{1}{2}(b-1)} (s-\tfrac{1}{2}(b-1)) \sum_{\varepsilon=\pm 1} (\delta,\varepsilon)\Gamma(\tfrac{1}{2}+s-\tfrac{1}{2}\varepsilon r)\xi_r^\varepsilon(s) = a(\mathrm{sgn} r,\delta) \text{ for some}$$

constant a. Hence for $\varepsilon = \pm 1$

(13.5.9) $$\lim_{s\to\frac{1}{2}(b-1)} (s-\tfrac{1}{2}(b-1))\Gamma(\tfrac{1}{2}+s-\tfrac{1}{2}\varepsilon r)\xi_r^\varepsilon(s) = \tfrac{1}{2}a(1+\varepsilon\,\mathrm{sgn} r).$$

If $\varepsilon r \geqslant b$ the Γ-factor gives a pole, and ξ_r^ε has to be holomorphic at $s = \frac{1}{2}(b-1)$. If $\varepsilon r \leqslant -b$, then the limit has to be zero, the Γ-factor is regular, so ξ_r^ε has to be holomorphic at $s = \frac{1}{2}(b-1)$. Checking the analogue of (13.4.9) is again just a computation. To obtain the conditions (13.3.32) and (13.3.33) one carries out similar computations, departing from (13.5.4) and (13.5.3).

13.5.5. Proposition. For $\xi \in H_\sigma^\tau$

(13.5.10) $$\Phi\xi = \frac{1}{2\pi i}\int_0^{i\infty} \frac{W(s)\xi(s)ds}{\Gamma(2s)\Gamma(-2s)} + \sum_{\substack{b\in Y_\tau \\ b>1}} \Gamma(b)^{-\frac{1}{2}}\Gamma(b-1)^{-\frac{1}{2}}D^b\xi_b$$

defines a map from H_σ^τ into the functions on G.

(13.5.11) $\Phi\xi(n(x)g) = e^{\frac{1}{2}ix}\,\Phi\xi(g)$;

(13.5.12) Φ is unitary for the scalar product given in (8.3.2)

(13.5.13) $$|\Phi\xi(a(y)k)| \ll \begin{cases} y^{\frac{1}{2}+\sigma} & \text{for } y \downarrow 0 \\ y^{\frac{1}{2}-\sigma} & \text{for } y \to \infty \end{cases}$$

uniformly for $k \in K$;

(13.5.14) $\quad \Phi\Xi_\sigma$ *is the identity on* ${}^{\frac12}S_\tau$;

(13.5.15) $\quad$ formula (13.4.1) converges for $f = \Phi\xi$ *and* $\omega(\Phi\xi)(s) = \xi(s)$ *if* $|\mathrm{Re}\, s| < \sigma$ *and* W *holomorphic at* $-\bar{s}$;

(13.5.16) $\quad$ *formula* (13.4.2) *converges for* $f = \Phi\xi$ *and* $b > 1-2\sigma$ *and* $\pi_b\Phi\xi = \xi_b$

Proof. It is sufficient to consider ξ of the form

$$\xi(s) = \frac{1}{\sqrt{2}}\,\Gamma(\tfrac12+s-\tfrac12\varepsilon r)\phi(s).\begin{cases} \frac{1}{\sqrt{2}}\,(\phi_r^1(s)+\varepsilon\phi_r^{-1}(s)) & \text{if } \tau = 0,1 \\ \phi_r^\eta(s) & \text{if } 0<\tau<1 \end{cases} \tag{13.5.17}$$

$$\xi_b = \begin{cases} \frac{1}{\sqrt{2}}\, a_b(-1)^{\frac12(\varepsilon r-b)}(\tfrac12(\varepsilon r-b))!^{-1}\phi(\tfrac12(b-1))\ \phi_r[b] & \text{if } b \equiv \varepsilon r(2),\\ & 1-2\sigma < b \leq \varepsilon r \\ 0 & \text{otherwise} \end{cases}$$

with $\eta,\varepsilon = \pm 1$, $\varepsilon r \equiv \eta\tau(2)$ and $\phi \in F_{\frac12\varepsilon r,\sigma}$. Then (13.5.10) amounts to

$$\begin{aligned}
\Phi\xi(n(x)a(y)j_\beta k(\theta)) &= \frac{1}{2\pi i}\int_0^{i\infty}\Gamma(\tfrac12+s-\tfrac12\varepsilon r)\phi(s)e^{\frac12 ix}\Gamma(\tfrac12-s-\tfrac12\beta r)W_{\frac12\beta r,s}(y)\,.\\
&\quad .\ e^{ir\theta}\,\delta_{\varepsilon,\beta}|\Gamma(2s)|^{-1}ds\\
&\quad + \sum_{\substack{b\equiv\varepsilon r(2)\\ 1<b\leq\varepsilon r}}\delta_{\varepsilon\beta}|b-1|^{\frac12}(-1)^{\frac12(\varepsilon r-b)}(\tfrac12(\varepsilon r-b))!^{-1}\phi(\tfrac12(b-1))e^{\frac12 ix}\\
&\quad .\ (-1)^{\frac12(\beta r-b)}\Gamma(b)\Gamma(\tfrac12(\beta r+b))^{-1}\ W_{\frac12\beta r,\frac12(b-1)}(y)\ e^{ir\theta}\Gamma(b)^{-\frac12}\Gamma(b-1)^{-\frac12}\\
&= \delta_{\varepsilon,\beta}\ e^{\frac12 ix+ir\theta}\Big(\frac{1}{2\pi i}\int_0^{i\infty}\Big|\frac{\Gamma(\frac12+s-\frac12\varepsilon r)}{\Gamma(2s)}\Big|^2\ \phi(s)\ W_{\frac12\varepsilon r,s}(y)ds\ +\\
&\quad + \sum_{\substack{1<b\leq\varepsilon r\\ b\equiv\varepsilon r(2)}}(b-1)(\tfrac12(\varepsilon r-b))!^{-1}\Gamma(\tfrac12(\varepsilon r+b))^{-1}\ \phi(\tfrac12(b-1))W_{\frac12\varepsilon r,\frac12(b-1)}(y)\Big) =\\
&= \delta_{\varepsilon,\beta}e^{\frac12 ix}\ f_\varphi(y)e^{ir\theta},
\end{aligned} \tag{13.5.18}$$

see proposition 13.3.10. So, indeed we get a map satisfying (13.5.11), and also (13.5.13) by corollary 13.3.12. Assertion (13.5.14) follows from lemma 13.4.3 and proposition 13.3.10. Assertions (13.5.15) and (13.5.16) are reformulations of proposition 13.3.13 . For (13.5.12) use propositions 13.3.10 and 13.3.5.

13.5.6. *Proposition.* Φ *maps* H_σ^τ *into the* K-*finite functions in* $C^\infty(G)$. *It intertwines the* g-K-*action on* H_σ^τ *with the pointwise right* g-K-*action on* $C^\infty(G)$.

Proof. The K-finiteness and the intertwining of the K-action is easy to see; one uses that $W(s)$ and D^b are intertwining operators. For the differentiability one has to justify taking the differentiation inside the integration over $(0,i\infty)$. Let $\underline{X} \in \mathfrak{g}_{\mathbb{R}}$ and $g \in G$ be given. For every $s \in (0,i\infty)$ and every $t \in \mathbb{R}$, $t \neq 0$, there is a number t_s between t and 0 such that

$$t^{-1}((W\xi)(s)(ge^{t\underline{X}}) - (W\xi)(s)(g)) = (W\underline{X}\xi)(s)(ge^{t_s\underline{X}}). \tag{13.5.19}$$

The estimate of $W_{\frac{1}{2}\varepsilon r,s}(y)$ which one needs to see that the integral in (13.5.10) converges, is uniform for y in compact sets. From this one derives an estimate of the expression in (13.5.19) valid for all small $t \neq 0$; then one applies dominated convergence.

13.5.7. Proposition. Let $0 \leqslant \sigma_1 \leqslant \sigma$, $2\sigma_1+1 \notin \pm\tau(2)$.

For all $\xi \in H_\sigma^\tau$:

$$\Phi\xi = \frac{1}{2\pi i}\int_{\mathrm{Re}\, s=\sigma_1} M(s)\xi(s)\Gamma(2s)^{-1}ds + \sum_{\substack{b\in Y_\tau\\ b>2\sigma_1+1}} \Gamma(b)^{-\frac{1}{2}}(b-1)^{-\frac{1}{2}}D^b\xi_b \tag{13.5.20}$$

$$= \frac{1}{4\pi i}\int_{\mathrm{Re}\, s=\sigma_1} W(s)\xi(s)\,\Gamma(2s)^{-1}\Gamma(-2s)^{-1}ds + \sum_{b\in Y_\tau, b>2\sigma_1+1} \Gamma(b)^{-\frac{1}{2}}(b-1)^{-\frac{1}{2}}D^b\xi_b + \sum_{\substack{b\in Y_\tau, b\notin\mathbb{Z}\\ 1-2\sigma_1<b<1+2\sigma_1}} \tfrac{1}{2}\Gamma(b-1)^{-1}|b-1|^{-\frac{1}{2}}D^b\xi_b. \tag{13.5.21}$$

Proof. This is just proposition 13.3.11 in another notation.

14. Convolution operators

14.1. Introduction

In subsection 6.4 I have defined convolution operators C_q. For the sum formula I need the case that $q \in {}_{\frac{1}{2}}Q_{\frac{1}{2}}$ and I let C_q act on ΦH_σ^τ with $\sigma > \frac{1}{2}$. With help of Bessel transforms, discussed in 14.2, one may translate the action of these operators to multiplication operators M_q in H_σ^τ, according to the scheme. This is shown in subsection 14.3.

$$\begin{array}{ccc} \Phi H_\sigma^\tau & \xrightarrow{\;C_q\;} & \Phi H_\sigma^\tau \\ \uparrow \Phi & & \uparrow \Phi \\ H_\sigma^\tau & \xrightarrow{\;M_q\;} & H_\sigma^\tau \end{array}$$

The operators M_q may occur as principal test functions in the sum formula. But in the proof I need more multiplication operators in H_σ^τ than those arising from compactly supported q. Therefore I define in 14.4 more spaces of multiplication operators in H_σ^τ. In 14.5 I give the Plancherel formula for the L^2-completion of ${}_{\frac{1}{2}}Q_{\frac{1}{2}}$. This Plancherel formula is an easy consequence of the results on Bessel transforms in 14.2.

These Bessel transforms have been studied by Kuznetsov and Proskurin, cf. [15] §2.5, [31], p. 121.

14.2. Bessel transforms

In proposition 6.8.1 we have seen that in studying the operators C_q with $q \in {}_{\frac{1}{2}}Q_{\frac{1}{2}}$ we meet a transform associating to $\phi \in C_c^\infty(0, \infty)$ the function

(14.2.1) $$s \mapsto \int_0^\infty \phi(y) J_{2s}^\varepsilon(y^{\frac{1}{2}}) y^{-3/2} dy$$

with J_{2s}^ε a Bessel function. In this section I need for this transform results like those derived for Whittaker transforms in 13.3. In this subsection these results are given. Actually, I do not consider the transform indicated above, but the transform in (14.2.9). In this way comparision with the results in [15] and [34] is easier. A general reference for Bessel functions is Ch. III of [25].

For $s \in \mathbb{C}$ and $\varepsilon = \pm 1$ we consider the differential equation

(14.2.2) $$y^2 f''(y) + y f'(y) + (\varepsilon y^2 - 4s^2) f(y) = 0.$$

For $\varepsilon = 1$ it is the standard Bessel equation, for $\varepsilon = -1$ the modified one. If $\mathrm{Re}\, s > 0$, then all solutions satisfying

(14.2.3) $$f(y) \ll y^{2\mathrm{Re}\, s} \quad \text{for } y \searrow 0$$

are multiples of the Bessel function

(14.2.4) $$J_{2s}^\varepsilon(y) = \sum_{n=0}^\infty \frac{(-\varepsilon)^n (\frac{1}{2}y)^{2s+2n}}{n!\,\Gamma(2s+1+n)} .$$

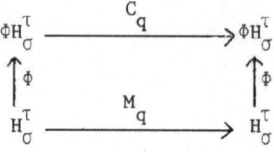

The notation J^{1}_{2s}, J^{-1}_{2s} is unusual, but convenient in our context. The relation with the standard notation is

(14.2.5) $\qquad J^{1}_{2s} = J_{2s}$

(14.2.6) $\qquad J^{-1}_{2s}(y) = I_{2s}(y) = 2^{-\frac{1}{2}-4s}\Gamma(2s+1)^{-1}\ y^{-\frac{1}{2}}\ M_{0,2s}(2y).$

For $a \in \mathbb{Z}$ we have

(14.2.7) $\qquad J^{\varepsilon}_{-a}(y) = J^{\varepsilon}_{a}(y)(-\varepsilon)^{a}.$

14.2.1. Definition. Let $\alpha \in \mathbb{R}$, $\varepsilon = \pm 1$. The Bessel transform b^{ε}_{α} is defined for distributions f with compact support on $(0, \infty)$ satisfying

(14.2.8) $\qquad |f\{\phi\}| \ll \sup_{0<y<\infty} |\phi(y)|$

for all $\phi \in C^{\infty}_{c}(0, \infty)$, by

(14.2.9) $\qquad b^{\varepsilon}_{\alpha}f(s) = 2(\sin 2\pi s)^{-1} e^{-\frac{1}{2}\pi i\ \varepsilon\alpha}.$

$\qquad\qquad . (\sin \pi(\tfrac{1}{2}-s+\tfrac{1}{2}\varepsilon\alpha)f\{J^{\varepsilon}_{-2s}\} - \sin \pi(\tfrac{1}{2}+s+\tfrac{1}{2}\alpha)\ f\{J^{\varepsilon}_{2s}\}).$

Remark that $f\{\phi\}$ means the value of the distribution f in the function ϕ. In definition 14.2.1 the distribution f may be of the form $g(y)y^{-1}dy$ with $g \in C^{\infty}_{c}(0, \infty)$; in this case I write $b^{\varepsilon}_{\alpha}g$ instead of $b^{\varepsilon}_{\alpha}f$. I also use the Bessel transform of delta distributions

(14.2.10) $\qquad \delta_{t}: \phi \to \phi(t),$

with $t > 0$ fixed. We obtain

(14.2.11) $\qquad b^{\varepsilon}_{\alpha}\delta_{t}(s) = \dfrac{2e^{-\frac{1}{2}\pi i\varepsilon\alpha}}{\sin 2\pi s}\ (\sin \pi(\tfrac{1}{2}-s+\tfrac{1}{2}\varepsilon\alpha)J^{\varepsilon}_{-2s}(t) - \sin \pi(\tfrac{1}{2}+s+\tfrac{1}{2}\alpha)J^{\varepsilon}_{2s}(t)).$

Remark that

(14.2.12) $\qquad b^{1}_{\alpha}f(-s) = b^{1}_{\alpha}f(s),$

but that $b^{-1}_{\alpha}f$ is not even for each f as in definition 14.2.1. Using (14.2.7) one shows that $b^{\varepsilon}_{\alpha}f$ is a holomorphic function on $\mathbb{C}$.

14.2.2. Lemma. Let f be as in definition 14.2.1. Let $\sigma > 0$,

(14.2.13) $\qquad |b^{\varepsilon}_{\alpha}f(s)| \ll (1+|\mathrm{Im}\ s|)^{2|\mathrm{Re}\ s|-\frac{1}{2}}$ for $|\mathrm{Re}\ s| \leq \sigma$

(14.2.14) $\qquad |f\{J^{\varepsilon}_{2s}\}| \ll C^{\mathrm{Re}\ s}|\Gamma(2s+1)|^{-1}$ for $\mathrm{Re}\ s \geq 0$

for some constant $C > 0$, depending on f.

Proof. The power series in (14.2.4) converges absolutely on the support of f, so we may write out $f\{J^{\varepsilon}_{2s}\}$ as a sum containing Mellin transforms of f. These are easily estimated, and so we obtain (14.2.14). Using Stirling's formula we obtain (14.2.13) from the same sum.

14.2.3. Lemma. Let $\sigma > 0$. For $g \in C^{\infty}_{c}(0, \infty)$

(14.2.15) $\qquad |b^{\varepsilon}_{\alpha}g(s)| \ll (1+|\mathrm{Im}\ s|)^{-a}$ for $|\mathrm{Re}\ s| \leq \sigma$

for each $a \in \mathbb{R}$;

(14.2.16) $\qquad |g\{J^{\varepsilon}_{2s}\}| \ll C^{\mathrm{Re}\ s}|\Gamma(2s+1)|^{-1}(1+|\mathrm{Im}\ s|)^{-a}$ for $\mathrm{Re}\ s \geq 0$

for each $a \in \mathbb{R}$, for some $C > 0$.

Proof. Use the estimates for Mellin transforms of compactly supported functions.

14.2.4. Lemma. For f as in definition 14.2.1 and $b > 1$, $b \equiv \varepsilon\alpha$ (2)

(14.2.17) $$b_\alpha^\varepsilon f(\tfrac{1}{2}(b-1)) = 2e^{-\frac{1}{2}\pi i b} f\{J_{b-1}^1\} \text{ if } \varepsilon = 1$$

(14.2.18) $$b_\alpha^\varepsilon f(\tfrac{1}{2}(b-1)) = 0 \text{ if } \varepsilon = -1.$$

Proof. Let us first consider the case $\alpha \notin \mathbb{Z}$; then

(14.2.19) $$b_\alpha^\varepsilon f(\tfrac{1}{2}(b-1)) =$$

$$\frac{2e^{-\frac{1}{2}\pi i\varepsilon\alpha}}{-\sin \pi b} (\sin \pi(1+\tfrac{1}{2}(\varepsilon\alpha-b)) f\{J_{1-b}^\varepsilon\} - \sin \pi \tfrac{1}{2}(b+\alpha) f\{J_{b-1}^\varepsilon\})$$

$$= \frac{2e^{-\frac{1}{2}\pi i\varepsilon\alpha}}{\sin \pi b} \sin \tfrac{1}{2}\pi(b+\alpha) \,.\, f\{J_{b-1}^\varepsilon\}$$

$$= \begin{cases} 0 & \text{if } \varepsilon = -1 \\ 2e^{-\frac{1}{2}\pi i\alpha} \dfrac{\sin \pi b}{\sin \pi b} (-1)^{\frac{1}{2}(\alpha-b)} f\{J_{b-1}^1\} & \text{if } \varepsilon = 1 \end{cases}$$

In the case $\alpha \in \mathbb{Z}$:

(14.2.20) $$b_\alpha^\varepsilon f(\tfrac{1}{2}(b-1)) =$$

$$\lim_{s\to\frac{1}{2}(b-1)} \frac{2e^{-\frac{1}{2}\pi i\varepsilon\alpha}}{\sin 2\pi s} (\sin \pi(\tfrac{1}{2}-s+\tfrac{1}{2}\varepsilon\alpha) f\{J_{-2s}^\varepsilon\}$$

$$- \sin \pi(\tfrac{1}{2}+s+\tfrac{1}{2}\alpha) f\{J_{2s}^\varepsilon\})$$

$$= \lim_{h\to 0} -2(-1)^b e^{-\frac{1}{2}\pi i\varepsilon\alpha} \Big(\frac{(-1)^{\frac{1}{2}(\varepsilon\alpha-b)} \sin \pi h}{\sin 2\pi h} f\{J_{-b+1-2h}^\varepsilon\}$$

$$- \frac{(-1)^{\frac{1}{2}(\alpha+b)} \sin \pi h}{\sin 2\pi h} f\{J_{b-1+2h}^\varepsilon\}\Big)$$

$$= \begin{cases} 2e^{-\frac{1}{2}\pi i b} f\{J_{b-1}^1\} & \text{if } \varepsilon = 1 \\ 0 & \text{if } \varepsilon = -1, \end{cases}$$

see (14.2.7).

14.2.5. Lemma. For f as in definition 14.2.1.

(14.2.21) $$|b_\alpha^1 f(\tfrac{1}{2}(b-1))| \ll C^b \Gamma(b)^{-1} \text{ for } b > 1,\ b \equiv \alpha(2)$$

for some $C > 0$.

Proof. Use the lemmas 14.2.2 and 14.2.4.

14.2.6. Proposition. Let $\alpha \in \mathbb{R}$ and $\varepsilon = \pm 1$. For $f, g \in C_c^\infty(0, \infty)$

(14.2.22) $$\int_0^\infty f(y)\overline{g(y)} y^{-1} dy =$$

$$\frac{1}{2\pi i} \int_0^{i\infty} b_\alpha^\varepsilon f(s)\, \overline{b_\alpha^\varepsilon g(s)}\, \frac{-2\pi s \sin 2\pi s}{\cos 2\pi s + \cos \pi\alpha}\, ds$$

$$+ \frac{1+\varepsilon}{4} \sum_{\substack{b>1 \\ b\equiv\alpha(2)}} (b-1) b_\alpha^\varepsilon f(\tfrac{1}{2}(b-1)) \overline{b_\alpha^\varepsilon g(\tfrac{1}{2}(b-1))}.$$

The map $f \mapsto b_\alpha^\varepsilon f$ extends to a unitary isomorphism between $L^2((0, \infty), y^{-1}dy)$ and

$L^2((0, i\infty), \frac{is \sin 2\pi s}{\cos 2\pi s+\cos \pi\alpha}ds) \oplus L^2(\{b > 1 | b \equiv \alpha(2)\}, \frac{1}{2}(b-1))$ _if_ $\varepsilon = 1$ _or_

$L^2((0, i\infty), \frac{is \sin 2\pi s}{\cos 2\pi s+\cos \pi\alpha}ds)$ _if_ $\varepsilon = -1$.

Remark. $\frac{is \sin 2\pi s}{\cos 2\pi s+\cos \pi\alpha} \cdot i$ is positive for $s \in (0, i\infty)$.

Remark. The sum in (14.2.22) only occurs in the case $\varepsilon = 1$.

Remark. The case $\varepsilon = -1$ amounts to the inversion of the Lebedev transform, see [25], p. 398.

Proof. By (14.2.6) the case $\varepsilon = -1$ is a special case of proposition 13.3.5. We have

(14.2.23) $\quad b_\alpha^{-1} f(s) = 4\pi^{-\frac{1}{2}} e^{\frac{1}{2}\pi i\alpha} \mathfrak{w}_0 f_1(2s) \sin \pi(\tfrac{1}{2}+s+\tfrac{1}{2}\alpha)$

with

$$f_1(y) = y^{\frac{1}{2}} f(\tfrac{1}{2}y).$$

Some computations lead to 14.2.22 for this case. The case $\varepsilon = 1$ follows directly from theorem 5, §2.5 of [15]; note that our α is $-\alpha$ there. In fact, Kuznetsov's proof has been my example in proving proposition 13.3.5.

To get the density of the image of $C_c^\infty(0, \infty)$ under b_α^ε one may use the same method as sketched in the proof of proposition 13.3.5.

14.2.7. _Definition. Let_ $\alpha \in \mathbb{R}$, $\sigma > 0$, $a \in \mathbb{R}$, $p = 1, 2$ _or_ ∞. ${}^p_1F^a_{\alpha,\sigma}$ _is the set of functions_ ϕ _on_ $\{s \in \mathbb{C} \,|\, |\mathrm{Re}\, s| \leq \sigma\} \cup \{\tfrac{1}{2}(b-1) \,|\, b \equiv \alpha(2),\ b > 2\sigma+1\}$ _satisfying:_

(14.2.24) $\quad \phi$ _is an even holomorphic function on_ $|\mathrm{Re}\, s| \leq \sigma$

(14.2.25) $\quad |\phi(s)| \ll (1+|\mathrm{Im}\, s|)^{-a}$ _for_ $|\mathrm{Re}\, s| \leq \sigma$

(14.2.26) $\quad \sum_{b>1,\, b\equiv\alpha(2)} (b-1)|\phi(\tfrac{1}{2}(b-1))|^p < \infty$ _if_ $p = 1$ _or_ 2

(14.2.27) $\quad |\phi(\tfrac{1}{2}(b-1))| \ll 1$ _for_ $b \equiv \alpha(2)$, $b > 1$ _if_ $p = \infty$.

${}_{-1}F^a_{\alpha,\sigma}$ _is the set of functions_ ϕ _on_ $\{s \in \mathbb{C} \,|\, |\mathrm{Re}\, s| \leq \sigma\}$ _satisfying:_

(14.2.28) $\quad \phi$ _is holomorphic on_ $|\mathrm{Re}\, s| \leq \sigma$, _with zero's in_ $\{\tfrac{1}{2}(b-1) \,|\, b \equiv -\alpha(2)\}$.

(14.2.29) $\quad s \mapsto \phi(s)(\sin \pi(\tfrac{1}{2}+s+\tfrac{1}{2}\alpha))^{-1}$ _is an even function._

(14.2.30) $\quad |\phi(s)| \ll (1+|\mathrm{Im}\, s|)^{-a}$ _for_ $|\mathrm{Re}\, s| \leq \sigma$.

If I write ${}^p_\varepsilon F^a_{\alpha,\sigma}$ with $\varepsilon = -1$, then I mean ${}_\varepsilon F^a_{\alpha,\sigma}$. Clearly ${}^1_1F^a_{\alpha,\sigma} \subset {}^2_1F^a_{\alpha,\sigma} \subset {}^\infty_1F^a_{\alpha,\sigma}$ and ${}^p_\varepsilon F^a_{\alpha,\sigma} \subset {}^pF^b_{\alpha,\sigma}$ if $a > b$. From the lemmas 14.2.2 and 14.2.4 and other remarks in this subsection follows that $b_\alpha^\varepsilon f \in {}^1_\varepsilon F^{-2\sigma+\frac{1}{2}}_{\alpha,\sigma}$ for each f satisfying the conditions in definition 14.2.1. By lemma 14.2.3 we see that for $g \in C_c^\infty(0, \infty)$ we have $b_\alpha^\varepsilon g \in {}^1_\varepsilon F^a_{\alpha,\sigma}$ for all $a \in \mathbb{R}$.

If $a > 1$ then the elements of ${}^2_\varepsilon F^a_{\alpha,\sigma}$ represent a dense subset of the Hilbert space mentioned in proposition 14.2.6.

14.2.8. Proposition. Let $\alpha \in \mathbb{R}$, $\varepsilon = \pm 1$, $\sigma > 0$, $a > +3/2$, then for each $\phi \in {}^2_\varepsilon F^a_{\alpha,\sigma}$ the following integral and sum converge absolutely and invert the unitary isomorphism mentioned in proposition 14.2.6:

$$(14.2.31) \qquad (b^\varepsilon_\alpha)^{\leftarrow}\phi(y) = \frac{1}{2\pi i}\int_0^{i\infty} \phi(s)\,\overline{b^\varepsilon_\alpha \delta_y(s)}\,\frac{-2\pi s \sin 2\pi s}{\cos 2\pi s + \cos \pi\alpha}\,ds$$

$$+ \frac{1+\varepsilon}{4}\sum_{\substack{b>1\\ b\equiv\alpha(2)}} (b-1)\phi(\tfrac{1}{2}(b-1))\,\overline{b^\varepsilon_\alpha \delta_y(\tfrac{1}{2}(b-1))}.$$

Proof. Clear from proposition 14.2.6 and definition 14.2.7.

In this way one obtains more functions on $(0, \infty)$ than compactly supported ones. If for a function $(b^\varepsilon_\alpha)^{\leftarrow}\phi$ the integrals in definition 14.2.1 converge absolutely, one gets back the function ϕ.

The functions $(b^\varepsilon_\alpha)^{\leftarrow}\phi$ with $\phi \in {}^1_\varepsilon F^a_{\alpha,\sigma}$ with some $\sigma > \frac{1}{2}$ and some $a > 2$ will be the test functions in the sum formula. I do not know exactly the relation with the class of test functions in Kuznetsov's and Proskurin's versions of the sum formula. I suspect that none of these classes is contained in the other one. In proposition 14.2.11 I give a class of functions satisfying the conditions of the sum formulas in [15], [34] as well as those of the sum formula in this paper.

14.2.9. Lemma. Let f be a continuous function on $[0, \infty)$ satisfying

$$(14.2.32) \qquad |f(x)| \ll x^2 \qquad \text{for } x \downarrow 0$$

$$(14.2.33) \qquad |f(x)| \ll x^{-\beta} \qquad \text{for } x \to \infty$$

for some $\beta \in (0, \frac{1}{2})$.

Let $\varepsilon = \pm 1$, $\alpha \in \mathbb{R}$; $\frac{1}{2} < \sigma < \frac{1}{2}(\beta+1)$. If $\varepsilon = 1$ then the integral $f\{J^1_{2s}\}$ converges absolutely for $\mathrm{Re}\, s \geq -\sigma$ and $b^1_\alpha f$ gives an element of ${}^2_1 F^{\frac{1}{2}-2\sigma}_{\alpha,\sigma}$.

If $\varepsilon = -1$ then $f\{(\sin 2\pi s)^{-1}(J^{-1}_{-2s} - J^{-1}_{2s})\}$ converges absolutely for $|\mathrm{Re}\, s| \leq \sigma$ and $b^{-1}_\alpha f$ gives an element of ${}_{-1}F^{+\frac{1}{2}-2\sigma}_{\alpha,\sigma}$.

Proof. The convergence of the integrals gives no problem near $x = 0$, as follows from (14.2.32) and (14.2.4). Near $x = \infty$ I use the well known estimate

$$(14.2.34) \qquad |J_{2s}(x)| \ll x^{-\frac{1}{2}} \qquad \text{for } x \to \infty$$

in the case $\varepsilon = 1$. For $\varepsilon = -1$ remark that

$$(14.2.35) \qquad (\sin 2\pi s)^{-1}(J^{-1}_{-2s}(x) - J^{-1}_{2s}(x)) = 2\pi^{-1}K_{2s}(x),$$

and that $K_{2s}(x)$ decreases exponentially for $x \to \infty$. To get the holomorphy of the integrals I have to justify carrying out the differentiation under the integral sign. On a compact interval $[0, x_1]$ with $x_1 > 1$, this is no problem. On $[x_1, \infty)$ I use

(14.2.36) $$J_{2s}(x) = \frac{1}{2\pi i}\int_{-\infty}^{(0+)} t^{-2s} e^{\frac{1}{2}x(t-t^{-1})} t^{-1} dt$$

and

(14.2.37) $$K_{2s}(x) = \tfrac{1}{2}\int_0^\infty t^{-2s} e^{-\frac{1}{2}x(t+t^{-1})} t^{-1} dt$$

to prove that $h^{-1}(J_{2s+2h}(x) - J_{2s}(x))$ and $h^{-1}(K_{2s+2h}(x) - K_{2s}(x))$ are bounded for small h and $x \geqslant x_1$. Then application of Lebesgue's theorem gives the holomorphy. To get (14.2.26) in the case $\varepsilon = 1$, remark that f represents an element of $L^2((0, \infty), x^{-1}dx)$ and that the J^1_{b-1} form an orthogonal system in this space. The difficult part of the proof are the estimates (14.2.25) and (14.2.30). We again split up the integrals at some $x_1 > 1$. For the first part we use the power series expansion (14.2.4) and get

(14.2.38) $$\left|\int_0^{x_1} f(x)\, J^{\varepsilon}_{2s}(x) x^{-1} dx\right| \ll x_1^{2\mathrm{Re}\, s} e^{x_1^2} |\Gamma(2s+1)|^{-1}$$
$$\ll e^{+\pi|\mathrm{Im}\, s|}(1+|\mathrm{Im}\, s|)^{+2\sigma-\frac{1}{2}}.$$

In the case $\varepsilon = -1$ we use the following formula, obtained from Basset's integral by partial integration:

(14.2.39) $$K_{2s}(x) = -i\pi^{\frac{1}{2}} x^{-1-2s} 2^{2s}\Gamma(2s+\tfrac{3}{2})\int_{-\infty}^{+\infty} u\, e^{-ixu}(u^2+1)^{-2s-\frac{3}{2}} du,$$

absolutely converging for $\mathrm{Re}\, s \geqslant 0$. (It is sufficient to consider the case $\mathrm{Re}\, s \geqslant 0$ only). From (14.2.39) follows:

(14.2.40) $$\left|\int_{x_1}^{\infty} f(x) K_{2s}(x) x^{-1} dx\right| \ll \int_{x_1}^{\infty} |f(x)| x^{-2\mathrm{Re}\, s-2} dx . 2^{\mathrm{Re}\, s}|\Gamma(\tfrac{3}{2}+2s)| . \int_{-\infty}^{+\infty} |u|(u^2+1)^{-\frac{3}{2}} du$$
$$\ll e^{-\pi|\mathrm{Im}\, s|}(1+|\mathrm{Im}\, s|)^{2\sigma-1}.$$

This completes the case $\varepsilon = -1$. In the other case I start with rewriting (14.2.9) with help of (1) and (2) on p. 74 of [43]:

(14.2.41) $$(\sin 2\pi s)^{-1}[\sin \pi(\tfrac{1}{2}-s+\tfrac{1}{2}\alpha)J_{-2s} - \sin \pi(\tfrac{1}{2}+s+\tfrac{1}{2}\alpha)J_{2s}]$$
$$= (\sin 2\pi s)^{-1}[-\cos \pi(\tfrac{1}{2}\alpha-s)\tfrac{1}{2}(H^{(1)}_{2s} + H^{(2)}_{2s})$$
$$+ \cos \pi(\tfrac{1}{2}\alpha+s)\tfrac{1}{2}(e^{2\pi i s}H^{(1)}_{2s} + e^{-2\pi i s}H^{(2)}_{2s})]$$

$$= +\tfrac{1}{2}i\ e^{\pi is-\frac{1}{2}\pi i\alpha}H_{2s}^{(1)} - \tfrac{1}{2}i\ e^{-\pi is+\frac{1}{2}\pi i\alpha}H_{2s}^{(2)}.$$

Now I take $x \geqslant x_1$ and $\mathrm{Re}\ s \geqslant 0$. Using formulas (3) and (4) of [43], p. 168 one sees that one has to estimate

$$(14.2.42)\qquad \left(\frac{2}{\pi x}\right)^{\frac{1}{2}} \frac{e^{\pm i(x-\frac{1}{4}\pi)\mp\frac{1}{2}\pi i\alpha}}{\Gamma(\frac{1}{2}+2s)} \int_0^{\infty e^{i\beta}} e^{-u}\ u^{2s-\frac{1}{2}}\left(1 \pm \frac{iu}{2x}\right)^{2s-\frac{1}{2}} du$$

with $|\beta| < \frac{1}{2}\pi$.

Let $|\mathrm{Im}\ s|$ be large. In the integral with $\mp\ \mathrm{Im}\ s < 0$ I take $\beta = 0$; the integral is estimated by

$$(14.2.43)\qquad \int_0^\infty e^{-u}\ u^{2\mathrm{Re}\ s-\frac{1}{2}}\left|1 \pm \tfrac{1}{2}iux^{-1}\right|^{2\mathrm{Re}\ s-\frac{1}{2}}\ e^{\mp 2\mathrm{Im}\ s\ \arctan\frac{u}{2x}}\ du,$$

this is bounded, uniformly in $x \geqslant x_1$ and $|\mathrm{Re}\ s| \leqslant \alpha$. In the other integral we have $\mp\ \mathrm{Im}\ s > 0$. I choose $\beta = \mp\ \pi/3$. Then we get, with $\tau = \frac{ue^{-i\beta}}{2x}$,

$$(14.2.44)\qquad \arg u\left(1 \pm \frac{iu}{2x}\right) = \beta + \arg(1 \mp \tau \sin\beta \pm i\tau\cos\beta)$$

$$= \beta + \arctan\ (\pm\ \tau\cos\beta/(1 \mp \tau\sin\beta));$$

hence

$$2i\ \mathrm{Im}\ s\ i\ \arg u\left(1 \pm \frac{iu}{2x}\right) =$$

$$\pm\ 2|\mathrm{Im}\ s|(\beta \ \pm\ \arctan\ (\tau\cos\beta/(1 \mp \tau\sin\beta))) =$$

$$2|\mathrm{Im}\ s|(-\ \pi/3 + \arctan\ (\tau\cos\beta/(1 + \tau\sin\pi/3))) \leqslant$$

$$2|\mathrm{Im}\ s|(-\ \pi/3 + \arctan\ (\tfrac{1}{2}\tau/\tau\tfrac{1}{2}\sqrt{3}))$$

$$= 2|\mathrm{Im}\ s|(-\ \pi/3 + \pi/6) < 0.$$

So for the other integral we also get a uniform estimate. So the quantity in (14.2.41) is estimated by

$$(14.2.45)\qquad x^{-\frac{1}{2}}e^{\pi|\mathrm{Im}\ s|}(1 + |\mathrm{Im}\ s|)^{0},$$

uniformly for $x \geqslant x_1$, $|\mathrm{Re}\ s| \leqslant \sigma$.

This completes the proof of lemma 14.2.9.

14.2.10. Lemma. Let ε, α, β and σ be as in lemma 14.2.9. Let f be twice continuously differentiable on $[0, \infty)$, let

$$(14.2.46)\qquad f(0) = f'(0) = 0$$

$$(14.2.47)\qquad |f(x)| \ll x^{-2-\beta},\quad |f''(x)| \ll x^{-2-\beta} \qquad \text{for } x \to \infty$$

$$(14.2.48)\qquad |f'(x)| \ll x^{-1-\beta} \qquad \text{for } x \to \infty$$

Let $f_1(x) = x^2 f''(x) + xf'(x) + \varepsilon x^2 f(x)$. Then f_1 satisfies the conditions of lemma 14.2.9. If $b_\alpha^\varepsilon f_1 \in {}^2_\varepsilon F^a_{\alpha,\sigma}$, then $b_\alpha^\varepsilon f \in {}^1_\varepsilon F^{a+2}_{\alpha,\sigma}$.

Proof. It is clear from (14.2.46) - (14.2.48) that f_1 satisfies (14.2.32) and (14.2.33). Further, by repeated partial integration,

$$(14.2.49)\qquad \int_0^\infty f_1(x)J_{2s}^\varepsilon(x)\frac{dx}{x} = 4s^2\int_0^\infty f(x)J_{2s}^\varepsilon(x)\frac{dx}{x}.$$

So it is clear that from an estimate of $b_\alpha^\varepsilon f_1$ a better estimate of $b_\alpha^\varepsilon f$ is obtained. As f satisfies the conditions in lemma 14.2.9 the lemma follows.

14.2.11. Proposition. Let $\varepsilon = \pm 1$, $\alpha \in \mathbb{R}$. Let $\beta \in (0, \frac{1}{2})$, let $\frac{1}{2} < \sigma < \frac{1}{2}(1+\beta)$. Let f be a four times continuously differentiable function on $[0, \infty)$ satisfying

(14.2.50) $\quad f(0) = f'(0) = 0$

(14.2.51) $\quad |f(x)|, \ |f''(x)|, \ |f''''(x)| \ll x^{-4-\beta} \quad$ for $x \to \infty$

(14.2.52) $\quad |f'(x)|, \ |f'''(x)| \ll x^{-3-\beta} \quad$ for $x \to \infty$;

then $b_\alpha^\varepsilon f$ converges absolutely and determines an element of ${}_\varepsilon^1F_{\alpha,\sigma}^{2\sigma+3/2}$.

Proof. Define f_1 as in lemma 14.2.10. Now f and f_1 both satisfy the conditions of lemma 14.2.10. Define $f_2(x) = x^2 f_1''(x) + x f_1'(x) + \varepsilon x^2 f_1(x)$. Then $b_\alpha^\varepsilon f_2 \in {}_\varepsilon^2F_{\alpha,\sigma}^{-2\sigma+3/2}$ by lemma 14.2.9. Applying lemma 14.2.10 two times we get $b_\alpha^\varepsilon f \in {}_\varepsilon^1F_{\alpha,\sigma}^{-2\sigma+9/2}$. But as $\sigma < \frac{1}{2}(1+\frac{1}{2})$ we have $-2\sigma + 9/2 > 2\sigma + 3/2$.

Remark that the conditions in proposition 14.2.11 are stronger than e.g. those in [15], p. 40, [34], p. 33.

14.2.12. Definition. Let $\alpha \in \mathbb{R}$, $\varepsilon = \pm 1$, $\sigma > 0$, $a + b > 2$; $p = q = 2$ or $p, q \in \{1, \infty\}$, $p \neq q$; for $\phi \in {}_\varepsilon^pF_{\alpha,\sigma}^a$, $\psi \in {}_\varepsilon^qF_{\alpha,\sigma}^b$

(14.2.53)
$$\langle\phi, \psi\rangle = \frac{1}{2\pi i}\int_0^{i\infty} \phi(s)\overline{\psi(s)}\,\frac{-2\pi s \sin 2\pi s\,ds}{\cos 2\pi s + \cos \pi\alpha} + \frac{1+\varepsilon}{4}\sum_{\substack{b>1\\ b\equiv\alpha(2)}} (b-1)\phi(\tfrac{1}{2}(b-1))\overline{\psi(\tfrac{1}{2}(b-1))}.$$

14.2.13. Proposition. Let $\alpha \in \mathbb{R}$, $\varepsilon = \pm 1$, $\sigma > 0$, $2\sigma + 1 \not\equiv \alpha(2)$, $a > 3/2$, $\phi \in {}_\varepsilon^2F_{\alpha,\sigma}^a$ and let f satisfy the conditions in definition 14.2.1. Then the integral in (14.2.53) for $\langle\phi, b_\alpha^\varepsilon f\rangle$ converges absolutely and

(14.2.54)
$$\langle\phi, b_\alpha^\varepsilon f\rangle = \frac{1}{2\pi i}\int_{\mathrm{Re}\, s = \sigma} \phi(s)\overline{f}\{J_{2s}^\varepsilon\}\,\frac{2\pi e^{\frac{1}{2}\pi i\varepsilon\alpha}\,s\,ds}{\sin \pi(\frac{1}{2}+s-\frac{1}{2}\varepsilon\alpha)} + \frac{\varepsilon+1}{4}\sum_{\substack{b>2\sigma+1\\ b\equiv\alpha(2)}} (b-1)\phi(\tfrac{1}{2}(b-1))\overline{b_\alpha^\varepsilon f(\tfrac{1}{2}(b-1))}.$$

Remark. (14.2.54) may be used to reformulate the inversion formula (14.2.31).

Proof. The convergence in (14.2.53) follows from (14.2.13) and (14.2.21). We have

(14.2.55)
$$\frac{1}{2\pi i}\int_0^{i\infty} \phi(s)\overline{b_\alpha^\varepsilon f(s)}\,\frac{-2\pi s \sin 2\pi s\,ds}{\cos 2\pi s + \cos \pi\alpha}$$
$$= \frac{1}{2\pi i}\int_0^{i\infty} \phi(s)\overline{f}\{J_{2s}^\varepsilon\}(-2)\,\frac{e^{\frac{1}{2}\pi i\varepsilon\alpha}\sin \pi(\frac{1}{2}+s+\frac{1}{2}\varepsilon\alpha)(-2\pi s)\sin 2\pi s\,ds}{\sin 2\pi s\ 2\sin \pi(\frac{1}{2}+s+\frac{1}{2}\alpha)\sin \pi(\frac{1}{2}+s-\frac{1}{2}\alpha)}$$
$$+ \frac{1}{2\pi i}\int_0^{i\infty} \phi(-s)\frac{\sin \pi(\frac{1}{2}-\varepsilon s+\frac{1}{2}\alpha)}{\sin \pi(\frac{1}{2}-s+\frac{1}{2}\alpha)}\, 2\,\frac{e^{\frac{1}{2}\pi i\varepsilon\alpha}\sin \pi(\frac{1}{2}-s+\frac{1}{2}\alpha)(-2\pi s)\sin 2\pi s}{\sin 2\pi s\ 2\sin \pi(\frac{1}{2}+s+\frac{1}{2}\alpha)\sin \pi(\frac{1}{2}+s-\frac{1}{2}\alpha)}$$
$$\cdot\ \overline{f}\{J_{-2s}^\varepsilon\}\,ds$$
$$= \frac{1}{2\pi i}\int_{\mathrm{Re}\, s=0} \phi(s)\overline{f}\{J_{2s}^\varepsilon\}(+2)e^{\frac{1}{2}\pi i\varepsilon\alpha}\,\frac{\pi s\,ds}{\sin \pi(\frac{1}{2}+s-\frac{1}{2}\varepsilon\alpha)}.$$

By (14.2.14) we may move the line of integration to Re $s = \sigma$. We have to take account of the residues. These turn out to cancel terms of the sum in (14.2.53); use lemma 14.2.4.

14.2.14. Corollary. Let $\phi \in {}^2_\varepsilon F^a_{\alpha,\sigma}$ as in proposition 14.2.13. Then

$$(14.2.56)\qquad |(b^\varepsilon_\alpha)^{\leftarrow}\phi(y)| \ll y^{2\sigma} \quad \text{for } y \searrow 0.$$

Proof. Combine propositions 14.2.13, 14.2.8 and (14.2.4).

14.2.15. Example. Take $\varepsilon = 1$, $p > \sigma$, $2p+1 \not\equiv \alpha(2)$, $\phi(s) = (s^2-p^2)^{-1}$ for all $s \in \mathbb{C}$, $s \neq \pm p$. Clearly $\phi \in {}^2_1F^2_{\alpha,\sigma}$. Compute $(b^\varepsilon_\alpha)^{-1}\phi(y) = \langle\phi, b^\varepsilon_\alpha\delta_y\rangle$ using (14.2.54) and move off the integral to the right. We obtain

$$(14.2.57)\qquad (b^1_\alpha)^{\leftarrow}\phi(y) = \frac{-\pi e^{\frac12\pi i\alpha}}{\sin\pi(\frac12+p-\frac12\alpha)}\, J^1_{2p}(y).$$

14.2.16. Lemma. Let $a > 1$, $\alpha \in \mathbb{R}$, $\sigma > 0$. Then $\langle\phi, 1\rangle$ is well defined for all $\phi \in {}^1_1F^a_{\alpha,\sigma}$. It is zero for $\phi \in b^1_\alpha C^\infty_c(0,\infty)$. It does not vanish for all $\phi \in {}^1_1F^a_{\alpha,\sigma}$.

Proof. See definition 14.2.12. If ϕ is positive on $(0, i\infty)$ and on $\{\frac12(b-1)\,|\,b \equiv \alpha(2)\}$, then $\langle\phi, 1\rangle > 0$. This is easily arranged; take e.g. $\phi(s) = (s^2 - p^2)^{-2}$, p suitable. If $\phi = b^1_\alpha f$, with $f \in C^\infty_c(0,\infty)$ then

$$(14.2.58)\qquad \langle\phi, 1\rangle = \frac{1}{2\pi i}\int_{\mathrm{Re}\, s=0} e^{-\frac12\pi i\alpha}\, f\{J^1_{2s}\}\, \frac{s\,ds}{\sin\pi(\frac12+s-\frac12\alpha)} + \frac{1}{2\pi}\sum_{\substack{b>1\\ b\equiv\alpha(2)}} (b-1)e^{-\frac12\pi i b} f\{J^1_{b-1}\}.$$

From lemma 14.2.3 follows that the path of integration may be moved off to the right. This absorbs the sum; the integral goes to zero.

Remark. With the methods leading to proposition 14.2.11 one might extend the class of $\phi \in {}^1_1F^a_{\alpha,\sigma}$ for which $\langle\phi, 1\rangle$ vanishes. I do not go into this question.

14.3. Convolution operators.

The set ${}_{\frac12}Q_{\frac12}$ of functions on G has been defined in definition 6.4.1. For each $q \in {}_{\frac12}Q_{\frac12}$ we have defined in 6.4.2 an operator C_q. It acts on functions f on G satisfying

$$(14.3.1)\qquad |f(na(y)k)| \ll y^{\frac12+\sigma} \qquad \text{for } y \searrow 0$$

uniformly in $n \in N$, $k \in K$, for some $\sigma > 0$.

$$(14.3.2)\qquad f(n(x)g) = e^{\frac12 ix} f(y).$$

The function $C_q f$ again satisfies (14.3.2), but not necessarily (14.3.1). We have described q in (6.4.4) by two functions $q_1, q_{-1} \in C^\infty_c(0,\infty)$. Now I define q^1 and $q^{-1} \in C^\infty_c(0,\infty)$ by

$$(14.3.3)\qquad q_\varepsilon(y) = y^{\frac12} q^\varepsilon(y^{\frac12}).$$

This I do to get in the situation of proposition 6.8.1 :

$$(14.3.4) \qquad v(q,\ s) = 2\pi^{-1} \sum_{\varepsilon=\pm 1} q^{\varepsilon}\{J^{\varepsilon}_{2s}\}\ \Gamma(2s+1)\ \sin \pi(\tfrac{1}{2}+s+\tfrac{1}{2}\tau)X^{\varepsilon}(s).$$

The operator ${}_{\frac{1}{2}}V^{\varepsilon}(t)$ defined in definition 6.2.3 may be obtained by replacing q by the distribution $d(t,\ \varepsilon)$, determined by

$$(14.3.5) \qquad d(t,\ \varepsilon)^{\varepsilon} = \pi t^{\frac{1}{2}}\ \delta_{t^{\frac{1}{2}}}$$

$$d(t,\ \varepsilon)^{-\varepsilon} = 0.$$

A computation like the one in (6.4.8) gives

$$(14.3.6) \qquad C_{d(t,\ \varepsilon)}f = {}_{\frac{1}{2}}V^{\varepsilon}(t)f.$$

In this subsection I take $q \in {}_{\frac{1}{2}}Q_{\frac{1}{2}}$ or $q = d(t,\ \varepsilon)$ and study C_q on H^{τ}_{σ} with $\sigma > 0$. Remark that in the notations of this section propositions 6.8.1 and 6.8.5 read:

$$(14.3.7) \qquad C_q M(s) = 2\pi^{-1}\Gamma(2s+1)\sin \pi(\tfrac{1}{2}+s+\tfrac{1}{2}\tau) \sum_{\varepsilon=\pm 1} q^{\varepsilon}\{J_{2s}\}W(s)X^{\varepsilon}(s)$$

for Re $s > 0$;

$$(14.3.8) \qquad C_q D^{b} = 2e^{-\frac{1}{2}\pi i b}q^{1}\{J^{1}_{b-1}\}D^{b}$$

for $b \in Y_{\tau}$, $b > 1$.

14.3.1. <u>Lemma</u>. <u>Let</u> $q \in {}_{\frac{1}{2}}Q_{\frac{1}{2}}$ <u>or</u> $q = d(t,\ \eta)$. <u>Let</u> $\xi \in H^{\tau}_{\sigma}$. <u>Then</u>

$$(14.3.9) \qquad C_q\Phi\xi = \frac{1}{2\pi i}\int_0^{i\infty} \Gamma(2s)^{-1}\Gamma(-2s)^{-1}W(s)$$

$$.\sum_{\varepsilon=\pm 1} \frac{2}{\sin 2\pi s}(\sin \pi(\tfrac{1}{2}-s+\tfrac{1}{2}\tau)q^{\varepsilon}\{J_{-2s}\}\iota(-s)X^{\varepsilon}(-s)\iota(s)-\sin \pi(\tfrac{1}{2}+s+\tfrac{1}{2}\tau).$$

$$q^{\varepsilon}\{J_{2s}\}X^{\varepsilon}(s))\xi(s)ds + \sum_{b>1,\, b\equiv\eta\tau(2)} \Gamma(b)^{-\frac{1}{2}}\Gamma(b-1)^{-\frac{1}{2}}b^{1}_{\eta\tau}q^{1}(\tfrac{1}{2}(b-1)D^{b}\xi_{b}$$

<u>Proof</u>. From (13.5.13) follows that $C_q\Phi\xi$ is well defined. I write $\Phi\xi$ as in (13.5.20) with $\sigma_1 = \sigma$. The integrals occurring in $C_q\Phi\xi$ converge absolutely; so one may take C_q inside the integral. As $C_qM(s)$ is holomorphic on Re $s > 0$ we may move the line of integration to the left. We pick up residues, use (14.3.8), and restore the terms with $1 < b < 2\sigma+1$. To see that the integrand decreases quickly at infinity, use (14.3.7). In this way we get

$$(14.3.10) \qquad C_q\Phi\xi = \frac{1}{2\pi i}\int_{\mathrm{Re}\ s=\sigma_1} C_qM(s)\xi(s)\Gamma(2s)^{-1}ds$$

$$+ \sum_{\substack{b\in Y_{\tau} \\ b>1}} \Gamma(b)^{-\frac{1}{2}}\Gamma(b-1)^{-\frac{1}{2}}C_qD^{b}\xi_{b},$$

with σ_1 between 0 and all positive nonregular points. $C_qM(s)$ is only defined for Re $s > 0$, but the expression in the right hand side of (14.3.7) is meromorphic on the whole complex plane. If we replace $C_qM(s)$ we see that the integral in (14.3.10) equals

$$(14.3.11) \qquad \frac{1}{2\pi i}\int_{\mathrm{Re}\ s=0} 2\pi^{-1}2s\ \sin \pi(\tfrac{1}{2}+s+\tfrac{1}{2}\tau) \sum_{\varepsilon=\pm 1} q^{\varepsilon}\{J_{2s}\}W(s)X^{\varepsilon}(s)\xi(s)ds$$

$$= \frac{1}{2\pi i}\int_0^{i\infty} 2\pi^{-1}[2s\ \sin \pi(\tfrac{1}{2}+s+\tfrac{1}{2}\tau) \sum_{\varepsilon=\pm 1} q^{\varepsilon}\{J^{\varepsilon}_{2s}\}W(s)X^{\varepsilon}(s)\xi(s) -$$

$$-2s \sin \pi(\tfrac{1}{2}-s+\tfrac{1}{2}\tau) \sum_{\varepsilon=\pm 1} q^{\varepsilon}\{J_{-2s}\}W(s)\iota(-s)X^{\varepsilon}(-s)\iota(s)\xi(s)]ds$$

use (13.5.4) and (5.5.3). This is the integral in (14.3.9); the sum is obtained from (14.3.10), (14.3.8) and (14.2.17).

14.3.2. Proposition. *Let* $\sigma > 0$, $2\sigma+1 \not\equiv \pm\tau(2)$. *Let* $q \in {}_{\frac{1}{2}}Q_{\frac{1}{2}}$ *or* $q = d(t, \beta)$ *for some* $t > 0$, $\beta = \pm 1$.

Define for $\xi \in H_{\sigma}^{\tau}$

(14.3.12) $$(M_q\xi)(s) = \begin{cases} \begin{pmatrix} \sum_{\varepsilon=\pm1} b_{\tau}^{\varepsilon}q^{\varepsilon}(s) & 0 \\ 0 & \sum_{\varepsilon=\pm1} \varepsilon b_{\tau}^{\varepsilon}q^{\varepsilon}(s) \end{pmatrix} \xi(s) & \text{if } \tau = 0,\ 1 \\ \begin{pmatrix} b_{\tau}^{1}q^{1}(s) & b_{\tau}^{-1}q^{-1}(s) \\ b_{-\tau}^{-1}q^{-1}(s) & b_{-\tau}^{1}q^{1}(s) \end{pmatrix} \xi(s) & \text{if } 0 < \tau < 1 \end{cases}$$

(14.3.13) $(M_q\xi)_b = b_{\eta\tau}^{1}q^{1}(\tfrac{1}{2}(b-1))\xi_b$ *with* $\eta = -1$ *if* $0 < \tau < 1$ *and* $b \equiv -\tau(2)$ *and* $\eta = 1$ *otherwise.*

Then $M_q\xi \in H_{\sigma}^{\tau}$ *and on* H_{σ}^{τ}:

(14.3.14) $C_q\Phi = \Phi M_q$.

Remark. The matrices refer to the decomposition (3.5.3).

Proof. (14.3.12) has been arranged to correspond to (14.3.9). So by lemma 14.3.1 we have only to check that $M_q\xi \in H_{\sigma}^{\tau}$.

The holomorphy of $(M_q\xi)(s)$ for general s is clear, for all coefficients of the matrices are holomorphic on $\mathbb{C}$. Condition (13.5.2) follows from lemma 14.2.4. As the $b_{\pm\tau}^{\varepsilon}q^{\varepsilon}$ have at most polynomial growth (13.5.3) is satisfied. To check (13.5.4) we have to show that for the matrices $M_q(s)$ in (14.3.12)

(14.3.15) $\iota(s)M_q(s)\iota(-s) = M_q(-s)$

This is an easy computation, using

(13.3.16) $b_{\eta\tau}^{\varepsilon}q^{\varepsilon}(s) = b_{\eta\tau}^{\varepsilon}q^{\varepsilon}(-s) \sin \pi(\tfrac{1}{2}+s+\tfrac{1}{2}\eta\tau)(\sin \pi(\tfrac{1}{2}+\varepsilon s+\tfrac{1}{2}\eta\tau))^{-1}$,

definition 3.5.2 and (3.3.15).

14.3.3. Corollary. *For* $t > 0$, $\varepsilon = \pm 1$ *we have on* H_{σ}^{τ}

(14.3.17) ${}_{\frac{1}{2}}V^{\varepsilon}(t)\Phi = \Phi M_{d(t,\varepsilon)}$.

Notation. I use the notation $\delta(t, \varepsilon)$ for the operator $M_{d(t, \varepsilon)}$.

14.4. Multiplication operators.

We have seen that the convolution operators C_q on ΦH_{σ}^{τ} correspond to multiplication operators M_q in H_{σ}^{τ}. I now define more general multiplication operators in H_{σ}^{τ}.

14.4.1. Proposition. $\sigma > 0$, $2\sigma + 1 \not\equiv \pm\tau(2)$, $a \in \mathbb{R}$, $p = 1$, 2 or ∞. *The operators*

μ *in* H_σ^τ *defined by*

(14.4.1) $\quad \mu\xi(s) = \mu(s)\xi(s)$

with

(14.4.2)
$$\mu(s) = \begin{cases} \begin{pmatrix} \mu^{1}(s) & 0 \\ 0 & \mu^{-1}(s) \end{pmatrix} & \text{if } \tau = 0, 1 \\ \begin{pmatrix} \mu^{11}(s) & \mu^{-11}(s) \\ \mu^{1-1}(s) & \mu^{-1-1}(s) \end{pmatrix} & \text{if } 0 < \tau < 1 \end{cases}$$

(14.4.3) $\quad (\mu\xi)_b = \mu_b \xi_b$

map H_σ^τ *into itself provided they satisfy*:

(14.4.4) *the coefficient functions of* $\mu(s)$ *are holomorphic on* $|\mathrm{Re}\ s| \leqslant \sigma$ *and are estimated by* $(1 + |\mathrm{Im}\ s|)^{-a}$

(14.4.5) $\quad \mu(\frac{1}{2}(b-1))|D_b = \mu_b \binom{10}{01}|D_b$ *for* $b \in Y_\tau$ *or* $b = \tau = 1$ *and* $1 - 2\sigma < b < 1 + 2\sigma$

(14.4.6) $\quad \iota(s)\mu(s)\iota(-s) = \mu(-s)$

(14.4.7) $\quad \sum\limits_{\substack{b>1 \\ b\in Y_\tau}} (b-1)|\mu_b|^p < \infty \qquad$ *if* $p = 1$ *or* 2

(14.4.8) $\quad |\mu_b| \ll 1 \quad$ *for* $b > 1,\ b \in Y_\tau \qquad$ *if* $p = \infty$.

The set of these operators is denoted ${}^pR_\sigma^a$.

Proof. As in the proof of proposition 14.3.2.

14.4.2. *Remark*. The elements of ${}^pR_\sigma^a$ are *g*-K-intertwining operators; ${}^1R_\sigma^a \subset {}^2R_\sigma^a \subset {}^\infty R_\sigma^a$ and ${}^pR_\sigma^a \subset {}^pR_\sigma^b$ if $b \leqslant a$. The identity operator $1 \in {}^\infty R_\sigma^0$.
Also $\pi_\varepsilon \in {}^\infty R_\sigma^0$ (see definition 9.4.1), with

(14.4.9)
$$\pi_\varepsilon(s) = \begin{cases} \binom{10}{01} & \text{if } \tau = 0, 1 \\ \binom{10}{00} & \text{if } 0 < \tau < 1,\ \varepsilon = 1 \\ \binom{00}{01} & \text{if } 0 < \tau < 1,\ \varepsilon = -1 \end{cases}$$

$$(\pi_\varepsilon)_b = \begin{cases} 1 & \text{if } b \equiv \varepsilon\tau(2) \\ 0 & \text{otherwise} \end{cases}$$

Under composition as operators in H_σ^τ:

(14.4.10)
$${}^\infty R_\sigma^a\ {}^\infty R_\sigma^b \subset {}^\infty R_\sigma^{a+b}$$

$${}^2R_\sigma^a\ {}^\infty R_\sigma^b \subset {}^2R_\sigma^{a+b},\quad {}^\infty R_\sigma^a\ {}^2R_\sigma^b \subset {}^2R_\sigma^{a+b}$$

$${}^1R_\sigma^a\ {}^pR_\sigma^b \subset {}^1R_\sigma^{a+b},\quad {}^pR_\sigma^a\ {}^1R_\sigma^b \subset {}^1R_\sigma^{a+b}$$

14.4.3. *Remark*. If $q \in {}_{\frac{1}{2}}Q_{\frac{1}{2}}$ then $M_q \in {}^1R_\sigma^a$ for all $a \in \mathbb{R}$.
If $t > 0$, $\varepsilon = \pm 1$, then $\delta(t, \varepsilon)$, defined in 14.3.3, is an element of ${}^1R_\sigma^{\frac{1}{2}-2\sigma}$.

14.4.4. Lemma. Let μ be in some ${}^pR^a_\sigma$. If $\tau = 0, 1$ then

(14.4.11) $\mu^\delta(-s) = \mu^{(-1)^\tau\delta}(s)$ for $\delta = \pm 1$.

If $0 < \tau < 1$ then

(14.4.12) $\mu^{\delta\varepsilon}(-s) = \dfrac{\sin \pi(\frac{1}{2}+s+\frac{1}{2}\delta\tau)}{\sin \pi(\frac{1}{2}+s+\frac{1}{2}\varepsilon\tau)}\, \mu^{\delta\varepsilon}(s)$ for $\delta,\varepsilon = \pm 1$.

Conversely, these formulas imply (14.4.6).

Proof. Write out condition (14.4.6).

14.4.5. Lemma. Let μ be in some ${}^pR^a_\sigma$. If $\tau = 0, 1$ and $b \equiv \tau(2)$, $1 \leqslant b < 2\sigma + 1$, then

(14.4.13) $\mu^\delta(\frac{1}{2}(b-1)) = \mu^\delta(\frac{1}{2}(1-b)) = \mu_b$ for $\delta = \pm 1$.

If $0 < \tau < 1$, $b \equiv \eta\tau(2)$ and $|b-1| < 2\sigma$, then

(14.4.14) $\mu^{\eta\eta}(\frac{1}{2}(b-1)) = \mu^{\eta\eta}(-\frac{1}{2}(b-1)) = \mu_b$

(14.4.15) $\mu^{\eta-\eta}(\frac{1}{2}(b-1)) = \mu^{-\eta\eta}(-\frac{1}{2}(b-1)) = 0.$

Proof. Use (14.4.5), the standard embedding of D_b in H^τ (see (3.5.6)) and the previous lemma.

14.4.6. Proposition. Let $\sigma > 0$, $2\sigma + 1 \not\equiv \pm\tau(2)$, $a \in \mathbb{R}$, $p = 1, 2$ or ∞. If $\mu \in {}^pR^a_\sigma$, then μ^*, defined by

(14.4.16) $\mu^*(s) = \overline{\mu(-\bar{s})}^t$ (t means the matrix-transpose)

(14.4.17) $(\mu^*)_b = \overline{\mu_b}$,

is an element of ${}^pR^a_\sigma$.

For $\xi, \eta \in H^\tau_\sigma$:

(14.4.18) $\{\mu\xi, \eta\} = \{\xi, \mu^*\eta\}$.

Proof: (14.4.16) has been chosen to satisfy (14.4.18). Then condition (14.4.5) forces (14.4.17), use lemma 14.4.5. To obtain (14.4.6) for μ^* use lemma 14.4.4. The other conditions in proposition 14.4.1 follow easily.

14.4.7. Remark. $1^* = 1$, $\pi^*_\varepsilon = \pi_\varepsilon$

14.4.8. Definition. If $\mu \in {}^pR^a_\sigma$ then $\mathrm{tr}\mu$ is the function on $|\mathrm{Re}\ s| \leqslant \sigma$ defined by

(14.4.19) $(\mathrm{tr}\ \mu)(s) = \begin{cases} \mu^1(s) + \mu^{-1}(s) & \text{if } \tau = 0, 1 \\ \mu^{11}(s) + \mu^{-1-1}(s) & \text{if } 0 < \tau < 1. \end{cases}$

Remark that from lemma 14.4.5 follows that for $b \in Y_\tau$ or $b = \tau = 1$ with $|\frac{1}{2}(b-1)| < \sigma$

(14.4.20) $\mathrm{tr}\ \mu(\frac{1}{2}(b-1)) = \begin{cases} 2\mu_b & \text{if } \tau = 0, 1 \\ \mu_b + \mu_{2-b} & \text{if } 0 < \tau < 1. \end{cases}$

14.5. Scalar products

On ${}_{\frac{1}{2}}Q_{\frac{1}{2}}$ we have a natural unitary structure; for $q, r \in {}_{\frac{1}{2}}Q_{\frac{1}{2}}$:

$$(14.5.1)\qquad \langle q, r\rangle = \int_{N\backslash G/N} q(g)\overline{r(g)}d\dot g$$

$$= \frac{1}{2\pi}\int_H q(hw)\overline{r(hw)}\alpha(h)^{-1}dh \qquad \text{see (2.4.10)}$$

$$= \sum_{\varepsilon=\pm 1} \frac{1}{2\pi}\int_0^\infty q_\varepsilon(y)\overline{r_\varepsilon(y)}y^{-2}dy \qquad \text{see (6.4.4)}$$

$$= \sum_{\varepsilon=\pm 1} \pi^{-1}\int_0^\infty q^\varepsilon(y)\overline{r^\varepsilon(y)}y^{-1}dy \qquad \text{see (14.3.3)}$$

Applying proposition 14.2.6 we see that in the case $\tau = 0, 1$ this equals

$$(14.5.2)\qquad \frac{1}{2\pi i}\int_0^{i\infty} (\mathrm{tr}\ M_r^* M_q)(s)\ \frac{-s\sin 2\pi s}{\cos 2\pi s + \cos\pi\tau}\,ds$$

$$+ \frac{1}{2\pi}\sum_{b\in Y_\tau} (b-1)(M_q)_b\overline{(M_r)_b}.$$

Working back from the integral in (14.5.2) we see that in the case $0 < \tau < 1$ we have

$$(14.5.3)\qquad \langle q, r\rangle = \frac{1}{2\pi i}\int_0^{i\infty} (\mathrm{tr}\ M_r^* M_q)(s)\ \frac{-s\sin 2\pi s}{\cos 2\pi s + \cos\pi\tau}\,ds$$

$$+ \frac{1}{4\pi}\sum_{\substack{b\in Y_\tau\\ b>1}} (b-1)(M_q)_b\overline{(M_r)_b}.$$

This leads to the following definition.

14.5.1. Definition. Let $\sigma > 0$, $2\sigma + 1 \not\equiv \pm\tau(2)$. Let $a, b \in \mathbb{R}$, $a + b > 2$. Let $p = q = 2$ or $p \neq q$, $p, q \in \{1, \infty\}$. Take

$$(14.5.4)\qquad \varepsilon_\tau = \begin{cases} 2 & \text{if } \tau = 0, 1 \\ 1 & \text{if } 0 < \tau < 1 \end{cases}$$

For $\mu \in {}^pR^a_\sigma$, $\nu \in {}^qR^b_\sigma$

$$(14.5.5)\qquad \langle \mu, \nu\rangle = \frac{1}{2\pi i}\int_0^{i\infty} (\mathrm{tr}\ \nu^*\mu)(s)\ \frac{-s\sin 2\pi s}{\cos 2\pi s + \cos\pi\tau}\,ds$$

$$+ \frac{1}{4\pi}\varepsilon_\tau \sum_{\substack{b\in Y_\tau\\ b>1}} (b-1)\mu_b\overline{\nu_b}.$$

Remark that $\langle q, r\rangle = \langle M_q, M_r\rangle$ for $q, r \in {}_{\frac12}Q_{\frac12}$. For λ, μ, ν in appropriate spaces we have

$$(14.5.6)\qquad \langle \lambda\mu, \nu\rangle = \langle \mu, \lambda^*\nu\rangle = \langle \lambda, \nu\mu^*\rangle.$$

In particular

$$(14.5.7)\qquad \langle \mu\pi_\varepsilon, \nu\rangle = \langle \mu, \nu\pi_\varepsilon\rangle$$

$$\langle \pi_\varepsilon\mu, \nu\rangle = \langle \mu, \pi_\varepsilon\nu\rangle.$$

14.5.2. Remark. The definition of $\langle \mu, \nu\rangle$ makes sense if the estimate of the matrix coefficients of μ and ν is only valid on the line Re $s = 0$. So $\langle \mu, \delta(t, \varepsilon)\rangle$ is well defined for $\mu \in {}^2R^{3/2+\beta}_\sigma$ with $\beta > 0$. One may define a function on G with support in NAwN $\cup$ NAjwN by

(14.5.8) $\quad n(x)a(y)j_{\varepsilon}wn(u) \to e^{\frac{1}{2}i(x+u)}\langle\mu,\ \delta(t,\ \varepsilon)\rangle.$

If $\mu = M_q$ with $q \in {}_{\frac{1}{2}}Q_{\frac{1}{2}}$, then one gets back q.

Now I apply proposition 14.2.13 to write $\langle\mu,\ \delta(t,\ \varepsilon)\rangle$ with an integral over $\mathrm{Re}\ s = \sigma$. This I shall need in the proof of the sum formula.

14.5.3. Proposition. Let $\sigma > 0$, $2\sigma + 1 \not\equiv \pm\tau(2)$. Let $\mu \in {}^2R^a_\sigma$ with $\alpha > 3/2$. Let $t > 0$, $\varepsilon = \pm 1$. Then

(14.5.9) $\quad \langle\mu,\ \delta(t,\ \varepsilon)\rangle =$

$$\frac{1}{2\pi i}\int_{\mathrm{Re}\ s=\sigma} s\left(\sin\pi(\tfrac{1}{2}+s-\tfrac{1}{2}\tau)\right)^{-1}\pi t^{\frac{1}{2}}J^{\varepsilon}_{2s}(t^{\frac{1}{2}})\,\mathrm{tr}(\overline{X^{\varepsilon}}(s)^{t}\mu(-s))\,ds$$

$$+\ \tfrac{1}{4}(1+\varepsilon)\pi^{-1}\varepsilon_{\tau}\sum_{\substack{b>2\sigma+1\\ b\in Y_\tau}}(b-1)e^{\frac{1}{2}\pi i b}\pi t^{\frac{1}{2}}J^{1}_{b-1}(t^{\frac{1}{2}})\mu_b$$

Proof.

Take q as in subsection 14.3. I compute $\langle\mu,\ M_q\rangle$. From (14.3.9) and proposition 14.3.2 follows that

(14.5.10) $\quad M_q(s) = 2(\sin 2\pi s)^{-1}[-\sin\pi(\tfrac{1}{2}+s+\tfrac{1}{2}\tau)\sum_{\delta=\pm 1} q^{\delta}\{J^{\delta}_{2s}\}X^{\delta}(s)$

$$+\ \sin\pi(\tfrac{1}{2}-s+\tfrac{1}{2}\tau)\sum_{\delta=\pm 1} q^{\delta}\{J^{\delta}_{-2s}\}\iota(-s)X^{\delta}(-s)\iota(s)].$$

For matrixvalued functions A and B with diagonal form if $\tau = 1$ one easily checks:

(14.5.11) $\quad \mathrm{trace}(\overline{A(+\bar{s})}^{t}\iota(+s)B(+s)\iota(-s)) =$

$$\mathrm{trace}(\overline{(\iota(\bar{s})A(\bar{s})\iota(-\bar{s}))}^{t}\ B(s)).$$

Using (14.4.6) we obtain

(14.5.12) $\quad (\mathrm{tr}\ M_q^{*}\mu)(s) =$

$$\overline{2(\sin 2\pi(-\bar{s}))}^{-1}[-\sin\pi(\tfrac{1}{2}-s+\tfrac{1}{2}\tau)\sum_{\delta=\pm 1}\overline{q^{\delta}}\{J^{\delta}_{-2s}\}\mathrm{trace}(\overline{X^{\delta}}(-s)^{t}\mu(s))$$

$$+\ \sin\pi(\tfrac{1}{2}+s+\tfrac{1}{2}\tau)\sum_{\delta=\pm 1}\overline{q^{\delta}}\{J^{\delta}_{2s}\}\mathrm{trace}(\overline{X^{\delta}}(s)^{t}\mu(-s))].$$

So by (14.5.5)

(14.5.13) $\quad \langle\mu,\ M_q\rangle = \dfrac{-1}{2\pi i}\displaystyle\int_{\mathrm{Re}\ s=0}\frac{2}{\sin 2\pi s}\sin\pi(\tfrac{1}{2}+s+\tfrac{1}{2}\tau)\sum_{\delta=\pm 1}\overline{q^{\delta}}\{J^{\delta}_{2s}\}$

$$.\ \mathrm{trace}(\overline{X^{\delta}}(s)^{t}\mu(-s))\,\frac{-s\sin 2\pi s}{\cos 2\pi s+\cos\pi\tau}\,ds$$

$$+\ \frac{\varepsilon\tau}{4\pi}\sum_{b>1,\,b\in Y_\tau}(b-1)\mu_b\ 2e^{\frac{1}{2}\pi i b}\ \overline{q^{1}}\{J^{1}_{b-1}\}$$

$$= \frac{1}{2\pi i}\int_{\mathrm{Re}\ s=0}\frac{s}{\sin\pi(\frac{1}{2}+s-\frac{1}{2}\tau)}\sum_{\delta=\pm 1}\overline{q^{\delta}}\{J^{\delta}_{2s}\}\mathrm{trace}(\overline{X^{\delta}}(s)^{t}\mu(-s))\,ds$$

$$+\ \frac{1}{2\pi}\varepsilon_{\tau}\sum_{\substack{b>1\\ b\in Y_\tau}}(b-1)\ e^{\frac{1}{2}\pi i b}\ \overline{q^{1}}\{J^{1}_{b-1}\}\mu_b$$

The integrand goes to zero quickly enough to admit moving the line of integration to Re s = σ. We have to take account of the poles. In the case $\tau = 0, 1$ it is easy to see that the residues cancel terms of the sum. In the case $0 < \tau < 1$ we have

(14.5.14) $$\sum_{\delta=\pm 1} \overline{q^{\delta}\{J^{\delta}_{2s}\}} \operatorname{trace} (\overline{X^{\delta}}(s)^{t}\mu(-s))$$
$$= \sum_{\delta,\eta=\pm 1} \overline{q^{\delta\eta}} \{J^{\delta\eta}_{2s}\} \frac{\sin \pi(\frac{1}{2}+s+\frac{1}{2}\eta\tau)}{\sin \pi(\frac{1}{2}+s+\frac{1}{2}\tau)} e^{\frac{1}{2}\pi i\tau\delta} \mu^{\delta\eta}(-s).$$

If $b > 1$, $b \equiv \alpha\tau(2)$ then the residue is

(14.5.15) $$\operatorname*{res}_{s\to\frac{1}{2}(b-1)} s \sum_{\delta,\eta=\pm 1} \overline{q^{\delta\eta}}\{J^{\delta\eta}_{2s}\}(\sin \pi(\tfrac{1}{2}-s+\tfrac{1}{2}\eta\tau))^{-1} e^{\frac{1}{2}\pi i\tau\delta} \mu^{\delta\eta}(-s)$$
$$= \tfrac{1}{2}(b-1) \sum_{\delta=\pm 1} \overline{q^{\alpha\delta}}\{J^{\alpha\delta}_{b-1}\}\pi^{-1}(-1)^{\frac{1}{2}(b-\alpha\tau)} e^{\frac{1}{2}\pi i\delta\tau} \mu^{\delta\alpha}(-\tfrac{1}{2}(b-1))$$
$$= \frac{1}{2\pi} (b-1)\, \overline{q^{1}}\{J^{1}_{b-1}\}e^{\frac{1}{2}\pi ib} \mu_{b},$$

as it should be to cancel the terms in the sum in (14.5.13). I have used lemma 14.4.5. Replacing q by $d(t, \varepsilon)$ we obtain

(14.5.16) $$\langle\mu, \delta(t, \varepsilon)\rangle =$$
$$= \frac{1}{2\pi i} \int_{\mathrm{Re}\ s=0} s(\sin \pi(\tfrac{1}{2}+s-\tfrac{1}{2}\tau))^{-1}\pi t^{\frac{1}{2}}J^{\varepsilon}_{2s}(t^{\frac{1}{2}})\operatorname{trace}(\overline{X^{\varepsilon}}(s)^{t}\mu(-s))ds$$
$$+ \frac{1}{2\pi} \varepsilon_{\tau} \sum_{\substack{b>2\sigma+1 \\ b\in Y_{\tau}}} (b-1)e^{\frac{1}{2}\pi ib}\tfrac{1}{2}(1+\varepsilon)\pi t^{\frac{1}{2}}J^{1}_{b-1}(t^{\frac{1}{2}})\mu_{b}.$$

<u>Remark</u>. One may prove proposition 14.5.3 directly from (14.5.5) and proposition 14.2.13. I work this out, for I shall need some of the intermediate results later on.

(14.5.17) In the case $\tau = 0$ or 1 I take $\mu_{\delta} = \mu^{1} + \delta\mu^{-1}$. Then $\mu_{\delta} \in {}^{2}_{\delta}F^{a}_{\tau,\sigma}$. We have

(14.5.18) $$\operatorname{tr}(\delta(t, \varepsilon)*\mu)(s) = \pi t^{\frac{1}{2}}\mu_{\varepsilon}(s)\overline{b^{\varepsilon}_{\tau}\delta_{t^{\frac{1}{2}}}(s)}$$

(14.5.19) $$\delta(t, \varepsilon)_{b} = \tfrac{1}{2}(\varepsilon+1)\pi t^{\frac{1}{2}}b^{1}_{\tau}\delta_{t^{\frac{1}{2}}}(\tfrac{1}{2}(b-1)).$$

So by (14.5.5)

(14.5.20) $$\langle\mu, \delta(t, \varepsilon)\rangle = \frac{1}{2\pi i}\int_{0}^{i\infty} \pi t^{\frac{1}{2}}\mu_{\varepsilon}(s)\overline{b^{\varepsilon}_{\tau}\delta_{t^{\frac{1}{2}}}(s)} \frac{-s \sin 2\pi s}{\cos 2\pi s+\cos \pi\tau} ds$$
$$+ \frac{1}{2\pi} \pi t^{\frac{1}{2}}\tfrac{1}{2}(\varepsilon+1) \sum_{\substack{b>1 \\ b\equiv\tau(2)}} (b-1)\ \tfrac{1}{2}\mu_{1}(\tfrac{1}{2}(b-1))\ \overline{b^{1}_{\tau}\delta_{t^{\frac{1}{2}}}(\tfrac{1}{2}(b-1))}$$
$$= \tfrac{1}{2}t^{\frac{1}{2}}\langle\mu_{\varepsilon}, b^{\varepsilon}_{\tau}\delta_{t^{\frac{1}{2}}}\rangle$$

by definition 14.2.12. On the other hand

(14.5.21) $\quad tr(\overline{X^{\varepsilon}}(s)\,{}^t\mu(-s)) = e^{\frac{1}{2}\pi i\tau\varepsilon}\mu_{\varepsilon}(-s)$

$= ((-1)^{\tau}, \varepsilon)\, e^{\frac{1}{2}\pi i\tau\varepsilon}\mu_{\varepsilon}(s),$

use lemma 14.4.4. Proposition 14.5.3 amounts to

(14.5.22) $\quad \langle\mu, \delta(t, \varepsilon)\rangle =$

$$\frac{1}{2\pi i}\int_{\mathrm{Re}\ s=\sigma} \pi t^{\frac{1}{2}} J^{\varepsilon}_{2s}(t^{\frac{1}{2}})((-1)^{\tau}, \varepsilon) e^{\frac{1}{2}\pi i\tau\varepsilon}\mu_{\varepsilon}(s)\,\frac{s\ ds}{\sin\pi(\frac{1}{2}+s-\frac{1}{2}\tau)}$$

$$+\ \tfrac{1}{2}(1+\varepsilon)t^{\frac{1}{2}} \sum_{\substack{b\equiv\tau(2)\\ b>2\sigma+1}} \tfrac{1}{2}(b-1)\tfrac{1}{2}\mu_1(\tfrac{1}{2}(b-1))\ \overline{b^{1}\delta_{t^{\frac{1}{2}}}(\tfrac{1}{2}(b-1))}$$

$$= \tfrac{1}{2}t^{\frac{1}{2}}\langle\mu_{\varepsilon}, b^{\varepsilon}_{\tau}\delta_{t^{\frac{1}{2}}}\rangle$$

by proposition 14.2.13.

In the case $0 < \tau < 1$ we have

(14.5.23) $\quad \mu^{\delta\eta} \in {}^{2}_{\delta\eta}F^{a}_{\eta\tau,\sigma}$

as follows from proposition 14.4.1 and the lemmas 14.4.4 and 14.4.5.

(14.5.24) $\quad tr(\delta(t, \varepsilon)*\mu)(s) = \pi t^{\frac{1}{2}} \sum_{\eta=\pm1} \mu^{\varepsilon\eta\ \eta}(s)\ \overline{b^{\varepsilon}_{\eta\tau}\delta_{t^{\frac{1}{2}}}(-\bar{s})}$

(14.5.25) $\quad \delta(t, \varepsilon)_b = \frac{1}{2}(\varepsilon+1)\pi t^{\frac{1}{2}} b^{1}_{\eta\tau}\delta_{t^{\frac{1}{2}}}(\frac{1}{2}(b-1)) \quad$ if $b \equiv \eta\tau(2)$.

By (14.5.5):

(14.5.26) $\quad \langle\mu, \delta(t, \varepsilon)\rangle =$

$$\pi t^{\frac{1}{2}} \sum_{\eta=\pm1} \frac{1}{2\pi i}\int_0^{i\infty} \mu^{\varepsilon\eta\ \eta}(s)\ \overline{b^{\varepsilon}_{\eta\tau}\delta_{t^{\frac{1}{2}}}(s)}\ \frac{-s\ \sin 2\pi s\ ds}{\cos 2\pi s+\cos\pi\tau}$$

$$+\ \frac{1}{4\pi}\pi t^{\frac{1}{2}}\tfrac{1}{2}(\varepsilon+1) \sum_{\eta=\pm1} \sum_{\substack{b>1\\ b\equiv\eta\tau(2)}} (b-1)\mu_b\ \overline{b^{1}_{\eta\tau}\delta_{t^{\frac{1}{2}}}(\tfrac{1}{2}(b-1))}$$

$$= \tfrac{1}{2}t^{\frac{1}{2}} \sum_{\eta=\pm1} \langle\mu^{\varepsilon\eta\ \eta}, b^{\varepsilon}_{\eta\tau}\delta_{t^{\frac{1}{2}}}\rangle,$$

see (14.2.53).

(14.5.27) $\quad tr(\overline{X^{\varepsilon}}(s)\,{}^t\mu(-s)) = \sum_{\eta=\pm1} c_{\varepsilon\eta}(-s)\ e^{\frac{1}{2}\pi i\eta\tau}\mu^{\eta\ \varepsilon\eta}(-s)$

$$= \sum_{\eta=\pm1} \frac{\sin\pi(\frac{1}{2}+s+\frac{1}{2}\varepsilon\eta\tau)}{\sin\pi(\frac{1}{2}+s+\frac{1}{2}\tau)}\ e^{\frac{1}{2}\pi i\varepsilon\eta\tau}\mu^{\varepsilon\eta\ \eta}(s)$$

(14.5.28) $\quad \langle\mu, \delta(t, \varepsilon)\rangle =$

$$\sum_{\eta=\pm1} \pi t^{\frac{1}{2}}\frac{1}{2\pi i}\int_{\mathrm{Re}\ s=\sigma} \mu^{\varepsilon\eta\ \eta}(s)\ J^{\varepsilon}_{2s}(t^{\frac{1}{2}})e^{\frac{1}{2}\pi i\varepsilon\eta\tau}\ \frac{s\ ds}{\sin\pi(\frac{1}{2}+s-\frac{1}{2}\varepsilon\eta\tau)}$$

$$+\ \pi t^{\frac{1}{2}}\tfrac{1}{4}(1+\varepsilon)\pi^{-1} \sum_{\eta=\pm1} \sum_{\substack{b>2\sigma+1\\ b\equiv\eta\tau(2)}} (b-1)\tfrac{1}{2}\ \overline{b^{1}_{\eta\tau}\delta_{t^{\frac{1}{2}}}}\ \mu^{\eta\ \eta}(\tfrac{1}{2}(b-1))$$

$$= \tfrac{1}{2}t^{\frac{1}{2}} \sum_{\eta=\pm 1} \langle \mu^{\varepsilon\eta\ \eta},\ b^{\varepsilon}_{\eta\tau}\delta_{t^{\frac{1}{2}}} \rangle.$$

We have seen:

14.5.4. Lemma. If $\mu \in {}^{p}R^{a}_{\sigma}$, then

(14.5.29) $\mu^{1} + \delta\mu^{-1} \in {}^{p}_{\delta}F^{a}_{\tau,\sigma}$ if $\tau = 0,\ 1,\ \delta = \pm 1$

(14.5.30) $\mu^{\delta\ \eta} \in {}^{p}_{\delta\eta}F^{a}_{\eta\tau,\sigma}$ if $0 < \tau < 1,\ \delta,\eta = \pm 1$.

14.5.5. Lemma. Let $a \in \mathbb{R}$, $p = 1$ or 2, $\delta,\eta = \pm 1$, $\sigma > \frac{1}{2}$; let

(14.5.31) $\phi \in {}^{p}_{\delta}F^{a}_{\tau,\sigma}$ if $\tau = 0,\ 1,$

$\phi \in {}^{p}_{\delta\eta}F^{a}_{\eta\tau,\sigma}$ if $0 < \tau < 1.$

Define μ in the case $\tau = 0,\ 1$ by

(14.5.32) $\mu(s) = \frac{1}{2}\begin{pmatrix} \phi(s) & \\ & \delta\phi(s) \end{pmatrix}$ $|\mathrm{Re}\ s| \leq \sigma$

$\mu_b = \begin{cases} \frac{1}{2}\phi(\frac{1}{2}(b-1)) & \text{if } \delta = 1 \\ 0 & \text{if } \delta = -1 \end{cases}$ $\Big\}$ $b \in Y_{\tau}$ or $b = \tau = 1;$

and in the case $0 < \tau < 1$ by

(14.5.33) $\mu^{\delta\eta}(s) = \phi(s)$
the other matrix elements of $\mu(s)$ are zero $\Big\}$ $|\mathrm{Re}\ s| \leq \sigma$

$\mu_b = \begin{cases} \phi(\frac{1}{2}(b-1)) & b \equiv \delta\tau(2),\ \delta = \eta \\ 0 & \text{otherwise} \end{cases}$ $\Big\}$ $b \in Y_{\tau},\ b > 0$

Then $\mu \in {}^{p}R^{a}_{\sigma}$.

Proof. This is checked in a straightforward way using definition 14.2.7, proposition 14.4.1 and lemma 14.4.4

14.5.6. Lemma. Let ϕ and μ be as in lemma 14.5.5. Let $t > 0$, $a > 2$, $p = 1$. Then in the case $\tau = 0,\ 1$:

(14.5.34) $\langle \mu,\ \delta(t,\ \delta) \rangle = \frac{1}{2}t^{\frac{1}{2}}(b^{\delta}_{\tau})^{\leftarrow}\ \phi(t^{\frac{1}{2}})$

and in the case $0 < \tau < 1$:

(14.5.35) $\langle \pi_{\eta}\mu,\ \delta(t,\ \delta\eta) \rangle = \frac{1}{2}t^{\frac{1}{2}}(b^{\delta\eta}_{\eta\tau})^{\leftarrow}\phi(t^{\frac{1}{2}})$

and if $\delta = 1$ in the case $\tau = 0,\ 1$ and $\eta = \delta$ in the case $0 < \tau < 1$

(14.5.36) $\langle \pi_{\delta}\mu,\ 1 \rangle = \frac{1}{2\pi} \langle \phi,\ 1 \rangle$

Proof. Work out (14.5.5) for this case. Use proposition 14.2.8 to obtain (14.5.34) and (14.5.35).

15. Scalar product of Poincaré series.

15.1. Introduction.

In section 8 we studied the operator Θ_α. For $\alpha \in \Lambda^1$ it acts on ΦH_σ^τ with $\sigma > \frac{1}{2}$. The resulting functions $\Theta_\alpha \Phi\xi$ with $\xi \in H_\sigma^\tau$ are square integrable. I call them Poincaré series; see subsection 15.2.

The purpose of section 15 is to compute in two ways $\langle \Theta_\alpha\Phi\xi, \Theta_\beta\Phi\theta\rangle$ with $\alpha,\beta \in \Lambda^1$, $\xi,\theta \in H_\sigma^\tau$. The equality so obtained is the basis of the sum formula to be derived in section 16. This equality turns up in all proofs of the sum formula known to me. In [1] it is disguised as a formula for Fourier coefficients of Poincaré series.

The first way of computing $\langle \Theta_\alpha\Phi\xi,\Theta_\beta\Phi\theta\rangle$ uses propositions 8.3.3 and 8.5.1. It gives an expression containing Kloosterman sums; see subsection 15.3.

The other way uses the spectral decomposition of $L^2(\Gamma\backslash G,\chi)$. To get the part due to ${}^cL^2$ I need the Eisenstein transform of Poincaré series. In 15.4 we see that this transform is described by the Fourier coefficients of the Eisenstein model. To describe the contribution of the discrete subspaces ${}^eL^2(s_0)$ and the cuspidal spaces, I introduce the notion of Poincaré elements in spaces of square integrable automorphic models. The Fourier coefficients $T \mapsto \alpha(U,T)$ are linear forms on $A(U)$; the restrictions to $A^{sq}(U)$ may be given by elements of $A^{sq}(U)$, for $A^{sq}(U)$ has a scalar product. These elements I call Poincaré elements. These Poincaré elements I study in 15.5. In 15.6 I give the second formula for $\langle \Theta_\alpha\Phi\xi,\Theta_\beta\Phi\theta\rangle$. This formula contains the $\underline{f}_\alpha$ discussed in definition 9.5.1 and proposition 11.7.16, and scalar products of Poincaré elements.

15.2. Poincaré series.

Take $\sigma > \frac{1}{2}$, $2\sigma+1 \not\equiv \pm\tau(2)$. Let $\alpha \in \Lambda^1$ and $\xi \in H_\sigma^\tau$ be given. In proposition 13.5.5 we see that

$$(15.2.1) \qquad \Phi\xi(n(x)g) = e^{\frac{1}{2}ix}\Phi\xi(g)$$

$$(15.2.2) \qquad |\Phi\xi(a(y)k)| \ll \begin{cases} y^{\frac{1}{2}+\sigma} & y \downarrow 0 \\ y^{\frac{1}{2}-\sigma} & y \to \infty \end{cases}$$

uniformly for $k \in K$. So by proposition 8.2.3 the function $\Theta_\alpha\Phi\xi$ is well defined and bounded. Hence $\Theta_\alpha\Phi$ maps H_σ^τ into $L^2(\Gamma\backslash G,\chi)$. It intertwines the $\underline{g}$-K-action in H_σ^τ and the pointwise $\underline{g}$-K-action in $C^\infty(G)$. By lemma 10.2.4 this pointwise action on $\Theta_\alpha\Phi H_\sigma^\tau$ coincides with the $\underline{r}(\underline{g})$-$\underline{r}(K)$-action in L^2. We have proved:

15.2.1. Proposition. Let $\sigma > \frac{1}{2}$, $2\sigma+1 \not\equiv \pm\tau(2)$; $\alpha \in \Lambda^1$. Then $\Theta_\alpha\Phi$ maps H_σ^τ into the bounded functions in $L^2(\Gamma\backslash G,\chi)$. It is a $\underline{g}$-K-operator.

I call the elements of $\Theta_\alpha \Phi H_\sigma^\tau$ Poincaré series. To justify this name take $b \in Y_\tau$, $b > 2$ and suppose that $2\sigma+1 < b$. Take

(15.2.3) $$\begin{aligned} \xi(s) &= 0 \\ \xi_b &= \phi_b[b] \\ \xi_{b_1} &= 0 \qquad \text{if } b_1 \neq b. \end{aligned}$$

Clearly $\xi \in H_\sigma^\tau$, and by proposition 13.5.5, (13.5.10)

(15.2.4) $$\Phi\xi = \Gamma(b)^{-\frac{1}{2}}\Gamma(b-1)^{-\frac{1}{2}} D^b \phi_b[b].$$

In (8.2.18)-(8.2.22) we have seen that $\Theta_\alpha D^b \phi_b[b]$ corresponds to classical holomorphic Poincaré series.

15.3. Expansion using Kloosterman sums.

15.3.1. Proposition. $\sigma > \frac{1}{2}$, $2\sigma+1 \not\equiv \pm\tau(2)$. Let $\alpha,\beta \in \Lambda^1$ and $\xi,\theta \in H_\sigma^\tau$.

(15.3.1) $$\begin{aligned} &\langle \Theta_\alpha \Phi\xi, \Theta_\beta \Phi\theta \rangle = \\ &4\pi|n_\alpha| \delta_{\alpha\beta} \langle \xi, \pi_{\varepsilon(\beta)}\theta \rangle \\ &+ e^{-\frac{1}{2}\pi i \tau(\varepsilon(\beta)+1)} \sum_{c \in C_{\alpha\beta}} S(\alpha,\beta;c) \ \langle \xi, \delta(16\pi^2|n_\alpha n_\beta|c^{-2}, \varepsilon(\alpha)\varepsilon(\beta))\pi_{\varepsilon(\beta)}\theta \rangle \end{aligned}$$

Proof. By (15.2.2) and the boundedness of Poincaré series we may apply proposition 8.3.3 to obtain

(15.3.2) $$\langle \Theta_\alpha \Phi\xi, \Theta_\beta \Phi\theta \rangle = 4\pi|n_\alpha| \langle \Phi\xi, F_\alpha \Theta_\beta \Phi\theta \rangle.$$

By proposition 8.5.1:

(15.3.3) $$\begin{aligned} F_\alpha \Theta_\beta \Phi\theta = {} & \delta_{\alpha\beta} P_{\varepsilon(\beta)} \Phi\theta + (4\pi|n_\alpha|)^{-1} e^{\frac{1}{2}\pi i \tau(\varepsilon(\beta)-1)} \sum_{c \in C_{\beta\alpha}} S(\beta,\alpha;c) \\ & \cdot {}_{\frac{1}{2}}V^{\varepsilon(\alpha)\varepsilon(\beta)}(16\pi^2|n_\alpha n_\beta|c^{-2}) P_{\varepsilon(\beta)} \Phi\theta. \end{aligned}$$

From the definitions of $P_\varepsilon, \pi_\varepsilon$ and Φ (see definition 8.2.1, definition 9.4.1, (14.4.9) and proposition 13.5.5) follows that

(15.3.4) $$P_\varepsilon \Phi = \Phi \pi_\varepsilon \qquad \varepsilon = \pm 1.$$

So the first term of the right hand side is obtained from the unitarity of Φ, see (13.5.12). The series in the right hand side of (15.3.3) converges absolutely and the series of absolute values gives a bounded function on G by proposition 8.2.3 and definition 7.3.2. So we obtain as second term in the expression for $\langle \Theta_\alpha \Phi\xi, \Theta_\beta \Phi\theta \rangle$:

$$(15.3.5)\qquad e^{\frac{1}{2}\pi i\tau(1-\varepsilon(\beta))}\sum_{c\in C_{\beta,\alpha}}\overline{S(\beta,\alpha;c)\langle\Phi\xi,{}_{\frac{1}{2}}V^{\varepsilon(\alpha)\varepsilon(\beta)}(16\pi^2|n_\alpha n_\beta|c^{-2})P_{\varepsilon(\beta)}\Phi\theta\rangle}$$

$$= e^{-\frac{1}{2}\pi i\tau(\varepsilon(\beta)+1)}\sum_{c\in C_{\alpha\beta}} S(\alpha,\beta;c)\langle\xi,\delta(16\pi^2|n_\alpha n_\beta|c^{-2},\varepsilon(\alpha)\varepsilon(\beta))\pi_{\varepsilon(\beta)}\theta\rangle$$

by proposition 8.4.5, the unitarity of Φ, (15.3.4) and corollary 14.3.3.

Remark that in general

$$(15.3.6)\qquad \sum_{c\in C_{\beta\alpha}} S(\beta,\alpha;c)\delta(16\pi^2|n_\alpha n_\beta|c^{-2},\varepsilon(\alpha)\varepsilon(\beta))\pi_{\varepsilon(\beta)}\theta$$

does not converge to give an element of H^τ_σ.

15.4. Eisenstein transform of Poincaré series.

Let $\alpha\in\Lambda^1$, $\xi\in H^\tau_\sigma$, $\sigma>\frac{1}{2}$, $2\sigma+1\not\equiv\pm\tau(2)$.

By the boundedness of $\Theta_\alpha\Phi\xi$ (see proposition 15.2.1) definition 12.4.1 of the Eisenstein transform $\eta\Theta_\alpha\Phi\xi(s)$ makes sense provided $|\mathrm{Re}\ s|<\frac{1}{2}$ and $\underline{E}$ is holomorphic at $-\bar{s}$.

15.4.1. Proposition. $\sigma>\frac{1}{2}$, $2\sigma+1\not\equiv\pm\tau(2)$; $\alpha\in\Lambda^1$, $\xi\in H^\tau_\sigma$.

The Eisenstein transform $\eta\Theta_\alpha\Phi\xi$ is for $|\mathrm{Re}\ s|<\frac{1}{2}$, $\underline{E}$ holomorphic at $-\bar{s}$, given by

$$(15.4.1)\qquad \eta\Theta_\alpha\Phi\xi(s) = 4\pi|n_\alpha|\pi_1(X^{\varepsilon(\alpha)})^*(s)\xi(s)\otimes\underline{f}_\alpha(-s).$$

Proof. Let ψ be in the standard basis of H^τ and $\lambda\in\Lambda^0$. Then by propositions 8.3.3 and 9.5.2:

$$(15.4.2)\qquad \langle\Theta_\alpha\Phi\xi,\underline{E}(-\bar{s})\psi(-\bar{s})\otimes\underline{e}_\lambda\rangle =$$

$$4\pi|n_\alpha|((\underline{f}_\alpha(-s),\underline{e}_\lambda))\langle\Phi\xi,W(-\bar{s})X^{\varepsilon(\alpha)}(-\bar{s})\pi_1\psi(-\bar{s})\rangle.$$

So by definition 12.4.1:

$$(15.4.3)\qquad \eta\Theta_\alpha\Phi\xi(s) =$$

$$4\pi|n_\alpha|\sum_\psi\langle\Phi\xi,W(-\bar{s})X^{\varepsilon(\alpha)}(-\bar{s})\pi_1\psi(-\bar{s})\rangle\psi(s)\otimes\underline{f}_\alpha(-s).$$

If $\tau=0,1$ then one easily checks from definition 6.6.2 that for each element ψ of the standard basis there is a complex number z_ψ such that:

$$(15.4.4)\qquad X^{\varepsilon(\alpha)}(-\bar{s})\psi(-\bar{s}) = \overline{z_\psi}\ \psi(-\bar{s})$$

$$(X^{\varepsilon(\alpha)})^*(s)\psi(s) = z_\psi\ \psi(s).$$

For $\tau=0,1$ the proposition follows from (13.5.15) and (13.4.1). In the case $0<\tau<1$ remark that

(15.4.5) $$X^{\varepsilon(\alpha)}(-\bar{s})\phi_r^1(-\bar{s}) = e^{-\frac{1}{2}\pi i\tau}c_{\varepsilon(\alpha)}(\bar{s})\phi_r^{\varepsilon(\alpha)}(-\bar{s})$$

$$(X^{\varepsilon(\alpha)})^*(s)\phi_r^{\delta}(s) = \begin{cases} e^{\frac{1}{2}\pi i\tau}c_{\varepsilon(\alpha)}(s)\phi_r^1(s) & \text{if } \delta = \varepsilon(\alpha) \\ \text{a multiple of } \phi_r^{-1}(s) & \text{if } \delta = -\varepsilon(\alpha). \end{cases}$$

So

(15.4.6) $$\begin{aligned} &\sum_{\psi} \langle \Phi\xi, W(-\bar{s})X^{\varepsilon(\alpha)}(-\bar{s})\pi_1\psi(-\bar{s})\rangle\, \psi(s) \\ &= \sum_{r\equiv\pm\tau(2)} \langle \Phi\xi, W(-\bar{s})\phi_r^{\varepsilon(\alpha)}(-\bar{s})e^{-\frac{1}{2}\pi i\tau}c_{\varepsilon(\alpha)}(\bar{s})\rangle\, \phi_r^1(s) \\ &= \sum_{r\equiv\pm\tau(2)} \langle \Phi\xi, W(-\bar{s})\phi_r^{\varepsilon(\alpha)}(-\bar{s})\rangle\, (X^{\varepsilon(\alpha)})^*(s)\phi_r^{\varepsilon(\alpha)}(s) \\ &= \pi_1(X^{\varepsilon(\alpha)})^*(s)\xi(s) \end{aligned}$$

15.4.2. Corollary. *$\eta\Theta_\alpha\Phi\xi$ has a meromorphic extension to* $\{s \in \mathbb{C} \mid |\mathrm{Re}\, s| \leqslant \sigma\}$; *it is holomorphic on* $\mathrm{Re}\ s = 0$, $s \neq 0$; *if* $s_0 \in S^e$ *and* $s_0 \neq \frac{1}{2}$, *then it has at most a first order pole at* $-s_0$.

Proof. Most of the assertions follow directly from proposition 11.7.16 and the definition of $(X^{\varepsilon})^*$ and H_σ^τ. The only thing to worry about is the behaviour at $-s_0 = \frac{1}{2}(\tau-1)$. By proposition 11.7.16 we know that $\underline{f}_\alpha(-s)$ is holomorphic at this point, but in the case $0 < \tau < 1$ both $(X^{\varepsilon(\alpha)})^*(s)$ and $\xi(s)$ might contribute a pole. But

(15.4.7) $$\pi_1(X^{\varepsilon(\alpha)})^*(s)\xi(s) = \begin{cases} e^{\frac{1}{2}\pi i\tau}\pi_1\xi(s) & \text{if } \varepsilon(\alpha) = 1 \\ e^{\frac{1}{2}\pi i\tau}c_{-1}(s)Y^{-1}\pi_{-1}\xi(s) & \text{if } \varepsilon(\alpha) = -1, \end{cases}$$

and $\pi_{-1}\xi$ is holomorphic at $\frac{1}{2}(\tau-1)$.

15.4.3. Remark. If $0 < \tau < 1$, $s_0 = \frac{1}{2}(1-\tau)$, $\varepsilon(\alpha) = -1$, we have even $\eta\Theta_\alpha\Phi\xi$ holomorphic at s_0, for $\underline{f}_\alpha(s_0) = 0$ as follows from corollary 11.7.14 and the fact that (9.5.3) stays valid after continuation.

15.5. Poincaré elements.

Let U be an irreducible $\underline{g}$-K-space with unitary structure. In proposition 10.3.4 we have seen that $A^{sq}(U)$ is a Hilbert space. Let $\alpha \in \Lambda^1$ then $\alpha(U,T)$, defined in proposition 7.4.2 is linear in T, so there is a unique element $p_\alpha(U) \in A^{sq}(U)$ such that

(15.5.1) $$\langle T, p_\alpha(U)\rangle = \alpha(U,T) \text{ for all } T \in A^{sq}(U).$$

I call $p_\alpha(U)$ the Poincaré element of U of order α. It is possible to define Poincaré elements for linear subspaces of $A^{sq}(U)$. If V and W are subspaces of $A^{sq}(U)$, $V \perp W$, then the Poincaré element of $V \oplus W$ is equal to the sum of the

Poincaré elements of V and W. This may be applied to

(15.5.2) $$A^{sq}(U) = A^{0}(U) \oplus A^{esq}(U);$$

with the corresponding decomposition of Poincaré elements

(15.5.3) $$p_{\alpha}(U) = p^{0}_{\alpha}(U) + p^{e}_{\alpha}(U).$$

By choosing an orthonormal basis $T_1, T_2, \ldots, T_l$ of $A^{sq}(U)$ we get

(15.5.4) $$A^{sq}(U) = \bigoplus_{j=1}^{l} \mathbb{C}T_j.$$

Clearly the Poincaré element of $\mathbb{C}T_j$ is $\overline{\alpha(U,T_j)T_j}$, so we get

(15.5.5) $$p_{\alpha}(U) = \sum_{j=1}^{l} \overline{\alpha(U,T_j)T_j}.$$

For $p^{0}_{\alpha}(U)$ and $p^{e}_{\alpha}(U)$ we have similar formulas.

15.5.1. <u>Lemma</u>. <u>Let</u> $\sigma > \frac{1}{2}$, $2\sigma+1 \not\equiv \pm\tau(2)$, $\xi \in H^{\tau}_{\sigma}$, $\alpha \in \Lambda^{1}$. <u>Let</u> U <u>be an irreducible g-K-space with unitary structure</u>. <u>The orthogonal projection onto the closure of</u> $\{T\phi \mid T \in A^{sq}(U),\ \phi \in U\}$ <u>maps</u> $\Theta_{\alpha}\Phi\xi$ <u>onto</u>

$4\pi\lvert n_{\alpha}\rvert p_{\alpha}(H^{\delta}_{\tau}(s))\xi^{\delta}(s)$	<u>if</u> $U = H^{\delta}_{\tau}(s)$, $\tau = 0,1$, s <u>general</u>
$4\pi\lvert n_{\alpha}\rvert p_{\alpha}(H_{\tau}(s))\pi_{\varepsilon(\alpha)}\xi(s)$	<u>if</u> $U = H_{\tau}(s)$, $0 < \tau < 1$, Re $s = 0$
$4\pi\lvert n_{\alpha}\rvert c_{\varepsilon(\alpha)}(s)p_{\alpha}(H_{\tau}(s))\pi_{\varepsilon(\alpha)}\xi(s)$	<u>if</u> $U = H_{\tau}(s)$, $0 < \tau < 1$, $0 < s < \frac{1}{2}(1-\tau)$
0	<u>if</u> $U = D^{\delta}_{0}$
$4\pi\lvert n_{\alpha}\rvert p_{\alpha}(D_1)\xi_1$	<u>if</u> $U = D_1$
$4\pi\lvert n_{\alpha}\rvert\Gamma(b)\lvert b-1\rvert^{-\frac{1}{2}}p_{\alpha}(D_b)\xi_b$	<u>if</u> $U = D_b$, $b \in Y_{\tau}$, $b \equiv \varepsilon(\alpha)\tau(2)$
0	<u>if</u> $U = D_b$, $b \in Y_{\tau}$, $b \not\equiv \varepsilon(\alpha)\tau(2)$

<u>Proof</u>. We take $T \in A^{sq}(U)$, $\phi \in U$. By proposition 8.3.3

(15.5.6) $$\langle \Theta_{\alpha}\Phi\xi, T\phi\rangle = 4\pi\lvert n_{\alpha}\rvert\langle \Phi\xi, F_{\alpha}T\phi\rangle.$$

We work this out in the cases of the lemma, using proposition 7.4.2, (13.5.15), (13.4.1), (13.5.16) and (13.4.2). From the results the statements concerning the projection of $\Theta_{\alpha}\Phi\xi$ will be clear.

If $U = H^{\delta}_{\tau}(s)$, $\tau = 0,1$, s general:

(15.5.7) $$\langle \Phi\xi, F_{\alpha}T\phi\rangle = \overline{\alpha(H^{\delta}_{\tau}(s),T)}\langle \Phi\xi, W^{\delta}_{\tau}(s)\phi\rangle =$$

$$(15.5.8)\qquad \langle p_\alpha(H_\tau^\delta(s)),T\rangle\{\xi^\delta(-\bar{s}),\phi\}_{-\bar{s}} =$$

$$\langle p_\alpha(H_\tau^\delta(s)),T\rangle\{\iota_\tau^\delta(\bar{s})\xi^\delta(\bar{s}),\phi\}_{-\bar{s}} =$$

$$(15.5.9)\qquad \langle p_\alpha(H_\tau^\delta(s)),T\rangle\{\xi^\delta(\bar{s}),\iota_\tau^\delta(s)\phi\}_{\bar{s}}$$

If Re $s = 0$ use (15.5.8) and (3.4.3); if $0 < s < \frac{1}{2}$ use (15.5.9) and (3.4.4). The case $0 < \tau < 1$ goes similarly. Now let $U = D_b^{(\delta)}$. For the cases $b = 0$ and $b \neq \varepsilon(\alpha)\pi(2)$ see propositions 7.4.2 and 7.4.3 to obtain the vanishing of $F_\alpha T$. In the other cases:

$$(15.5.10)\qquad \langle \Phi\xi, F_\alpha T\phi\rangle = \overline{\alpha(D_b,T)}\langle \Phi\xi, D^b\phi\rangle$$

$$= \langle p_\alpha(D_b),T\rangle\Gamma(b)a_b^{-1}\langle \xi_b,\phi\rangle_b.$$

In the sum formula the $p_\alpha(U)$ will occur. The term $p_\alpha^e(U)$ in the decomposition (15.5.3) may be expressed using the $\underline{f}_\alpha(s)$, as is shown in proposition 15.5.2. If one substitutes this into the sum formula, only the cuspidal Poincaré elements $p_\alpha^0(U)$ are left.

15.5.2. Proposition. Let $s_0 \in S^e$, $\alpha \in \Lambda^1$.
If $\tau = 0$, $0 < s_0 < \frac{1}{2}$, $\delta = \pm 1$, then

$$(15.5.11)\qquad p_\alpha^e(H_0^\delta(s_0)) = (\delta,\varepsilon(\alpha))(\underline{E}_{s_0}(\underline{N}_{s_0} \operatorname{res}_{s\to s_0} \underline{f}_\alpha(s)))|H_0^\delta(s_0).$$

If $0 < \tau < 1$, $0 < s_0 < \frac{1}{2}(1-\tau)$, then

$$(15.5.12)\qquad p_\alpha^e(H_\tau(s_0)) = c_{\varepsilon(\alpha)}(-s_0)e^{\frac{1}{2}\pi i\tau}(\underline{E}_{s_0}(\underline{N}_{s_0} \operatorname{res}_{s\to s_0} \underline{f}_\alpha(s)))|\pi_1 H^\tau(s_0)$$

If $s_0 = \frac{1}{2}(1-\tau)$, $0 < \tau < 1$, $\varepsilon(\alpha) = 1$, then

$$(15.5.13)\qquad p_\alpha^e(D_\tau)\iota^*_{2-\tau} = -\Gamma(1-\tau)e^{\frac{1}{2}\pi i\tau}(\underline{E}_{s_0}(\underline{N}_{s_0}\underline{f}_\alpha(s_0)))|\pi_1 H^\tau(s_0)$$

15.5.3. Lemma. $s_0 \in S^e$, $\underline{e}$, $\underline{e}_1 \in E(s_0)$. Let

$$(15.5.14)\qquad U = \begin{cases} H_0^\delta(s_0) & , \delta = \pm 1, \text{ if } \tau = 0,\ 0 < s_0 < \frac{1}{2} \\ H_\tau(s_0) & \text{if } 0 < \tau < 1,\ 0 < s_0 < \frac{1}{2}(1-\tau) \\ D_\tau & \text{if } 0 < \tau < 1,\ s_0 = \frac{1}{2}(1-\tau). \end{cases}$$

Let

$$(15.5.15)\qquad T = \underline{E}_{s_0}\underline{e}|H_\tau^{(\delta)}(s_0) \qquad \text{if } 0 < s_0 < \frac{1}{2}(1-\tau)$$

$$T\iota^*_{2-\tau} = \underline{E}_{s_0}\underline{e} \qquad \text{if } s_0 = \frac{1}{2}(1-\tau),$$

and let T_1 correspond to $\underline{e}_1$ in the same way. Then

(15.5.16) $$\langle T,T_1\rangle = \begin{cases} ((\underline{M}_{s_0}\underline{e},\underline{e}_1)) & \text{if } 0 < s_0 < \tfrac12(1-\tau) \\ -\pi^{-1}\sin\pi\tau((\underline{M}_{s_0}\underline{e},\underline{e}_1)) & \text{if } s_0 = \tfrac12(1-\tau) \end{cases}$$

Proof. Let $\phi_1 \in \pi_1 H^{\tau}(s_0)$. By (12.2.10) we have

(15.5.17) $$\langle \underline{E}_{s_0}\underline{e}\phi_1, \underline{E}_{s_0}\underline{e}_1\phi_1\rangle = \{\phi_1, \iota_{s_0}\phi_1\}_{s_0}((\underline{M}_{s_0}\underline{e},\underline{e}_1)).$$

To compute $\langle T,T_1\rangle$ in the case $0 < s_0 < \frac12(1-\tau)$ we take $\phi_1 \in H^{\delta}_{\tau}(s_0)$, resp. $\pi_1 H^{\tau}(s_0)$ such that

(15.5.18) $$1 = [\phi_1,\phi_1]_{s_0} = \{\phi_1, \iota_{s_0}\phi_1\}_{s_0}$$

and use proposition 10.3.4:

(15.5.19) $$\langle T,T_1\rangle = \langle \underline{E}_{s_0}\underline{e}\phi_1, \underline{E}_{s_0}\underline{e}_1\phi_1\rangle.$$

This gives the lemma in this case. In the other case we take $\phi \in D_{\tau}$ with $\langle\phi,\phi\rangle_{\tau} = 1$. Choose $\phi_1 \in \pi_1 H^{\tau}(s_0)$ such that $\iota^*_{2-\tau}\phi_1 = \phi$. Then

(15.5.20) $$1 = B(\phi,\phi) = B(\iota^*_{2-\tau}\phi_1, \iota^*_{2-\tau}\phi_1) = \{\phi_1, \iota^*_{2-\tau}\phi_1\}_{\frac12(1-\tau)}$$
$$= -\pi(\sin\pi\tau)^{-1}\{\phi_1, \iota_{s_0}\phi_1\}_{s_0}.$$

We have

(15.5.21) $$\langle T,T_1\rangle = \langle T\phi, T_1\phi\rangle$$
$$= \langle \underline{E}_{s_0}\underline{e}\phi_1, \underline{E}_{s_0}\underline{e}_1\phi_1\rangle$$
$$= -\pi^{-1}\sin\pi\tau((\underline{M}_{s_0}\underline{e},\underline{e}_1)).$$

Proof of proposition 15.5.2; case $0 < s_0 < \frac12(1-\tau)$.
We try to find $\underline{e}_1 \in E(s_0)$, such that

(15.5.22) $$p^{e}_{\alpha}(H^{(\delta)}_{\tau}(s_0)) = \underline{E}_{s_0}\underline{e}_1 | H^{\delta}_0(s_0) \text{ or } \pi_1 H^{\tau}(s_0).$$

For $\underline{e} \in E(s_0)$ we find, with $U = H^{\delta}_{\tau}(s_0)$ or $\pi_1 H^{\tau}(s_0)$:

(15.5.23) $$\langle \underline{E}_{s_0}\underline{e}|U, p^{e}_{\alpha}(U)\rangle = \alpha(U, \underline{E}_{s_0}\underline{e}|U)$$

and on the other hand, by lemma 15.5.3,

$$\langle \underline{E}_{s_0}\underline{e}|U, p^{e}_{\alpha}(U)\rangle = ((\underline{M}_{s_0}\underline{e},\underline{e}_1)).$$

Application of corollary 11.7.12 completes the proof.
In the case $s_0 = \frac12(1-\tau)$ we try to find $\underline{e}_1 \in E(s_0)$ such that

$$p^{e}_{\alpha}(D_{\tau})\iota^*_{2-\tau} = \underline{E}_{s_0}\underline{e}_1 | \pi_1 H^{\tau}(s_0).$$

We have for $\underline{e} \in E(s_0)$ and $T \in A^{esq}(D_\tau)$ such that $T\iota^*_{2-\tau} = \underline{E}_{s_0}\underline{e}|\pi_1 H^\tau(s_0)$

(15.5.24) $\langle T, p^e_\alpha(D_\tau)\rangle = \alpha(D_\tau, T)$

and

(15.5.25) $\langle T, p^e_\alpha(D_\tau)\rangle = -\pi^{-1}\sin\pi\tau((\underline{M}_{s_0}\underline{e},\underline{e}_1))$.

By corollary 11.7.14

(15.5.26) $\alpha(H_\tau(s_0), T\iota^*_{2-\tau}) = e^{-\frac{1}{2}\pi i\tau}((\underline{e},\underline{f}_\alpha(s_0)))$.

To finish the proof we have to relate $\alpha(D_\tau,T)$ and $\alpha(H_\tau(s_0),T\iota^*_{2-\tau})$. By proposition 7.4.2

(15.5.27) $\alpha(D_\tau,T)D^\tau\iota^*_{2-\tau} = \alpha(H_\tau(s_0),T\iota^*_{2-\tau}) \operatorname{res}_{s\to s_0} W^1_\tau(s)$.

This amounts to

(15.5.28) $$\alpha(D_\tau,T)\Gamma(1-\tau)^{-1}W^1_\tau(\tfrac{1}{2}(\tau-1))\pi(\sin\pi(2-\tau))^{-1} \operatorname{res}_{s\to\frac{1-\tau}{2}} \iota^1_\tau(s)$$

$$= \alpha(H_\tau(s_0),T\iota^*_{2-\tau}) \operatorname{res}_{s\to s_0} W^1_\tau(-s)\iota^1_\tau(s)$$

see definition 5.4.4, (3.3.17), (3.3.18) and (5.5.3).

So we get

(15.5.29) $$-\pi^{-1}\sin\pi\tau((\underline{M}_{s_0}\underline{e},\underline{e}_1)) =$$

$$\alpha(H_\tau(s_0),T\iota^*_{2-\tau})\Gamma(1-\tau)\pi^{-1}\sin\pi\tau =$$

$$\pi^{-1}\sin\pi\tau\,\Gamma(1-\tau)e^{-\frac{1}{2}\pi i\tau}((\underline{e},\underline{f}_\alpha(s_0))).$$

This completes the proof.

Notations.

(15.5.30) $S = S^e \cup S^0$.

The spaces $U(s)$ and $A^0(s)$ have been defined in (10.6.4) - (10.6.6). We may define $A^{sq}(s)$ in the same way, using $A^{sq}(.)$ instead of $A^0(.)$.

For $\alpha \in \Lambda^1$ and $s \in S$ I define

(15.5.31) $$p_\alpha(s) = \begin{cases} p_\alpha(H^1_\tau(s)) & \text{if } \tau = 0,1,\ s \text{ general} \\ p_\alpha(H_\tau(s)) & \text{if } 0 < \tau < 1,\ s \text{ general} \\ p_\alpha(D_b) & \text{if } s = \tfrac{1}{2}(b-1),\ b \equiv \pm\tau(2),\ b > 0. \end{cases}$$

Remark that $p_\alpha(s)$ can be considered as an element of $A^{sq}(s)$.

In the case $\tau = 0,1$, s general the spaces $A^{sq}(H^1_\tau(s))$ and $A^{sq}(H^{-1}_\tau(s))$ are isomorphic as Hilbert spaces, see propositions 4.5.6 and 10.4.1. In (7.4.4) we see that under this correspondence the linear form $\alpha(H_\tau^{-1}(s),.)$ on $A^{sq}(H_\tau^{-1}(s))$ corresponds to $\varepsilon(\alpha)\alpha(H_\tau^1(s),.)$ on $A^{sq}(H_\tau^1(s))$. Hence

(15.5.32) $$\langle p_\alpha(H_\tau^{-1}(s)),p_\beta(H_\tau^{-1}(s))\rangle = \varepsilon(\alpha)\varepsilon(\beta)\langle p_\alpha(s),p_\beta(s)\rangle.$$

For this reason it is sufficient to consider only $A^{sq}(H_\tau^1(s))$.

15.5.4. Lemma. Let $\tau = 0,1$, $b \geq 1$, $b \equiv \tau(2)$. If $\alpha,\beta \in \Lambda^1$ and $\varepsilon(\alpha) \neq \varepsilon(\beta)$, then

(15.5.33) $$\langle p_\alpha(\tfrac12(b-1)),p_\beta(\tfrac12(b-1))\rangle = 0.$$

Proof. This follows easily from the Fourier series expansions (7.4.5) and (7.4.6), and proposition 10.4.2.

15.6. Spectral decomposition.

15.6.1. Proposition. $\sigma > \frac12$, $2\sigma+1 \not\equiv \pm\tau(2)$. Let $\alpha,\beta \in \Lambda^1$, and $\xi,\theta \in H_\sigma^\tau$. Denote for $\delta,\varepsilon = \pm1$, s general:

(15.6.1) $$t_{\delta\varepsilon}(s) = \begin{cases} e^{\frac12\pi i\tau(\varepsilon-\delta)} & \text{if } \tau = 0,1 \\ c_\delta(-s)c_\varepsilon(s) & \text{if } 0<\tau<1,\ \mathrm{Re}\, s = 0 \\ c_\varepsilon(s) & \text{if } 0<\tau<1,\ 0<s<\frac12(1-\tau). \end{cases}$$

Then

(15.6.2) $$\langle \Theta_\alpha\Phi\xi,\Theta_\beta\Phi\theta\rangle =$$
$$\frac{1}{2\pi i}\int_0^{i\infty} 16\pi^2|n_\alpha n_\beta|\,((\underline{f}_\alpha(s),\underline{f}_\beta(s)))\{\pi_1 X^{\varepsilon(\alpha)}*\xi,\pi_1 X^{\varepsilon(\beta)}*\theta\}(s)\,ds$$
$$+ \sum_{\substack{s\in S\\ s \text{ general}}} 16\pi^2|n_\alpha n_\beta|\,t_{\varepsilon(\alpha),\varepsilon(\beta)}(s)\langle p_\alpha(s),p_\beta(s)\rangle.$$
$$\cdot\{\pi_1 X^{\varepsilon(\alpha)}*\xi,\pi_1 X^{\varepsilon(\beta)}*\theta\}(s)$$
$$+ \sum_{\substack{b>0,\frac12(b-1)\in S\\ b\equiv\varepsilon(\alpha)\tau(2)\\ \varepsilon(\alpha)=\varepsilon(\beta)}} 16\pi^2|n_\alpha n_\beta|\,\Gamma(b)^2 a_b^{-2}\langle p_\alpha(\tfrac12(b-1)),p_\beta(\tfrac12(b-1))\rangle\,.$$
$$.\,\langle \xi_b,\theta_b\rangle_b$$

Remark. The sum over b only occurs if $\varepsilon(\alpha) = \varepsilon(\beta)$.

Proof. For the projections onto ${}^cL^2$ we have, by proposition 12.5.1, remark 12.5.4, proposition 15.4.1, (11.7.31) and (11.6.1):

$$(15.6.3)\qquad \langle(\Theta_\alpha\Phi\xi)_c,(\Theta_\beta\Phi\theta)_c\rangle =$$

$$\frac{1}{2\pi i}\int_0^{i\infty}\langle\eta\Theta_\alpha\Phi\xi(s),\eta\Theta_\beta\Phi\theta(s)\rangle_s ds$$

$$= \frac{16\pi^2|n_\alpha n_\beta|}{2\pi i}\int_0^{i\infty}((\underline{f}_\alpha(-s),\underline{f}_\beta(-s)))\{\pi_1 X^{\varepsilon(\alpha)}*\xi(s),\pi_1 X^{\varepsilon(\beta)}*\theta(s)\}_s ds$$

$$= \frac{16\pi^2|n_\alpha n_\beta|}{2\pi i}\int_0^{i\infty}((\underline{f}_\alpha(s),\underline{f}_\beta(s)))\{\pi_1 X^{\varepsilon(\alpha)}*\xi,\pi_1 X^{\varepsilon(\beta)}*\theta\}(s)ds.$$

Lemma 15.5.1 gives the projections of $\Theta_\alpha\Phi\xi$ and $\Theta_\beta\Phi\theta$ onto the various discrete subspaces of L^2. Let us first consider the case $\tau = 0,1$. If $s \in S$ is general the subspaces of type $H^1_\tau(s)$ and $H^{-1}_\tau(s)$ give the contribution

$$(15.6.4)\qquad \sum_{\delta=\pm1} 16\pi^2|n_\alpha n_\beta|\langle p_\alpha(H^\delta_\tau(s))\xi^\delta(s),p_\beta(H^\delta_\tau(s))\theta^\delta(s)\rangle$$

$$= 16\pi^2|n_\alpha n_\beta|\sum_{\delta=\pm1}\langle p_\alpha(H^\delta_\tau(s)),p_\beta(H^\delta_\tau(s))\rangle\ .$$

$$\cdot\begin{cases}\{\xi^\delta(s),\theta^\delta(s)\}_s & \text{if}\ \ \mathrm{Re}\, s = 0\\ \{\xi^\delta(s),\iota^\delta_\tau(s)\theta^\delta(s)\}_s & \text{if}\ \ 0 < s < \tfrac12,\ \tau = 0\end{cases}$$

$$= 16\pi^2|n_\alpha n_\beta|\langle p_\alpha(s),p_\beta(s)\rangle\sum_\delta(\delta,\varepsilon(\alpha)\varepsilon(\beta))\{\xi^\delta(s),\theta^\delta(-\bar{s})\}_s,$$

see (15.5.32). This equals the desired expression; see (6.6.3) and (14.4.16). The contribution in type D^δ_0 vanishes. Let $b \equiv \tau(2)$, $b \geq 1$. Take

$$(15.6.5)\qquad a_b = \begin{cases}1 & \text{if}\ \ b = 1\\ |b-1|^{\frac12} & \text{if}\ \ b > 1.\end{cases}$$

Then the contribution in type D_b is

$$(15.6.6)\qquad 16\pi^2|n_\alpha n_\beta|\Gamma(b)^2 a_b^{-2}\langle p_\alpha(D_b)\xi_b,p_\beta(D_b)\theta_b\rangle$$

$$= 16\pi^2|n_\alpha n_\beta|\Gamma(b)^2 a_b^{-2}\langle p_\alpha(\tfrac12(b-1)),p_\beta(\tfrac12(b-1))\rangle.$$

$$.\ \langle\xi_b,\theta_b\rangle_b\ .$$

This vanishes if $\varepsilon(\alpha) \neq \varepsilon(\beta)$, see lemma 15.5.4.

In the case $0 < \tau < 1$ we get for $\mathrm{Re}\, s = 0$, $s \in S$, the contribution

$$(15.6.7)\qquad 16\pi^2|n_\alpha n_\beta|\langle p_\alpha(H_\tau(s))\pi_{\varepsilon(\alpha)}\xi(s),p_\beta(H_\tau(s))\pi_{\varepsilon(\beta)}\theta(s)\rangle$$

$$= 16\pi^2|n_\alpha n_\beta|\langle p_\alpha(s),p_\beta(s)\rangle\{\pi_{\varepsilon(\alpha)}\xi(s),\pi_{\varepsilon(\beta)}\theta(s)\}_s.$$

From (6.6.3) and (14.4.16) follows that

(15.6.8) $$\pi_1 X^{\delta}*(s) = e^{\frac{1}{2}\pi i\tau} c_{\delta}(s) Y^{\delta} \pi_{\delta}.$$

So

(15.6.9) $$\{\pi_{\delta}\xi(s), \pi_{\varepsilon}\theta(s)\}_s =$$

$$(c_{\delta}(s)c_{\varepsilon}(-s))^{-1}\{\pi_1 X^{\delta}*\xi, \pi_1 X^{\varepsilon}*\theta\}(s)$$

This gives the desired expression. For $s \in S$, $0 < s < \frac{1}{2}(1-\tau)$:

(15.6.10) $$16\pi^2|n_{\alpha}n_{\beta}|c_{\varepsilon(\alpha)}(s)c_{\varepsilon(\beta)}(s)\langle p_{\alpha}(H_{\tau}(s))\pi_{\varepsilon(\alpha)}\xi(s), p_{\beta}(H_{\tau}(s))\pi_{\varepsilon(\beta)}\theta(s)\rangle$$

$$= 16\pi^2|n_{\alpha}n_{\beta}|c_{\varepsilon(\alpha)}(s)c_{\varepsilon(\beta)}(s)\langle p_{\alpha}(s), p_{\beta}(s)\rangle$$

$$\cdot \{\pi_{\varepsilon(\alpha)}\xi(s), \iota_{\tau}^{1}(s)\pi_{\varepsilon(\beta)}\theta(s)\}_s$$

$$= 16\pi^2|n_{\alpha}n_{\beta}|c_{\varepsilon(\alpha)}(s)\langle p_{\alpha}(s), p_{\beta}(s)\rangle\{\pi_{\varepsilon(\alpha)}\xi(s), \pi_{\varepsilon(\beta}(\iota\theta)(-s)\}_s$$

$$= 16\pi^2|n_{\alpha}n_{\beta}|c_{\varepsilon(\beta)}(s)\langle p_{\alpha}(s), p_{\beta}(s)\rangle\{\pi_1 X^{\varepsilon(\alpha)}*\xi, \pi_1 X^{\varepsilon(\beta)}*\theta\}(s)$$

Finally, let $b > 0$, $b \equiv \pm\tau(2)$. In lemma 15.5.1 we see that the contribution of type D_b is zero, unless $b \equiv \varepsilon(\alpha)\tau = \varepsilon(\beta)\tau$. The further computations are the same as those in the case $\tau = 0,1$.

16. Sum formula

16.1. Introduction

Propositions 15.3.1 and 15.6.1 give two ways to compute $\langle\Theta_\alpha\Phi\xi, \Theta_\beta\Phi\eta\rangle$. The resulting equality is the basis for the sum formula. The ξ and η are only auxiliary test functions. I replace ξ by $\mu\xi$ with $\mu \in {}^1R_\sigma^{2+\varepsilon}$. The sum formula is obtained by letting ξ and η tend to suitable limits. This is done in the subsections 16.2 and 16.3.

The sum formula (16.3.26) relates values of the multiplication operator $\pi_{\varepsilon(\alpha)}\mu$ in the various components of the spectral decomposition of $L^2(\Gamma\backslash G, \chi)$ to a sum in which occur Kloosterman sums and values of the kernel of the convolution operator associated to $\pi_{\varepsilon(\alpha)}\mu$.

In subsection 16.4 I reformulate the sum formula in two ways: the operator μ is described more explicitly by its matrix coefficients, and the scalar products of Poincaré elements are expressed by Fourier coefficients of automorphic forms on the upper half plane. This enables me to compare this sum formula with those derived by Kuznetsov and Proskurin.

16.2. Preparatory lemmas

This subsection serves to derive a formula for $\langle \xi, \delta(t, \varepsilon)\eta\rangle$, with $\xi, \eta \in H_\sigma^\tau$; see lemma 16.2.3. The importance of this formula is that the integrand and summand are $O(t^{\sigma+\frac{1}{2}})$.

16.2.1. <u>Lemma</u>. <u>Let</u> q <u>be as in lemma</u> 14.3.1; <u>let</u> $\xi, \eta \in H_\sigma^\tau$. <u>Take</u>

$$(16.2.1)\qquad \zeta(s) = \frac{2}{\sin 2\pi s} \sin \pi(\tfrac{1}{2}-s+\tfrac{1}{2}\tau) \sum_{\beta=\pm 1} q^\beta\{J_{-2s}^\beta\}\iota(-s)X^\beta(-s)\eta(-s).$$

<u>Then</u> $s \mapsto \{\xi, \zeta\}(s)\Gamma(2s)^{-1}\Gamma(-2s)^{-1}$ <u>is meromorphic on</u> $|\mathrm{Re}\ s| \leq \sigma$. <u>The only poles in</u> $\mathrm{Re}\ s \geq 0$ <u>are situated in</u> $s = \frac{1}{2}(b-1)$, $b \in Y_\tau$, <u>have at most first order and</u>

$$(16.2.2)\qquad \operatorname*{res}_{s\to\frac{1}{2}(b-1)} \{\xi, \zeta\}(s)\Gamma(2s)^{-1}\Gamma(-2s)^{-1} =$$

$$2e^{\frac{1}{2}\pi ib}\,\overline{q^1}\,\{J_{b-1}^1\}\langle\xi_b, \eta_b\rangle_b.$$

<u>Proof</u>. I write out ξ and η as in lemma 13.5.4.

$$(16.2.3)\qquad \xi(s) = \frac{1}{\sqrt{2}} \sum_{\substack{r\equiv\tau(2)\\ \varepsilon=\pm 1}} \Gamma(\tfrac{1}{2}+s-\tfrac{1}{2}\varepsilon r)\xi_r^\varepsilon(s),\quad \begin{cases} \frac{1}{\sqrt{2}}(\phi_r^1(s)+\varepsilon\phi_r^{-1}(s)) & \text{if } \tau = 0, 1\\[2ex] \phi_r^\alpha(s) & \text{if } 0<\tau<1,\ \varepsilon r \equiv \alpha\tau(2)\end{cases}$$

$$(16.2.4)\qquad \zeta(s) = \sqrt{2}(\sin 2\pi s)^{-1} \sum_{\beta=\pm 1} q^\beta\{J_{-2s}\} \sum_{\substack{r\equiv\tau(2)\\ \varepsilon=\pm 1}} \eta_r^\varepsilon(-s)\Gamma(\tfrac{1}{2}-s-\tfrac{1}{2}\varepsilon r).$$

$$\cdot\begin{cases} e^{-\frac12\pi i\tau\beta}\sin\pi(\tfrac12-s+\tfrac12\tau)\Gamma(\tfrac12+s-\tfrac12 r)\Gamma(\tfrac12-s-\tfrac12 r)^{-1}((-1)^\tau,\beta\varepsilon)\frac{1}{\sqrt2}(\phi_r^1(s)+\varepsilon\beta\phi_r^{-1}(s)) & \text{if } \tau=0,\,1\\ e^{-\frac12\pi i\tau\alpha}\sin\pi(\tfrac12+s-\tfrac12\alpha\beta\tau)\Gamma(\tfrac12+s-\tfrac12\varepsilon\beta r)\Gamma(\tfrac12-s-\tfrac12\varepsilon\beta r)^{-1}\phi_r^{\alpha\beta}(s) & \text{if } 0<\tau<1,\ \varepsilon r\equiv\alpha\tau(2).\end{cases}$$

Hence

(16.2.5)
$$\{\xi,\zeta\}(s)\Gamma(2s)^{-1}\Gamma(-2s)^{-1} =$$

$$-2s\pi^{-1}\sin 2\pi s\sum_{\beta=\pm1}\sum_{\substack{r\equiv\pm\tau\\ \varepsilon=\pm1}}\frac{1}{\sqrt2}\Gamma(\tfrac12+s-\tfrac12\varepsilon\beta r)\xi_r^{\varepsilon\beta}(s)\sqrt2(-\sin 2\pi s)^{-1}\,.$$

$$.\,\overline{q^\beta\{J^\beta_{2s}\}\eta_r^\varepsilon}(s)\Gamma(\tfrac12+s-\tfrac12\varepsilon r)\,.$$

$$\cdot\begin{cases} e^{\frac12\pi i\tau\beta}\sin\pi(\tfrac12+s+\tfrac12\tau)\Gamma(\tfrac12-s-\tfrac12 r)\Gamma(\tfrac12+s-\tfrac12 r)^{-1}((-1)^\tau,\beta\varepsilon)\\ e^{\frac12\pi i\tau\alpha}\sin\pi(\tfrac12-s-\tfrac12\alpha\beta\tau)\Gamma(\tfrac12-s-\tfrac12\varepsilon\beta r)\Gamma(\tfrac12+s-\tfrac12\varepsilon\beta r)^{-1}\end{cases}$$

$$= 2s\sum_{\beta=\pm1}\overline{q^\beta\{J_{2s}\}}\sum_{r,\varepsilon}\Gamma(\tfrac12+s-\tfrac12\varepsilon\beta r)\Gamma(\tfrac12+s-\tfrac12\varepsilon r)\xi_r^{\varepsilon\beta}(s)\,\overline{\eta_r^\varepsilon}(s)\,.$$

$$\cdot\begin{cases} e^{\frac12\pi i\tau\beta}((-1)^\tau,\varepsilon\beta)(-1)^{\frac12(\tau-r)}\Gamma(\tfrac12+s+\tfrac12 r)^{-1}\Gamma(\tfrac12+s-\tfrac12 r)^{-1}\\ e^{\frac12\pi i\tau\alpha}(-1)^{\frac12\beta(\alpha\tau-\varepsilon r)}\Gamma(\tfrac12+s+\tfrac12\varepsilon\beta r)^{-1}\Gamma(\tfrac12+s-\tfrac12\varepsilon\beta r)^{-1}\end{cases}$$

$$= 2s\sum_{\beta=\pm1}\overline{q^\beta\{J^\beta_{2s}\}}\sum_{r,\varepsilon}\Gamma(\tfrac12+s-\tfrac12\varepsilon r)\Gamma(\tfrac12+s+\tfrac12\varepsilon\beta r)^{-1}\,.$$

$$.\,\xi_r^{\varepsilon\beta}(s)\overline{\eta_r^\varepsilon}(s).e^{\frac12\pi i\varepsilon r}$$

The meromorphy is clear. Possible poles have first order, and occur only for $\beta = 1$. The factor s takes care of holomorphy at $s = 0$. Let $b > 1$, $b \in Y_\tau$.

(16.2.6)
$$\operatorname*{res}_{s\to\frac12(b-1)}\{\xi,\zeta\}(s)\Gamma(2s)^{-1}\Gamma(-2s)^{-1} =$$

$$(b-1)\overline{q^1\{J^1_{b-1}\}}\sum_{\substack{\varepsilon r\equiv b(2)\\ \varepsilon r\geq b}}(\tfrac12(\varepsilon r-b))!^{-1}(-1)^{\frac12(\varepsilon r-b)}\Gamma(\tfrac12(b+\varepsilon r))^{-1}\,.$$

$$.\,\xi_r^\varepsilon(\tfrac12(b-1))\overline{\eta_r^\varepsilon(\tfrac12(b-1))}.e^{\frac12\pi i\varepsilon r} = 2(b-1)e^{\frac12\pi i b}\,\overline{q^1}\{J_{b-1}\}.$$

$$.\sum_{\substack{\varepsilon r\equiv b(2)\\ \varepsilon r\geq b}}\tfrac12(\tfrac12(\varepsilon r-b))!^{-1}\Gamma(\tfrac12(b+\varepsilon r))^{-1}\xi_r^\varepsilon(\tfrac12(b-1))\overline{\eta_r^\varepsilon(\tfrac12(b-1))}$$

Comparison of lemma 13.5.4, (13.4.9) and (3.4.5) gives the lemma.

16.2.2. Lemma. ξ, η, ζ and q as in lemma 16.2.1.

(16.2.7)
$$\langle\xi, M_q\eta\rangle =$$

$$\frac{1}{2\pi i}\int_{\mathrm{Re}\ s=\sigma}\{\xi,\ \zeta\}(s)\Gamma(2s)^{-1}\Gamma(-2s)^{-1}ds$$

$$+\sum_{\substack{b>2\sigma+1\\ b\in Y_\tau}}\langle\xi_b,\ (M_q\eta)_b\rangle_b$$

Proof. According to (14.3.11) and definition 14.3.2:

(16.2.8) $$M_q\eta(s) = \iota(-s)\zeta(-s) + \zeta(s)$$

So we get

(16.2.9) $$\frac{1}{2\pi i}\int_0^{i\infty}\{\xi,\ M_q\eta\}(s)\Gamma(2s)^{-1}\Gamma(-2s)^{-1}ds$$

$$= \frac{1}{2\pi i}\int_{\mathrm{Re}\ s=0}\{\xi,\ \zeta\}(s)\Gamma(2s)^{-1}\Gamma(-2s)^{-1}ds,$$

use $\iota\xi = \xi$ and (3.5.17). The integrand goes to zero quickly enough to admit moving the line of integration to Re $s = \sigma$; see (13.5.3) and (14.4.4) applied to M_q. Lemma 16.2.1 gives the contribution of the poles. Comparison of (14.3.13), (14.2.17) and (13.5.6) gives the lemma.

16.2.3. Lemma. $\sigma > \frac{1}{2}$, $2\sigma + 1 \equiv \pm\tau(2)$; $\xi,\ \eta \in H_\sigma^\tau$; $t > 0$, $\varepsilon = \pm 1$. Then

(16.2.10) $$\langle\xi,\ \delta(t,\ \varepsilon)\eta\rangle =$$

$$\frac{1}{2\pi i}\int_{\mathrm{Re}\ s=\sigma}\frac{4}{\pi}\ s\ \sin\pi(\tfrac{1}{2}+s+\tfrac{1}{2}\tau)\pi t^{\frac{1}{2}}J_{2s}^{\varepsilon}(t^{\frac{1}{2}})\{\xi,\ X^{\varepsilon}\eta\}(-s)ds$$

$$+\ (\varepsilon+1)\pi t^{\frac{1}{2}}\sum_{\substack{b\in Y_\tau\\ b>2\sigma+1}} e^{\frac{1}{2}\pi i b}J_{b-1}^{1}(t^{\frac{1}{2}})\langle\xi_b,\ \eta_b\rangle_b.$$

Proof. $\delta(t,\ \varepsilon) = M_{d(t,\ \varepsilon)}$, with $d(t,\ \varepsilon)$ as defined in (14.3.5). Define ζ by taking $q = d(t,\ \varepsilon)$ in (16.2.1). Then lemma 16.2.2 gives

(16.2.11) $$\langle\xi,\ \delta(t,\ \varepsilon)\eta\rangle =$$

$$\frac{1}{2\pi i}\int_{\mathrm{Re}\ s=\sigma}\pi^{-1}4s\ \sin\pi(\tfrac{1}{2}+s+\tfrac{1}{2}\tau)\overline{d(t,\ \varepsilon)\{J_{2s}^{\varepsilon}\}}$$

$$\cdot\ \{\xi(s),\ \iota(\bar{s})X^{\varepsilon}(\bar{s})\eta(\bar{s})\}_s ds$$

$$+\sum_{b\in Y_\tau,\ b>2\sigma+1}\langle\xi_b,\ \eta_b\rangle_b\overline{2e^{-\frac{1}{2}\pi i b}d(t,\ \varepsilon)^1\{J_{b-1}^1\}}.$$

16.3. Proof of the sum formula

I fix $\sigma \in (\frac{1}{2},\ \frac{1}{2}(\tau+1))$ if $\tau \neq 0$ and $\sigma \in (\frac{1}{2},\ 3/2)$ if $\tau = 0$. So the only non-general points in $|\mathrm{Re}\ s| \leq \sigma$ are $\pm\frac{1}{2}(1-\tau)$. Propositions 15.3.1 and 15.6.1 together give an equality by expressing $\langle\Theta_\alpha\Phi\xi,\ \Theta_\beta\Phi\eta\rangle$ in two different ways. In the sum formula the test functions are elements of ${}^1R_\sigma^a$ with $a > 2$. So I replace in the equality ξ by $\mu\xi$. Now I want to get rid of ξ and η. This I do by taking ξ and η in a suitable way, parametrized by a small positive numer v and take the limit for $v \searrow 0$. In the expression in proposition 15.3.1 I shall move the limit inside the sum and the integral. After that I know that the limit of the expression in proposition 15.6.1 exists; this knowledge permits moving the limit inside these sums and integrals as well. The equality of both expressions for the limit is the sum formula.

Now I fix ξ_v and η_v. This choice turns out to work; probably other choices work as well. Let $0 < v < (2\sigma+1)^{-2}$. Fix for general s:

$$(16.3.1) \qquad \eta_v(s) = \begin{cases} e^{vs^2}\Gamma(-\tfrac12+s+\tfrac12\tau)(\phi^1_{2-\tau}(s)+\phi^{-1}_{2-\tau}(s)) & \text{if } \tau = 0,\ 1 \\ e^{vs^2}(\Gamma(\tfrac12+s-\tfrac12\tau)\phi^1_\tau(s) + \Gamma(-\tfrac12+s+\tfrac12\tau)\phi^{-1}_{2-\tau}(s)) & \text{if } 0 < \tau < 1 \end{cases}$$

$$\xi_v(s) = \begin{cases} \frac{1}{4\pi}\Gamma(\frac32 -s-\tfrac12\tau)\Gamma(\frac32 +s-\tfrac12\tau)\eta_v(s) & \text{if } \tau = 0,\ 1 \\ \frac14 e^{vs^2}\left(\frac{\Gamma(\frac12+s+\frac12\tau)}{\sin \pi(\frac12+s-\frac12\tau)}\phi^1_\tau(s) - \frac{\Gamma(\frac32 +s-\frac12\tau)}{\sin \pi(\frac12+s+\frac12\tau)}\phi^{-1}_{2-\tau}(s)\right) & \text{if } 0<\tau<1 \end{cases}$$

For $b \equiv \pm\tau(2)$, $b \geqslant 1$ if $\tau = 0,\ 1$, $b > 1-2\sigma$ if $0 < \tau < 1$ I take

$$(16.3.2) \qquad (\eta_v)_b = \begin{cases} \left.\begin{array}{ll} e^{\frac14 v(b-1)^2} a_b(\phi^1_b(\tfrac12(b-1)) + \phi^{-1}_b(\tfrac12(b-1))) & \text{if } \tau = 0,\ 1 \\ e^{\frac14 v(b-1)^2} a_b\phi^\gamma_b(\tfrac12(b-1)) & \text{if } 0 < \tau < 1,\ b \equiv \gamma\tau(2) \end{array}\right\} & \text{if } b \leqslant \frac{1}{\sqrt{v}} \\ 0 & \text{if } b > \frac{1}{\sqrt{v}} \end{cases}$$

$$(\xi_v)_b = \frac{1}{4\pi}\Gamma(b)(\eta_v)_b.$$

The following lemmas state properties of ξ_v and η_v. Of course I started with those properties and adapted ξ_v and η_v.

16.3.1. Lemma. $\xi_v,\ \eta_v \in H^\tau_\sigma$.

If $A(s)$ is a matrix with meromorphic elements on $|\mathrm{Re}\ s| \leqslant \sigma$ and if $A(s)$ is a diagonal matrix in the case $\tau = 0,\ 1$, then

$$(16.3.3) \qquad \{A\xi_v,\eta_v\}(s) = \frac{\pi}{4} e^{2vs^2}(\sin \pi(\tfrac12+s+\tfrac12\tau)\sin \pi(\tfrac12+s-\tfrac12\tau))^{-1} \text{ trace } A(s)$$

If $b \in Y_\tau$ or $b = \tau = 1$, then

$$(16.3.4) \qquad \langle(\xi_v)_b,\ (\eta_v)_b\rangle_b = \begin{cases} \frac{1}{4\pi}\varepsilon_\tau a_b^2 e^{\frac12 v(b-1)^2} & \text{if } b \leqslant v^{-\frac12} \\ 0 & \text{if } b > v^{-\frac12} \end{cases}$$

Proof. Most of the properties in definition 13.5.1 of H^τ_σ are easily checked. I shall check the singularities and (13.5.4).

In the case $\tau = 0,\ 1$ the even factor $\Gamma(\frac32 -s-\frac12\tau)\Gamma(\frac32+s-\frac12\tau)$ is holomorphic on $|\mathrm{Re}\ s| \leqslant \sigma$ and has the value $\Gamma(2-\tau)$ at $\frac12(2-\tau-1)$; so we need only consider η_v. The only pole of η_v in $|\mathrm{Re}\ s| \leqslant \sigma$ is at $s = \frac12(1-\tau)$ and has first order.

(16.3.5) $$\operatorname*{res}_{s\to\frac12(1-\tau)} \eta_v(s) = e^{\frac14 v(1-\tau)^2}(\phi^{1}_{2-\tau}(\tfrac12(1-\tau)) + \phi^{-1}_{2-\tau}(\tfrac12(1-\tau)))$$

$$= (\eta_v)_{2-\tau} \in D_{2-\tau},$$

as it should be. (Remark that $a_2 = (2-1)^{\frac12} = 1$ and $a_1 = 1$.)

(16.3.6) $$\iota(-s)\eta_v(-s) =$$

$$= e^{v(-s)^2}\Gamma(-\tfrac12-s+\tfrac12\tau)\Gamma(\tfrac12+s-\tfrac12(2-\tau))\Gamma(\tfrac12-s-\tfrac12(2-\tau))^{-1}(\phi^{(-1)^\tau}_{2-\tau}(s)+\phi^{-(-1)^\tau}_{2-\tau}(s))$$

$$= e^{vs^2}\Gamma(-\tfrac12+s+\tfrac12\tau)(\phi^{1}_{2-\tau}(s)+\phi^{-1}_{2-\tau}(s)) = \eta_v(s).$$

In the case $0 < \tau < 1$ I only consider ξ_v; the treatment of η_v is easier.

(16.3.7) $$\xi_v(s) = \tfrac14 e^{vs^2} \sum_{\delta=\pm1} \delta\Gamma(1+s-\tfrac12\delta(1-\tau))\sin\pi(\tfrac12+s-\tfrac12\delta\tau)^{-1}\phi^{\delta}_{1-\delta(1-\tau)}(s)$$

Let $\varepsilon = \pm1$.

(16.3.8) $$\operatorname*{res}_{s\to\frac12\varepsilon(1-\tau)} \xi_v(s) = \tfrac14 e^{\frac14 v(1-\tau)^2}$$

$$.\ (\varepsilon\Gamma(1) \operatorname*{res}_{s\to\frac12\varepsilon(1-\tau)} \sin\pi(\tfrac12+s-\tfrac12\varepsilon\tau).\phi^{\varepsilon}_{1-\varepsilon(1-\tau)}(\tfrac12\varepsilon(1-\tau))$$

$$-\varepsilon\Gamma(1+\varepsilon(1-\tau)) \operatorname*{res}_{s\to\frac12\varepsilon(1-\tau)} \sin\pi(\tfrac12+s+\tfrac12\varepsilon\tau).\phi^{-\varepsilon}_{1+\varepsilon(1-\tau)}(\tfrac12\varepsilon(1-\tau)))$$

$$= \tfrac14 e^{\frac14 v(1-\tau)^2}(0-\varepsilon\Gamma(\varepsilon(1-\tau)+1)(-\varepsilon)\pi^{-1}\phi^{-\varepsilon}_{1+\varepsilon(1-\tau)}(\tfrac12\varepsilon(1-\tau)))$$

$$= \begin{cases} \frac{1}{4\pi}e^{\frac14(\tau-1)^2}\Gamma(\tau)\phi^{1}_{\tau}(\tfrac12(\tau-1)) = a_\tau^{-1}(\xi_v)_\tau \in D_\tau, & \varepsilon = -1 \\ \frac{1}{4\pi}e^{\frac14(1-\tau)^2}\Gamma(2-\tau)\phi^{-1}_{2-\tau}(\tfrac12(1-\tau)) = a^{-1}_{2-\tau}(\xi_v)_{2-\tau} \in D_{2-\tau}, & \varepsilon = 1 \end{cases}$$

(16.3.9) $$\iota(-s)\xi_v(-s) =$$

$$\tfrac14 e^{v(-s)^2} \sum_{\delta=\pm1} \frac{\delta\Gamma(1-s-\frac12\delta(1-\tau))}{\sin\pi(1-s-\frac12\delta\tau)} \frac{\Gamma(\frac12+s-\frac12+\frac12\delta(1-\tau))}{\Gamma(\frac12-s-\frac12+\frac12\delta(1-\tau))}\phi^{\delta}_{1-\delta(1-\tau)}(s)$$

$$= \tfrac14 e^{vs^2} \sum_{\delta=\pm1} \frac{\delta\pi \qquad \phi^{\delta}_{1-\delta(1-\tau)}(s)}{\sin\pi(s+\frac12\delta(1-\tau))\delta\ \sin\pi(-s+\frac12\delta(1-\tau))\Gamma(-s+\frac12\delta(1-\tau))}$$

$$= \tfrac14 e^{vs^2} \sum_{\delta=\pm1} \frac{\Gamma(1+s-\frac12\delta(1-\tau))}{\sin\pi(\frac12+s-\frac12\delta\tau)\delta}\phi^{\delta}_{1-\delta(1-\tau)}(s) = \xi_v(s)$$

The computation leading to (16.3.3) is of a similar nature. (16.3.4) is easy.

16.3. 2 . Lemma. Let $\mu \in {}^1R^a_\sigma$ with $a > 2$; $\delta = \pm1$. Then

(16.3.10) $$\lim_{v\downarrow 0} \langle\mu\xi_v,\ \pi_\delta\eta_v\rangle = \langle\pi_\delta\mu,\ 1\rangle.$$

Proof. Using definition 13.5.3:

(16.3.11) $$\langle\mu\xi_v,\ \pi_\delta\eta_v\rangle =$$

$$\frac{1}{2\pi i} \int_0^{i\infty} \{\pi_\delta \mu \xi_v, \ \eta_v\}(s) |\Gamma(2s)|^{-2} ds$$

$$+ \sum_{\substack{b>1 \\ b\in Y_\tau}} (\pi_\delta \mu)_b \stackrel{<}{\,} (\xi_v)_b, \ (\eta_v)_b \stackrel{>}{\,}_b$$

$$= \frac{1}{2\pi i} \int_0^{i\infty} \ \mathrm{tr}(\pi_\delta \mu(s)) \ \frac{-4s \sin 2\pi s}{\pi (\cos 2\pi s + \cos \pi \tau)} \ \frac{\pi}{4} \ e^{2vs^2} ds$$

$$+ \sum_{\substack{b\in Y_{\tau} \\ 1 < b \leqslant v^{-\frac{1}{2}}}} \frac{1}{4\pi} \varepsilon_\tau \ e^{\frac{1}{2}v(b-1)^2} (b-1)(\pi_\delta \mu)_b,$$

see (16.3.3) and (16.3.4). To get the expression in (14.5.5) we have to take the limit for $v \downarrow 0$. As $|e^{2vs^2}| \leqslant 1$ for $s \in (0, i\infty)$ and $e^{\frac{1}{2}v(b-1)^2}$ is bounded for $1 < b \leqslant v^{-\frac{1}{2}}$, we may use Lebesgue's theorem.

16.3.3. <u>Lemma</u>. Let $a > 2$, $\mu \in {}^1R_\sigma^a$ <u>and</u> α, $\beta \in \Lambda^1$. <u>The following limit exists and is equal to the expression in the right hand side</u>:

(16.3.12)
$$\lim_{v \downarrow 0} \stackrel{<}{\,} \Theta_\alpha \Phi \mu \xi_v, \ \Theta_\beta \Phi \eta_v \stackrel{>}{\,} =$$

$$4\pi |n_\alpha| \delta_{\alpha\beta} \stackrel{<}{\,} \pi_{\varepsilon(\beta)} \mu, \ 1 \stackrel{>}{\,}$$

$$+ e^{-\frac{1}{2}\pi i \tau (\varepsilon(\beta)+1)} \sum_{c \in C_{\alpha\beta}} S(\alpha, \ \beta; \ c) \stackrel{<}{\,} \pi_{\varepsilon(\alpha)} \mu, \ \delta(16 \ \pi^2 |n_\alpha n_\beta| c^{-2}, \ \varepsilon(\alpha)\varepsilon(\beta)) \stackrel{>}{\,}$$

<u>Proof</u>. Using proposition 15.3.1 and the lemmas 16.2.3 and 16.3.1 we obtain

(16.3.13)
$$\stackrel{<}{\,} \Theta_\alpha \Phi \mu \xi_v, \ \Theta_\beta \Phi \eta_v \stackrel{>}{\,} =$$

$$4\pi |n_\alpha| \delta_{\alpha\beta} \stackrel{<}{\,} \mu \xi_v, \ \pi_{\varepsilon(\beta)} \eta_v \stackrel{>}{\,}$$

$$+ e^{-\frac{1}{2}\pi i \tau \varepsilon(\beta)} \sum_{c \in C_{\alpha\beta}} S(\alpha, \ \beta; \ c)[$$

$$\frac{1}{2\pi i} \int_{\mathrm{Re}\ s=\sigma} 4\pi^{-1}s \sin \pi(\tfrac{1}{2}+s+\tfrac{1}{2}\tau)\pi(16 \ \pi^2 |n_\alpha n_\beta| c^{-2})^{\frac{1}{2}} J_{2s}^{\varepsilon(\alpha)\varepsilon(\beta)}(4\pi |n_\alpha n_\beta|^{\frac{1}{2}}c^{-1}).$$

$$\cdot \tfrac{1}{4}\pi e^{2vs^2} \sin \pi(\tfrac{1}{2}+s+\tfrac{1}{2}\tau)^{-1} \sin \pi(\tfrac{1}{2}+s-\tfrac{1}{2}\tau)^{-1} \mathrm{trace}(\pi_{\varepsilon(\beta)} \overline{x^{\varepsilon(\alpha)\varepsilon(\beta)}}(s) \ {}^t\mu(-s))ds$$

$$+(\varepsilon(\alpha)\varepsilon(\beta)+1)\pi(4\pi |n_\alpha n_\beta|^{\frac{1}{2}}c^{-1}) \sum_{\substack{b\in Y_\tau \\ v^{-\frac{1}{2}} \geqslant b > 2\sigma+1}} e^{\frac{1}{2}\pi i b} J_{b-1}^1(4\pi |n_\alpha n_\beta|^{\frac{1}{2}}c^{-1})(\pi_{\varepsilon(\beta)} \mu)_b \ .$$

$$\cdot \frac{1}{4\pi} \varepsilon_\tau(b-1)e^{\frac{1}{2}v(b-1)^2}]$$

The limit of the first term in the right hand side of (16.3.13) is taken care of by lemma 16.3.2. In the other term the limit may be taken inside the sum and the integral. For the integrand we get the estimate

(16.3.14) $\quad c^{-1-2\sigma}|s|(1+|\text{Im } s|)^{-2\sigma-\frac{1}{2}-a}$

see (14.4.4) and (14.2.4). In the sum the terms are bounded by a multiple of

(16.3.15) $\quad c^{-b}\Gamma(b-1)^{-1}.$

Using proposition 8.5.3 we see that we have a majorant enabling us to apply Lebesgue's theorem. The limit equals

(16.3.16)
$$4\pi|n_\alpha|\delta_{\alpha\beta}\langle\pi_{\varepsilon(\beta)}\mu,\ 1\rangle +$$
$$e^{-\frac{1}{2}\pi i\tau(\varepsilon(\beta)+1)}\sum_{c\in C_{\alpha\beta}} S(\alpha,\ \beta;\ c)\pi(4\pi|n_\alpha n_\beta|^{\frac{1}{2}}c^{-1}).$$
$$[\frac{1}{2\pi i}\int_{\text{Re } s=\sigma} s\ \sin\pi(\tfrac{1}{2}+s-\tfrac{1}{2}\tau)^{-1}J_{2s}^{\varepsilon(\alpha)\varepsilon(\beta)}(4\pi|n_\alpha n_\beta|^{\frac{1}{2}}c^{-1}).$$
$$.\text{trace}(\pi_{\varepsilon(\beta)}\overline{X^{\varepsilon(\alpha)\varepsilon(\beta)}}(s)^t\mu(-s))ds$$
$$+\tfrac{1}{4}(\varepsilon(\alpha)\varepsilon(\beta)+1)\pi^{-1}\varepsilon_\tau\sum_{\substack{b\in Y_\tau\\ b>2\sigma+1}} a_b^2 e^{\frac{1}{2}\pi i b}J_{b-1}(4\pi|n_\alpha n_\beta|^{\frac{1}{2}}c^{-1})(\pi_{\varepsilon(\beta)}\mu)_b]$$

The lemma is obtained by comparison with proposition 14.5.3. Remark that

(16.3.17) $\quad \pi_{\varepsilon(\beta)}\overline{X^{\varepsilon(\alpha)\varepsilon(\beta)}}(s)^t = \overline{X^{\varepsilon(\alpha)\varepsilon(\beta)}}(s)^t\pi_{\varepsilon(\alpha)}$

and that $\varepsilon(\beta) = \varepsilon(\alpha)$ if the sum occurs at all.

In proposition 15.6.1 we have another expression for $\langle\Theta_\alpha\Phi\mu\xi_v,\ \Theta_\beta\Phi\eta_v\rangle$. Working this out with help of lemma 16.3.1 we obtain

(16.3.18)
$$\langle\Theta_\alpha\Phi\mu\xi_v,\ \Theta_\beta\Phi\eta_v\rangle =$$
$$\frac{1}{2\pi i}\int_0^{i\infty}((\underline{f}_\alpha(s),\ \underline{f}_\beta(s)))e^{2vs^2}(\cos 2\pi s+\cos\pi\tau)^{-1}8\pi^3|n_\alpha n_\beta|$$
$$.\ \text{trace}(X^{\varepsilon(\beta)}(s)\pi_1\overline{X^{\varepsilon(\alpha)}}(-s)^t\mu(s))ds$$
$$+\sum_{\substack{s\in S\\ s\text{ general}}} 16\,\pi^2|n_\alpha n_\beta|t_{\varepsilon(\alpha),\varepsilon(\beta)}(s)\langle p_\alpha(s),\ p_\beta(s)\rangle\ .$$
$$.\ \tfrac{1}{2}\pi e^{2vs^2}(\cos 2\pi s+\cos\pi\tau)^{-1}\text{trace}(X^{\varepsilon(\beta)}(s)\pi_1\overline{X^{\varepsilon(\alpha)}}(-s)^t\mu(s))$$
$$+\ \tfrac{1}{2}(\varepsilon(\alpha)\varepsilon(\beta)+1)\sum_{\substack{v^{-\frac{1}{2}}\geq b>0,\ \frac{1}{2}(b-1)\in S\\ b\equiv\varepsilon(\alpha)\tau(2)}} 16\,\pi^2|n_\alpha n_\beta|\Gamma(b)^2\ .$$
$$.\langle p_\alpha(\tfrac{1}{2}(b-1)),\ p_\beta(\tfrac{1}{2}(b-1))\rangle\mu_b\ \frac{1}{4\pi}\ \varepsilon_\tau\ e^{\frac{1}{2}v(b-1)^2}$$

We know that the limit for $v\downarrow 0$ exists, and would like to take it inside the integral and sums. In general we have no a priori estimate of $((\underline{f}_\alpha(s),\ \underline{f}_\beta(s)))$

and the norms of the Poincaré elements; so we cannot use Lebesgue's theorem directly. We first consider a special situation in which everything is positive; so here we may apply Fatou's lemma.

16.3.4. Lemma.

$$(16.3.19)\qquad \operatorname{trace}(X^{\varepsilon(\beta)}(s)\pi_1\overline{X^{\varepsilon(\alpha)}}(-s)^t\mu(s)) =$$

$$= \begin{cases} t_{\varepsilon(\alpha),\ \varepsilon(\beta)}(s)^{-1}(\mu^1(s)+\varepsilon(\beta)\varepsilon(\alpha)\mu^{-1}(s)) & \text{if } \tau = 0,\ 1 \\ c_{\varepsilon(\alpha)}(s)c_{\varepsilon(\beta)}(-s)\mu^{\varepsilon(\beta)\varepsilon(\alpha)}(s) & \text{if } 0 < \tau < 1 \end{cases}$$

Proof. In the case $\tau = 0,\ 1$

$$(16.3.20)\qquad X^{\varepsilon}(s)\pi_1\overline{X^{\delta}}(-s)^t\mu(s) =$$

$$e^{-\frac{1}{2}\pi i\tau\varepsilon}\begin{pmatrix}1 & 0\\ 0 & \varepsilon\end{pmatrix}1\ e^{\frac{1}{2}\pi i\tau\delta}\begin{pmatrix}1 & 0\\ 0 & \delta\end{pmatrix}\begin{pmatrix}\mu^1(s) & 0\\ 0 & \mu^{-1}(s)\end{pmatrix},$$

so the trace has the desired form. In the case $0 < \tau < 1$:

$$(16.3.21)\qquad X^{\varepsilon}(s)\pi_1\overline{X^{\delta}}(-s)^t =$$

$$= \begin{pmatrix}1 & 0\\ 0 & c_{-1}(s)\end{pmatrix}Y^{\varepsilon}\begin{pmatrix}e^{-\frac{1}{2}\pi i\tau} & 0\\ 0 & e^{\frac{1}{2}\pi i\tau}\end{pmatrix}\begin{pmatrix}1 & 0\\ 0 & 0\end{pmatrix}\begin{pmatrix}e^{+\frac{1}{2}\pi i\tau} & 0\\ 0 & e^{-\frac{1}{2}\pi i\tau}\end{pmatrix}Y^{\delta}\begin{pmatrix}1 & 0\\ 0 & c_{-1}(s)\end{pmatrix}$$

$$= \begin{pmatrix}1 & 0\\ 0 & c_{-1}(-s)\end{pmatrix}Y^{\varepsilon\delta}c_{\delta}(s)\pi_{\delta}$$

$$= \begin{pmatrix}1 & 0\\ 0 & c_{-1}(-s)\end{pmatrix}\pi_{\varepsilon}Y^{\varepsilon\delta}c_{\delta}(s)$$

$$= c_{\varepsilon}(-s)c_{\delta}(s)Y^{\varepsilon\delta}\pi_{\delta}$$

and

$$(16.3.22)\qquad \operatorname{trace} Y^{\varepsilon\delta}\pi_{\delta}\mu(s) = \mu^{\varepsilon\delta}(s).$$

16.3.5. Lemma. $\alpha,\ \beta \in \Lambda^1$, $\mu \in {}^1R^a_{\sigma}$, $\delta = \varepsilon(\alpha)$ or $\delta = \varepsilon(\beta)$. There exists $\nu_{\delta} \in {}^1R^a_{\sigma}$ such that

$$(16.3.23)\qquad |\operatorname{trace}(X^{\varepsilon(\beta)}(s)\pi_1\overline{X^{\varepsilon(\alpha)}}(-s)^t\mu(s))| \leq \operatorname{trace}(X^{\delta}(s)\pi_1\overline{X^{\delta}}(-s)^t\nu_{\delta}(s))$$

for all general s with $\operatorname{Re} s = 0$ or $0 < s < \frac{1}{2}(1-\tau)$,
$|\mu_b| \leq (\nu_{\delta})_b$ for all $b > 0$ with $b \equiv \delta\tau(2)$.

Proof. Take $p > \sigma$, $p \neq \pm\tau(2)$ and choose in

$$(16.3.24)\qquad f(s) = (p^2-s^2)^{-\frac{1}{2}a}$$

the branch of the logarithm in such a way that $f(s)$ is positive for $\operatorname{Re} s = 0$ and for $s \in (-p,\ p)$. From (14.4.4) follows that all elements of the matrix $\mu(s)$ are estimated by a multiple of $f(s)$ for $\operatorname{Re} s = 0$ or $0 < s < \frac{1}{2}(1-\tau)$. For $b = 2-\tau$ or τ we can also arrange that $|\mu_b| \ll f(\frac{1}{2}(b-1))$. Now define $\nu_{\delta}(s)$ as in lemma 14.5.5

using a multiple of f; this is possible, for $f \in {}^1_1F^a_{\underline{+}\tau,\sigma}$. For $b > 2\sigma+1$ we just take $(\nu_\delta)_b = |\mu_b|$.

16.3.6. Lemma. *Let $\alpha = \beta$ and $\nu \in {}^1R^a_\sigma$ as in lemma 16.3.5. Then the integrand and the terms in the right hand side of (16.3.18) are non-negative, the limit for $v \downarrow 0$ can be taken inside the sums and the integral.*

Proof. The non-negativity is easily checked. By lemma 16.3.3 the limit exists, so by the lemma of Fatou the right hand side of (16.3.18) with $v = 0$ converges. Further, for $v \downarrow 0$, the factors e^{2vs^2} and $e^{\frac{1}{2}v(b-1)^2}$ stay bounded. So a multiple of the expression with $v = 0$ may be used as majorant in the application of Lebesgue's theorem. This gives the continuity at $v = 0$.

Now we can take the limit inside the integral and the sums in (16.3.18) in the general situation. Remark that

(16.3.25) $$|\langle p_\alpha(s), p_\beta(s)\rangle| \leqslant \langle p_\alpha(s), p_\alpha(s)\rangle + \langle p_\beta(s), p_\beta(s)\rangle,$$

and that for $((\underline{f}_\alpha(s), \underline{f}_\beta(s)))$ a similar relation holds. The $t_{\varepsilon(\alpha),\varepsilon(\beta)}(s)$ are bounded. Together with the boundedness of e^{2vs^2} and $e^{\frac{1}{2}v(b-1)^2}$ we see that we may majorize by a multiple of the sum of the corresponding expressions for $\nu_{\varepsilon(\alpha)}$ and for $\nu_{\varepsilon(\beta)}$ with $v = 0$.

Thus we have proved the sum formula:

16.3.7. Theorem. *Let $a > 2$, $\frac{1}{2} < \sigma$. Let $\alpha, \beta \in \Lambda^1$ and $\mu \in {}^1R^a_\sigma$, then the following sums and integrals converge absolutely and satisfy the equality*

(16.3.26)
$$\frac{1}{2\pi i}\int_0^{i\infty} ((\underline{f}_\alpha(s), \underline{f}_\beta(s)))\ 8\pi^3|n_\alpha n_\beta|\ (\cos 2\pi s+\cos \pi\tau)^{-1} . \operatorname{trace}(X^{\varepsilon(\beta)}(s)\pi_1 \overline{X^{\varepsilon(\alpha)}}(-s)\,{}^t\mu(s))ds$$
$$+ \sum_{\substack{s\in S\\ s \text{ general}}} 8\pi^3|n_\alpha n_\beta|t_{\varepsilon(\alpha),\varepsilon(\beta)}(s)\langle p_\alpha(s), p_\beta(s)\rangle . (\cos 2\pi s+\cos\pi\tau)^{-1}\operatorname{trace}(X^{\varepsilon(\beta)}(s)\pi_1\overline{X^{\varepsilon(\alpha)}}(-s)\,{}^t\mu(s))$$
$$+ \tfrac{1}{2}(\varepsilon(\alpha)\varepsilon(\beta)+1) \sum_{\substack{b>0,\ \frac{1}{2}(b-1)\in S\\ b\equiv\varepsilon(\alpha)\tau(2)}} 4\pi\varepsilon_\tau|n_\alpha n_\beta|\Gamma(b)^2 . \langle p_\alpha(\tfrac{1}{2}(b-1)),p_\beta(\tfrac{1}{2}(b-1))\rangle\ \mu_b$$
$$= 4\pi\delta_{\alpha,\beta}|n_\alpha|\ \langle\pi_{\varepsilon(\beta)}\mu, 1\rangle$$
$$+ e^{-\frac{1}{2}\pi i\tau(\varepsilon(\beta)+1)} \sum_{c\in S_{\alpha\beta}} S(\alpha,\beta;c)\langle\pi_{\varepsilon(\alpha)}\mu,\ \delta(16\pi^2|n_\alpha n_\beta|c^{-2},\ \varepsilon(\alpha)\varepsilon(\beta))\rangle$$

Notations. $\alpha, \beta \in \Lambda^1$ index Fourier coefficients and cusps, see definition 7.2.1. The operators $\mu \in {}^1R^a_\sigma$ have been defined in proposition 14.4.1.

$\underline{f}_\alpha$ is connected with a Fourier coefficient of the Eisenstein model, see definition

9.5.1, proposition 9.5.2 and proposition 11.7.16.
The operators $X^{\varepsilon}(s)$ have been defined in definition 6.6.2; the t denotes the transpose matrix and $\overline{X^{\varepsilon}}(-s) = X^{\varepsilon}(-\bar{s})$. See also lemma 16.3.4.

(16.3.27) $$t_{\delta,\varepsilon}(s) = \begin{cases} e^{\frac{1}{2}\pi i\tau(\varepsilon-\delta)} & \text{if } \tau = 0,\ 1 \\ \sin\pi(\frac{1}{2}+s+\frac{1}{2}\delta\tau)(\sin\pi(\frac{1}{2}+s+\frac{1}{2}\varepsilon\tau))^{-1} & \text{if } 0 < \tau < 1. \end{cases}$$

S parametrizes the classes of irreducible g-K-spaces occurring in $L^2(\Gamma\backslash G, \chi)$. The Poincaré elements are square integrable automorphic models, discussed in 15.5.

(16.3.28) $$\varepsilon_{\tau} = \begin{cases} 2 & \text{if } \tau = 0,\ 1 \\ 1 & \text{if } 0 < \tau < 1 \end{cases}$$

π_{ε} is a projection operator; see (14.4.9).
$S(\alpha, \beta; c)$ is a generalized Kloosterman sum; $C_{\alpha\beta}$ is a discrete set of positive numbers; see subsection 8.4. $\delta(t, \varepsilon) = M_{d(t,\varepsilon)}$, see (14.3.5) and also proposition 14.5.3.

16.4. Reformulations of the sum formula

16.4.1. Proposition. Let $\alpha, \beta \in \Lambda^1$. Let $\phi \in {}_{\varepsilon(\alpha)\varepsilon(\beta)}{}^1F^a_{\tau,\sigma}$ if $\tau = 0, 1$ and $\phi \in {}_{\varepsilon(\alpha)\varepsilon(\beta)}{}^1F^a_{\varepsilon(\alpha)\tau,\sigma}$ if $0 < \tau < 1$, with $a > 2$, $\sigma > \frac{1}{2}$. Then

(16.4.1) $$\frac{1}{2\pi i}\int_0^{i\infty} ((\underline{f}_{\alpha}(s), \underline{f}_{\beta}(s)))\ 8\pi^3|n_{\alpha}n_{\beta}|(\cos 2\pi s+\cos\pi\tau)^{-1}t_{\varepsilon(\alpha),\varepsilon(\beta)}(s)^{-1}\phi(s)ds$$
$$+8\pi^3|n_{\alpha}n_{\beta}| \sum_{\substack{s\in S \\ s \text{ general}}} u_{\varepsilon(\alpha)}(s)\langle p_{\alpha}(s), p_{\beta}(s)\rangle(\cos 2\pi s+\cos\pi\tau)^{-1}\phi(s)$$
$$+\tfrac{1}{2}(\varepsilon(\alpha)\varepsilon(\beta)+1) \sum_{\substack{b>0,\ \frac{1}{2}(b-1)\in S \\ b\equiv\varepsilon(\alpha)\tau\,(2)}} 4\pi|n_{\alpha}n_{\beta}|\Gamma(b)^2\langle p_{\alpha}(\tfrac{1}{2}(b-1)), p_{\beta}(\tfrac{1}{2}(b-1))\rangle\phi(\tfrac{1}{2}(b-1))$$
$$= 2\delta_{\alpha\beta}|n_{\alpha}|\langle\phi, 1\rangle$$
$$+ e^{-\frac{1}{2}\pi i\tau(\varepsilon(\beta)+1)}2\pi|n_{\alpha}n_{\beta}|^{\frac{1}{2}} \sum_{c\in C_{\alpha\beta}} c^{-1}S(\alpha, \beta; c)f(4\pi|n_{\alpha}n_{\beta}|^{\frac{1}{2}}c^{-1})$$

with $u_{\varepsilon}(s) = c_{\varepsilon}(s)$ if $0 < s < \frac{1}{2}(1-\tau)$, $0 < \tau < 1$, and $u_{\varepsilon}(s) = 1$ otherwise,

(16.4.2) $$f = \begin{cases} (b_{\tau}^{\varepsilon(\alpha)\varepsilon(\beta)})^{\leftarrow}\phi & \text{if } \tau = 0,\ 1 \\ (b_{\varepsilon(\alpha)\tau}^{\varepsilon(\alpha)\varepsilon(\beta)})^{\leftarrow}\phi & \text{if } 0 < \tau < 1 \end{cases}$$

For the spaces ${}_{\cdot\,\cdot}{}^1F^a$ see definition 14.2.7. An expression for $\langle\phi, 1\rangle$ may be found in definition 14.2.12 and for $(b^{\cdot}_{\cdot})^{\leftarrow}\phi$ in proposition 14.2.8.
Proof. Take μ in theorem 16.3.7 as indicated in lemma 14.5.5. Use lemma 16.3.4 and lemma 14.5.6 to obtain (16.4.1).

16.4.2. Corollary. Let $b > 2$, $b \in Y_\tau$ be given. Let $\alpha, \beta \in \Lambda^1$, $\varepsilon(\alpha) = \varepsilon(\beta)$. Put $\tau_1 = \tau$ if $\tau = 0, 1$ and $\tau_1 = \varepsilon(\alpha)\tau$ if $0<\tau<1$. Suppose $b \equiv \tau_1 (2)$. Then

$$(16.4.3) \qquad \langle p_\alpha(\tfrac{1}{2}(b-1)), p_\beta(\tfrac{1}{2}(b-1))\rangle = \frac{1}{4\pi}\delta_{\alpha\beta}\Gamma(b)^{-1}\Gamma(b-1)^{1}|n_\beta|^{-1}$$
$$+ \frac{\tfrac{1}{2}e^{-\frac{1}{2}\pi i\tau}(-1)^{\frac{1}{2}(b-\tau_1)}}{|n_\alpha n_\beta|^{\frac{1}{2}}\Gamma(b)\Gamma(b-1)} \sum_{c\in C_{\alpha\beta}} c^{-1} S(\alpha, \beta; c) J^1_{b-1}(4\pi(n_\alpha n_\beta)^{\frac{1}{2}}c^{-1}).$$

Proof. Take ϕ in proposition 16.4.1 equal to 0 except at $\frac{1}{2}(b-1)$; take $\phi(\frac{1}{2}(b-1))=1$ Then almost everything in (16.4.1) vanishes. The computation of $\langle\phi,1\rangle$ and f is also easy.

Remark. Corollary 16.4.2 is a classical result. It gives an expression for $\beta(D_b, p_\alpha(D_b))$. If we take ξ as in (15.2.3) then lemma 15.5.1 implies that

$$(16.4.4) \qquad p_\alpha(D_b)\phi_b[b] =$$
$$(4\pi|n_\alpha|)^{-1}\Gamma(b)^{-1}(b-1)^{\frac{1}{2}}\Gamma(b)^{-1}(b-1)^{\frac{1}{2}}\Theta_\alpha D^b\phi_b[b] =$$
$$(4\pi|n_\alpha|)^{-1}\Gamma(b)^{-1}\Gamma(b-1)^{-1}\Theta_\alpha D^b\phi_b[b]$$

In (8.2.21), (8.2.22) we have seen that for $\varepsilon(\alpha) = \pm 1$

$$(p_\alpha(D_b)\phi_b[b])^{\pm}(z) =$$
$$(4\pi|n_\alpha|)^{-1}\Gamma(b)^{-1}\Gamma(b-1)^{-1}\sqrt{2}(4\pi|n_\alpha|)^{\frac{1}{2}b}.$$
$$\sum_{M\in\bar{\Delta}_\alpha\backslash\bar{\Gamma}_\alpha} v^{\pm}(M)^{-1}(c_M z+d_M)^{-b} e^{-2\pi i b \underline{w}(A,M)} e^{2\pi i|n_\alpha|AMz}$$
$$= \sqrt{2}\,\Gamma(b)^{-1}\Gamma(b-1)^{-1}(4\pi|n_\alpha|)^{\frac{1}{2}b-1}\tfrac{1}{2}G_{-b}(z; v^{\pm}; A, \Gamma; R_{[|n_\alpha|]})v^{\pm}(A)$$

in the notation of Petersson in [28], p. 184. Here $v^{\pm}$ has been extended to $A\bar{\Gamma}$ and satisfies

$$(16.4.5) \qquad v^{\pm}(A\gamma) = v^{\pm}(A)v^{\pm}(\gamma)e^{2\pi i b\underline{w}(A,\gamma)}.$$

To compare corollary 16.4.2 with Petersson's result I suppose that $g_\beta \in H_0$; so the cusp c_β correspond to $i\infty$ in the upper half plane. This means that $\bar{g}_\beta^{-1} = \begin{pmatrix} \cdot & \cdot \\ 0 & d\end{pmatrix}$; I suppose further that $g^{-1}_{\beta\pm} = \begin{pmatrix} \cdot & \cdot \\ 0 & 1\end{pmatrix}$. Then the term corresponding to β in the Fourier series of $(p_\alpha(D_b)\phi_b[b])^{\pm}$ is by (7.4.5), (7.4.6):

$$(16.4.6) \qquad \pi^{\frac{1}{2}b}2^{b+\frac{1}{2}}|n_\beta|^{\frac{1}{2}b}e^{2\pi i|n_\beta|z}\cdot\begin{cases}\langle p_\alpha(D_b), p_\beta(D_b)\rangle & \text{if } \varepsilon(\alpha) = 1\\ \overline{\langle p_\alpha(D_b), p_\beta(D_b)\rangle} & \text{if } \varepsilon(\alpha) = -1\end{cases}$$

This means that the coefficient of $e^{2\pi i|n_\beta|z}$ in the Fourier series expansion of $G_{-b}(z; v^{\pm}; A, \Gamma; R_{[|n_\alpha|]})$ equals

(16.4.7) $$v^{\pm}(A)^{-1}(2\delta_{\alpha\beta}+4\pi\ e^{-\frac{1}{2}\pi ib}(\frac{n_\alpha}{n_\beta})^{\frac{1}{2}-\frac{1}{2}b}\sum_{c\in C_{\alpha\beta}} c^{-1}S(\alpha,\ \beta;\ c)^*J^1_{b-1}(4\pi\sqrt{n_\alpha n_\beta}c^{-1}))$$

with $S(\alpha,\ \beta;\ c)^* = S(\alpha,\ \beta;\ c)$ if $\varepsilon(\alpha) = 1$ and $S(\alpha,\ \beta;\ c)^* = \overline{S(\alpha,\ \beta;\ c)}$ if $\varepsilon(\alpha) = -1$. Let us compare this with Petersson's result in [28], p. 187. Comparison of notations:

(16.4.8)

Petersson	here
$\nu + \kappa$	$\lvert n_\alpha\rvert$
$n + \eta$	$\lvert n_\beta\rvert$
r	b
$N,\ N^1$	1
e_0	2

We have supposed that $\overline{g_\beta} \in \overline{HN}$, so we may assume $g_\beta = 1$. This means that Petersson's $\delta = 1$, and $\xi = 1$. If $\alpha = \beta$ then $A = 1$, $\lvert n_\alpha\rvert = \lvert n_\beta\rvert$ and the first term of Petersson's formula becomes

(16.4.9) $$2e^{2\pi i(z+1)\lvert n_\beta\rvert}\chi(n(1))^{-\varepsilon(\alpha)} = 2e^{2\pi i\lvert n_\beta\rvert z}$$

As $v(A) = v(1) = 1$, this is in accordance with (16.4.7). In the other term $m_1 = \varepsilon c$ with $c \in C_{\alpha\beta}$, $\varepsilon = \underline{+}1$. We have

(16.4.10) $$W_{m_1}(n+\eta,\ R_\nu) = \sum_{\substack{\gamma\in\Delta_\alpha\backslash\Gamma/\Delta_\beta \\ c_{\alpha\beta}(\gamma)=c}} e^{2\pi i(\lvert n_\beta\rvert x''_{\alpha\beta}(\gamma)+\lvert n_\alpha\rvert x'_{\alpha\beta}(\gamma))}v(A\overline{\gamma})^{-1},$$

with all γ chosen such that $(-1)^{1_{\alpha\beta}(\gamma)} = \varepsilon$. We may even arrange $1_{\alpha\beta}(\gamma) = \frac{1}{2}(1-\varepsilon)$. This implies $\widetilde{A\overline{\gamma}} = g_\alpha^{-1}\gamma$ and hence

(16.4.11) $$k(-2\pi\underline{w}(A,\overline{\gamma})) = \overline{\gamma}\tilde{\gamma}^{-1}$$

(16.4.12) $$v^{\pm}(A\overline{\gamma})^{-1} = v^{\pm}(A)^{-1}\chi(\gamma)^{-\varepsilon(\alpha)}$$

(16.4.13) $$W_{m_1}(\lvert n_\beta\rvert,\ R_{[\lvert n_\alpha\rvert]}) = S(\alpha,\ \beta;\ c)^*v^{\pm}(A)^{-1}e^{-\frac{1}{2}\pi i(1-\varepsilon)\varepsilon(\alpha)\tau}$$

One gets easily

(16.4.14) $$K_b(\lvert n_\beta\rvert,\ \lvert n_\alpha\rvert c^{-2}) = (n_\alpha n_\beta^{-1})^{\frac{1}{2}-\frac{1}{2}b}c^{b-1}2\pi\ e^{-\frac{1}{2}\pi ib}J_{b-1}(4\pi(n_\alpha n_\beta)^{\frac{1}{2}}c^{-1}).$$

With (16.4.13) and (16.4.14) substituted in Petersson's formula we get the second term in (16.4.7).

Of course the formula of Petersson could have been derived directly from proposition 8.5.1. I like to put it here to emphasize that the sum formula is a generalization of Petersson's formula for the Fourier coefficients of Poincaré series.

For the reformulation of the sum formula with Fourier coefficients of automorphic forms of weight $r \equiv \pm\tau(2)$ I need the Eisenstein series $E(z, \frac{1}{2}+s, 1; \overline{g_\lambda^{-1}}, \overline{\Gamma}, \tau, \mathbb{C}, v^+)$, see [37], (10.1) or (9.3.2), (9.3.4), an orthonormal basis of the square integrable forms in $F_{\tau, \frac{1}{4}-s^2}(\overline{\Gamma}, \mathbb{C}, v^+)$, see definition 4.5.3, and an orthonormal basis for the holomorphic square integrable forms in $\{\overline{\Gamma}, -b, v^\pm\}$. I give some lemmas relating the Fourier coefficients to the quantities occurring in proposition 16.4.1.

16.4.3. Lemma. Let, in the notation of (7.4.2), (7.4.3) the Fourier series expansion at the cusp c of the Eisenstein series be

(16.4.15) $$E(z, \tfrac{1}{2}+s, 1; \overline{g_\lambda^{-1}}, \overline{\Gamma}, \tau, \mathbb{C}, v^+)\, e^{i\tau \arg(cz+d)} = \rho_0^\lambda(y_A) + \sum_{\substack{\alpha\in\Lambda^1 \\ c_\alpha = c}} \rho_\lambda(s,\alpha)\, W_{\frac{1}{2}\varepsilon(\alpha)\tau, s}(4\pi|n_\alpha| y_A)\, e^{2\pi i n_\alpha x_A}$$

Then for $\alpha, \beta \in \Lambda^1$ and $\mathrm{Re}\, s = 0$

(16.4.16) $$8\pi^3 |n_\alpha n_\beta| ((\underline{f}_\alpha(s), \underline{f}_\beta(s)))\, (\cos 2\pi s + \cos \pi\tau)^{-1} t_{\varepsilon(\alpha),\varepsilon(\beta)}(s)^{-1} =$$
$$4\pi|n_\alpha n_\beta| \Gamma(\tfrac{1}{2}+s+\tfrac{1}{2}\varepsilon(\alpha)\tau)\Gamma(\tfrac{1}{2}-s+\tfrac{1}{2}\varepsilon(\beta)\tau) \sum_{\lambda\in\Lambda^0} \rho_\lambda(s, \beta)\overline{\rho_\lambda(s, \alpha)}$$

Proof. By (9.3.2), (9.3.4)

(16.4.17) $$E(z, \tfrac{1}{2}+s, 1; g_\lambda^{-1}, \overline{\Gamma}, \mathbb{C}, v^+) = \begin{cases} \underline{E}(s)\phi_\tau^1(s) \otimes \underline{e}_\lambda(p(z)) & \tau = 0, 1 \\ \frac{1}{\sqrt{2}} \underline{E}(s)\phi_\tau(s) \otimes \underline{e}_\lambda(p(z)) & 0 < \tau < 1. \end{cases}$$

In (7.4.4) and (7.4.8) we see that

(16.4.18) $$\rho_\lambda(s,\alpha) = \Gamma(\tfrac{1}{2}-s-\tfrac{1}{2}\varepsilon(\alpha)\tau)\alpha(H_\tau^{(1)}(s), \underline{E}(s)\underline{e}_\lambda | H_\tau^{(1)}(s))$$

So by corollary 9.5.3

(16.4.19) $$\rho_\lambda(s,\alpha) = \Gamma(\tfrac{1}{2}-s-\tfrac{1}{2}\varepsilon(\alpha)\tau)((\underline{e}_\lambda, \underline{f}_\alpha(\overline{s}))). \begin{cases} e^{-\frac{1}{2}\pi i\tau\varepsilon(\alpha)} \\ e^{-\frac{1}{2}\pi i \iota} c_{\varepsilon(\alpha)}(-s) \end{cases}$$

So we obtain, using the functional equation (11.7.31):

(16.4.20) $$((\underline{f}_\alpha(s),\ \underline{f}_\beta(s))) = \sum_{\lambda\in\Lambda^0}((\underline{e}_\lambda,\ \underline{f}_\beta(-s)))\overline{((\underline{e}_\lambda,\ \underline{f}_\alpha(-s)))}$$

$$= \sum_{\lambda\in\Lambda^0}\Gamma(\tfrac12-s-\tfrac12\varepsilon(\beta)\tau)^{-1}\Gamma(\tfrac12+s-\tfrac12\varepsilon(\alpha)\tau)^{-1}\rho_\lambda(s,\ \beta)\overline{\rho_\lambda(s,\ \alpha)}$$

$$\cdot\begin{cases} e^{\frac12\pi i\tau(\varepsilon(\beta)-\varepsilon(\alpha))} \\ c_{\varepsilon(\beta)}(s)c_{\varepsilon(\alpha)}(-s)\end{cases}$$

To get (16.4.16) use (15.6.1) and

(16.4.21) $$\Gamma(\tfrac12-s-\tfrac12\varepsilon\tau)^{-1}\Gamma(\tfrac12+s-\tfrac12\delta\tau)^{-1}\tfrac12\pi(\cos 2\pi s + \cos\pi\tau)^{-1} =$$

$$\frac{\pi}{2\Gamma(\frac12-s-\frac12\varepsilon\tau)\Gamma(\frac12+s-\frac12\delta\tau)2\sin\pi(\frac12+s+\frac12\tau)\sin\pi(\frac12+s-\frac12\tau)}$$

$$= \frac{1}{4\pi}\Gamma(\tfrac12-s+\tfrac12\varepsilon\tau)\Gamma(\tfrac12+s+\tfrac12\delta\tau).$$

For each general $s \in S$ there are nonzero square integrable automorphic forms in $F_{\tau,\frac14-s^2}(\overline{\Gamma},\ \mathbb{C},\ v^+)$; see proposition 4.5.6 and proposition 10.4.1. Let $\{f_{s,j}\}$ be an orthonormal basis. Suppose that at the cusp c the following expansion holds:

(16.4.22) $$f_{s,j}(z)e^{i\tau\arg(cz+d)} =$$

$$\rho_{0,j}(y_A) + \sum_{\substack{\alpha\in\Lambda^1\\ c_\alpha=c}}\rho(s,\alpha)_j W_{\frac12\varepsilon(\alpha)\tau,s}(4\pi|n_\alpha|y_A)e^{2\pi i n_\alpha x_A}$$

The zero order term $\rho_{0,j}(y_A)$ is a multiple of $y_A^{\frac12-s}$, only occurring if $0 < s < \frac12(1-\tau)$.

16.4.4. <u>Lemma</u>. <u>Let</u> $s \in S$ <u>be general</u>, <u>let</u> $\alpha,\ \beta \in \Lambda^1$. <u>Then</u>

(16.4.23) $$8\pi^3|n_\alpha n_\beta|u_{\varepsilon(\alpha)}(s)\langle p_\alpha(s),\ p_\beta(s)\rangle(\cos 2\pi s + \cos\pi\tau)^{-1} =$$

$$4\pi|n_\alpha n_\beta|\Gamma(\tfrac12+s+\tfrac12\varepsilon(\alpha)\tau)\Gamma(\tfrac12-s+\tfrac12\varepsilon(\beta)\tau)\sum_j\overline{\rho(s,\alpha)_j}\,\rho(s,\beta)_j.$$

<u>Proof</u>. Define $T_j \in A^{sq}(H_\tau^{(1)}(s))$ by

(16.4.24) $$^R(T_j\phi_\tau^{(1)}(s))^+ = \sqrt{2}\,\varepsilon_\tau^{-\frac12}f_j\cdot\begin{cases}1 & \text{if } \mathrm{Re}\ s = 0\\ (\Gamma(\frac12+s-\frac12\tau)\Gamma(\frac12-s-\frac12\tau)^{-1})^{-\frac12} & \text{if } 0 < s < \frac12(1-\tau)\end{cases}$$

From proposition 10.4.1 follows that $\{T_j\}$ is an orthonormal basis of $A^{sq}(H_\tau^{(1)}(s))$, and by (7.4.4), (7.4.8):

(16.4.25) $$\alpha(H_\tau^{(1)}(s),\ T_j) =$$
$$\Gamma(\tfrac12-s-\tfrac12\varepsilon(\alpha)\tau)^{-1}\rho(s,\alpha)_j\cdot\begin{cases}1 & \text{if } \mathrm{Re}\ s = 0\\ (\Gamma(\frac12+s-\frac12\tau)\Gamma(\frac12-s-\frac12\tau)^{-1})^{-\frac12} & \text{if } 0 < s < \frac12(1-\tau)\end{cases}$$

Formula (15.5.5) gives

(16.4.26) $$\langle p_\alpha(s),\ p_\beta(s)\rangle =$$

$$\sum_j \Gamma(\tfrac12-s-\tfrac12\varepsilon(\alpha)\tau)^{-1}\Gamma(\tfrac12-s-\tfrac12\varepsilon(\beta)\tau)^{-1}\overline{\rho(s,\alpha)_j}\ \rho(s,\beta)_j \cdot \begin{cases} 1 \\ \dfrac{\Gamma(\frac12-s-\frac12\tau)}{\Gamma(\frac12+s-\frac12\tau)} \end{cases}$$

$$= \sum_j \overline{\rho(s,\alpha)_j}\ \rho(s,\beta)_j \Gamma(\tfrac12+s-\tfrac12\varepsilon(\alpha)\tau)^{-1}\Gamma(\tfrac12-s-\tfrac12\varepsilon(\beta)\tau)^{-1} \ .$$

$$\cdot \begin{cases} 1 & \text{if } \mathrm{Re}\, s = 0 \\ c_{\varepsilon(\alpha)}(-s) & \text{if } 0 < s < \tfrac12(1-\tau) \end{cases}$$

This completes the proof.

Let $b \equiv \eta\tau(2)$, $b > 0$, $\eta = \pm 1$ and suppose $\frac12(b-1) \in S$. Then there are nonzero holomorphic square integrable automorphic forms in $\{\Gamma, -b, v^{\pm}\}$. Let $f^{\eta}_{b,j}$ be an orthonormal basis. Suppose the Fourier series expansion at the cusp c is

(16.4.27) $$f^{\eta}_{b,j}(z)(cz+d)^{-b} =$$
$$p^0_b + \sum_{\substack{\alpha\in\Lambda^1,\, c_\alpha = c \\ \varepsilon(\alpha)=\eta^\alpha}} \rho(\tfrac12(b-1),\ \alpha)^{\eta}_j e^{2\pi i|n_\alpha| z_A} .$$

16.4.5. <u>Lemma</u>. <u>Let</u> $\frac12(b-1) \in S$, α, $\beta \in \Lambda^1$, $\varepsilon(\alpha) = \varepsilon(\beta)$, $b \equiv \varepsilon(\alpha)\tau(2)$. <u>Then</u>

(16.4.28) $$4\pi|n_\alpha n_\beta|\Gamma(b)^2 \langle p_\alpha(\tfrac12(b-1)),\ p_\beta(\tfrac12(b-1))\rangle =$$

$$= (4\pi)^{1-b}|n_\alpha n_\beta|^{1-\frac12 b}\Gamma(b)\ . \begin{cases} \sum_j \overline{\rho(\frac12(b-1),\alpha)^1_j}\ \rho(\frac12(b-1),\beta)^1_j & \underline{\text{if}}\ \varepsilon(\alpha) = 1 \\ \sum_j \rho(\frac12(b-1),\alpha)^{-1}_j\ \overline{\rho(\frac12(b-1),\beta)^{-1}_j} & \underline{\text{if}}\ \varepsilon(\alpha) = -1 \end{cases}$$

<u>Proof</u>. Define T^{+1}_j and $T^{-1}_j \in A^{sq}(D_b)$ by

(16.4.29) $$(T^{\pm1}_j \phi_b[b])^{\pm} = \sqrt{2}\,\Gamma(b)^{-\frac12} f^{\pm1}_j$$
$$(T^{\pm1}_j \phi_b[b])^{\mp} = 0$$

According to propositions 10.4.2 and 10.4.3 an orthonormal basis of $A^{sq}(D_b)$ is given by

(16.4.30) $$\{T^{+1}_j\} \cup \{T^{-1}_j\} \qquad \text{if } b \in \mathbb{Z}$$
$$\{T^{+1}_j\} \qquad \text{if } b \notin \mathbb{Z},\ b \equiv \tau(2)$$
$$\{T^{-1}_j\} \qquad \text{if } b \notin \mathbb{Z},\ b \equiv -\tau(2)$$

From (7.4.5) and (7.4.6) follows:

(16.4.31) $$a(D_b,\ T^{\eta}_j) = \delta_{\varepsilon(\alpha),\eta}(4\pi)^{-\frac12 b}|n_\alpha|^{-\frac12 b}\Gamma(b)^{-\frac12} . \begin{cases} \overline{\rho(\frac12(b-1),\alpha)^1_j} & \text{if } \eta = 1 \\ \rho(\frac12(b-1),\alpha)^{-1}_j & \text{if } \eta = -1 \end{cases}$$

Application of (15.5.5) gives (16.4.28).

16.4.6. Proposition. Let $\alpha, \beta \in \Lambda^1$. Let $\phi \in {}_{\varepsilon(\alpha)\varepsilon(\beta)}^{1}F^a_{\tau,\sigma}$ if $\tau = 0, 1$, $\phi \in {}_{\varepsilon(\alpha)\varepsilon(\beta)}^{1}F^a_{\varepsilon(\alpha)\tau,\sigma}$ if $0 < \tau < 1$, with $\sigma > \frac{1}{2}$, $a > 2$. Then

(16.4.32)
$$\frac{1}{2\pi i}\int_0^{i\infty} \phi(s)4\pi|n_\alpha n_\beta|\Gamma(\tfrac{1}{2}+s+\tfrac{1}{2}\varepsilon(\alpha)\tau)\Gamma(\tfrac{1}{2}-s+\tfrac{1}{2}\varepsilon(\beta)\tau)\sum_{\lambda\in\Lambda^0}\rho_\lambda(s,\beta)\overline{\rho_\lambda(s,\alpha)}ds$$
$$+\sum_{\substack{s\in S\\ s\text{ gen.}}}\phi(s)4\pi|n_\alpha n_\beta|\Gamma(\tfrac{1}{2}+s+\tfrac{1}{2}\varepsilon(\alpha)\tau)\Gamma(\tfrac{1}{2}-s+\tfrac{1}{2}\varepsilon(\beta)\tau)\sum_j\rho(s,\beta)_j\overline{\rho(s,\alpha)_j}$$
$$+\delta_{\varepsilon(\alpha),\varepsilon(\beta)}\sum_{\substack{b>0,\, b\equiv\varepsilon(\alpha)\tau(2)\\ \frac{1}{2}(b-1)\in S}}\phi(\tfrac{1}{2}(b-1))(4\pi)^{1-b}|n_\alpha n_\beta|^{1-\frac{1}{2}b}\Gamma(b)\cdot\begin{cases}\sum_j\rho(\frac{1}{2}(b-1),\beta)^1_j\overline{\rho(\frac{1}{2}(b-1),\alpha)^1_j} & \text{if } \varepsilon(\alpha)=1\\ \sum_j\rho(\frac{1}{2}(b-1),\alpha)^{-1}_j\overline{\rho(\frac{1}{2}(b-1),\beta)^{-1}_j} & \text{if } \varepsilon(\alpha)=-1\end{cases}$$
$$= e^{-\frac{1}{2}\pi i\tau(1+\varepsilon(\beta))}2\pi|n_\alpha n_\beta|^{\frac{1}{2}}\sum_{c\in C_{\alpha\beta}}c^{-1}S(\alpha,\beta;c)f(4\pi|n_\alpha n_\beta|^{\frac{1}{2}}c^{-1})$$
$$+ 2\delta_{\alpha\beta}|n_\alpha|\langle\phi, 1\rangle,$$

with f as in (16.4.2).

Proof. It is a reformulation of proposition 16.4.1 with help of the lemmas 16.4.3 - 16.4.5.

Remark. After changing to the notations of subsection 1.2 one obtains (1.2.12) and (1.2.15).

16.4.7. Comparison with other sum formulas.

Let us take $\tau = 0$, $\overline{\Gamma} = Sl_2(\mathbb{Z})$, $\chi = 1$. If we take $A_1, A_2, \ldots, A_q \in \mathbb{C}$ such that $\sum_r A_r = 0$ and some different numbers u_r with $|\mathrm{Re}\, u_r| > \sigma$, $u_r \not\equiv \pm\tau(2)$, then

(16.4.33)
$$\phi(s) = \sum_r A_r(s^2 - u_r^2)^{-1}$$

satisfies the conditions in proposition 16.4.6 with $\varepsilon(\alpha) = \varepsilon(b) = 1$. Let $m, n \in \mathbb{N}$, take $\alpha = (\infty, m)$, $\beta = (\infty, n)$; then (16.4.32) becomes

(16.4.34)
$$\frac{1}{2\pi i}\int_0^{i\infty}\phi(s)(\cos\pi s)^{-1}\rho_{(\infty,\ 0)}(s,\ \beta)\overline{\rho_{(\infty,\ 0)}(s,\ \alpha)}ds$$
$$+\sum_{\substack{s\in S\\ s\text{ general}}}\phi(s)(\cos\pi s)^{-1}\sum_j\phi(s,\ \beta)_j\overline{\rho(s,\ \alpha)_j}$$
$$+\sum_{\substack{b>2\\ b\text{ even}}}\pi^{-1}\phi(\tfrac{1}{2}(b-1))(b-1)!(16\pi^2 mn)^{-\frac{1}{2}b}\sum_j\rho(\tfrac{1}{2}(b-1),\beta)^1_j\overline{\rho(\tfrac{1}{2}(b-1),\alpha)^1_j}$$
$$= \frac{1}{4\pi m}\sum_r A_r(\delta_{mn}\tan\pi u_r -$$

$$-(\cos \pi u_r)^{-1} 2\pi (m/n)^{\frac{1}{2}} \sum_{c=1}^{\infty} S(\alpha,\ \beta;\ c) c^{-1} J_{2u_r} (4\ \pi c^{-1} \sqrt{mn}).$$

This is just the formula in (5.11) of [1], after correction of two mistakes, one slipped in in section 3 in the factor of the Eisenstein term, and one in (5.9) in the discrete series term.

Take in the case $\varepsilon(\alpha) = \varepsilon(\beta)$ in proposition 16.4.6

(16.4.35) $\quad \phi(\frac{1}{2}(b-1)) = 0 \qquad$ for $b > 2\sigma+1$

We may use proposition 14.2.13 to express $f(.)$ by an integral over Re $s = \sigma$. So we get a generalisation of sum formulas like [18], theorem 1 and [1], theorem (3.29). (The sum formula in [1] should have a factor $(4\pi i)^{-1}$ in front of the Eisenstein term).

The test functions in these sum formulas are essentially multiplication operators on the spectrum of $L^2(\Gamma\backslash G,\ \chi)$. In [15] and [34] the test functions are functions on $(0,\ \infty)$, describing the kernels of convolution operators, as discussed in 14.3. Bessel transforms give the connection. To get Proskurin's formula (94) in [34], for the weight $k = \tau$, take $c_\alpha = c_\beta$ and $\varepsilon(\alpha) = \varepsilon(\beta) = 1$. Suppose $\phi \in {}^1_1F^a_{\tau,\sigma}$ satisfies $\langle\phi,\ 1\rangle = 0$. (This is the case for $f = (b^1_\tau)^{\leftarrow}\phi \in C^\infty_c(0,\ \infty)$, see lemma 14.2.16. I have not proved that $\langle b^1_\tau f,\ 1\rangle = 0$ for all f satisfying the conditions in Proskurin's theorem). Suppose the integrals in $b^1_\tau f$ converge absolutely, then the function $\hat{f}$ defined in (84) of [34] satisfies

(16.4.36) $$\phi(\ s) = \frac{2e^{-\pi i\tau}\hat{f}(-is)}{\cos \pi s \Gamma(\frac{1}{2}+s+\frac{1}{2}\tau)\Gamma(\frac{1}{2}-s+\frac{1}{2}\tau)}\ .$$

Further $\check{f}$, defined in (86) of [34] satisfies

(16.4.37) $$\check{f}(b) = \tfrac{1}{2}e^{+\frac{1}{2}\pi ib}\phi(\tfrac{1}{2}(b-1)).$$

With $n_\alpha = m-x$, $n_\beta = n-x$ we get for the Fourier coefficients of the Eisenstein series, with Re $s = 0$:

(16.4.38) $$\overline{\rho_\lambda(s,\ \alpha)}\rho_\lambda(s,\ \beta) =$$

$$\pi(m-x)^{-\frac{1}{2}-s}(n-x)^{-\frac{1}{2}+s}\Gamma(\tfrac{1}{2}+s+\tfrac{1}{2}\tau)^{-1}\Gamma(\tfrac{1}{2}-s+\tfrac{1}{2}\tau)^{-1}\overline{\phi_{\lambda m}(\tfrac{1}{2}+s)}\phi_{\lambda n}(\tfrac{1}{2}+s),$$

see (13) of [34]. Substituting all this into (16.4.32) we obtain

(16.4.39) $$\sum_{c\in C_{\alpha\beta}} c^{-1}S(\alpha,\ \beta;\ c)f(4\pi|n_\alpha n_\beta|^{\frac{1}{2}}c^{-1}) =$$

$$\sum_{\lambda\in\Lambda^0} \int_{-\infty}^{+\infty} \left(\frac{n-x}{m-x}\right)^{ir} \hat{f}(r) \frac{\overline{\phi_{\lambda,m}(\frac{1}{2}+ir)}\phi_{\lambda n}(\frac{1}{2}+ir)}{\mathrm{ch}\pi r\,|\Gamma(\frac{1}{2}+ir-\frac{1}{2}\tau)|^2}\,dr$$

$$+ 4((m-x)(n-x))^{\frac{1}{2}} \sum_{\substack{s\in S\\ s \text{ gen}}} \frac{\rho(s,\beta)_j\,\overline{\rho(s,\alpha)_j}}{\cos \pi s} \hat{f}(-is)$$

$$+ \sum_{\substack{b>0\\ b\equiv\tau(2)}} \frac{4\Gamma(b)e^{\frac{1}{2}\pi ib}}{(4\pi)^b((m-x)(n-x))^{\frac{1}{2}b-\frac{1}{2}}} \check{f}(b) \sum_j \rho(\tfrac{1}{2}(b-1),\beta)^1_j\,\overline{\rho(\tfrac{1}{2}(b-1),\alpha)^1_j}$$

This is almost Proskurin's formula (94), [34]; the difference concerns the square integrable automorphic forms of weight τ. In (16.4.39) these occur in the sum over b, whereas Proskurin puts them in the other sum. With help of (2.3.6) and (13.2.9) one may check that the corresponding terms are equal. Remark that Proskurin's theorem gives more in two respects: the class of test functions is precisely described in terms of the function f, and it is valid for more weights.

If one derives the sum formula working in weight zero only, as is done in [15], [31], [1], then it is natural that the terms corresponding to irreducible subspaces of $L^2(\Gamma\backslash G,\chi)$ of the type D_b ($b \geq 2$) do not occur explicitly in the sum formula, for these subspaces have no vectors of weight zero. I shall discuss the case $\tau = 0$, $\bar{\Gamma} = Sl_2(\mathbb{Z})$, $\chi = 1$, considered by Kuznetsov, [15], theorem 2, p. 40. In the proof of the sum formula in subsection 16.3 I have supposed that $\sigma > \frac{1}{2}$. This I needed in the proof of lemma 16.3.3, to get absolute convergence for $\mathrm{Re}\, s \geq \sigma$ of a series containing Kloosterman sums. In the case of the modular group we have Weil's estimates of the Kloosterman sums, implying

$$(16.4.40) \qquad \sum_{c\in C_{\alpha\beta}} |S(\alpha,\beta;c)|\,c^{-\frac{3}{2}-\varepsilon} < \infty$$

for each $\varepsilon > 0$. This means that for the modular case everything goes well with $\frac{1}{4} < \sigma < \frac{1}{2}$. Using this I show that Kuznetsov's transformation $f \mapsto f_H$ in (4.31) of [15] arises naturally:

Let $\phi \in {}^1_1F^a_{0,\sigma}$, $a > 2$ and let $f = (b^1_0)^{\leftarrow}\phi$. Working in weight 0 means ignoring the values of ϕ at $\frac{1}{2}(b-1)$, $b \geq 2$, so define

$$(16.4.41) \qquad \phi_H(s) = \begin{cases} \phi(s) & \text{if } |\mathrm{Re}\, s| \leq \sigma \\ 0 & \text{if } s = \frac{1}{2}(b-1),\ b \geq 2,\ b \text{ even.} \end{cases}$$

Then $\phi_H \in {}^1_1F^a_{0,\sigma}$; let $f_H = (b^1_0)^{\leftarrow} \phi_H$.

By (14.2.31) and (14.2.17)

$$(16.4.42) \qquad f_H(y) = f(y) - \sum_{\substack{b\geq 2\\ b \text{ even}}} 2(b-1)J^1_{b-1}(y)\, f\{J^1_{b-1}\}.$$

Proposition 16.4.6 applied to ϕ_H gives for $m,n \in \mathbb{Z}$, $mn > 0$:

$$(16.4.43)\qquad \sum_{c=1}^{\infty} c^{-1} S(n,m;c)\, f_H\,(4\pi\sqrt{mn}\, c^{-1}) + \pi^{-1}\delta_{mn}\langle \phi_H, 1\rangle =$$
$$= \tfrac{1}{2}\int_{-\infty}^{+\infty} (n/m)^{ir}\,\phi(ir)\sigma_{2ir}(m)\sigma_{-2ir}(n)\,|\zeta(1+2ir)|^{-2} dr$$
$$+ 2\pi\sqrt{mn} \sum_{\substack{s\in S\\ s \text{ general}}} \phi(s)(\cos\pi s)^{-1} \sum_j \rho(s,(\infty,m))_j\, \overline{\rho(s,(\infty,n))_j}$$

Use (9.5.12). Kuznetsov's $\hat{f}$ is given by

$$(16.4.44)\qquad \hat{f}(r) = \frac{i\pi}{2\operatorname{sh}\pi r}(f\{J_{2ir}\}-f\{J_{-2ir}\}) = \tfrac{1}{2}\pi\,\phi(ir),$$

and his Fourier coefficients $\rho_j(n)$ are $2|n|^{\frac{1}{2}}$ times our ones. So the right hand side of (16.4.43) is in accordance with (4.35) of [15]. By definition 14.2.12

$$(16.4.45)\qquad \langle \phi_H, 1\rangle = \frac{1}{2\pi i}\int_0^{i\infty} \phi(s)\,\frac{-2\pi s \sin 2\pi s ds}{\cos 2\pi s+1} =$$
$$= \frac{1}{2\pi i}\int_{\operatorname{Re} s=0} f\{J^1_{2s}\}(\cos\pi s)^{-1}\, 2\pi s ds.$$

If $f\{J^1_{2s}\}$ satisfies the right estimates, we may move off the line of integration to the right. This is anyhow the case if $f\in C_c^\infty(0,\infty)$, see lemma 14.2.3. Under this assumption we obtain by the method of (4.40)-(4.43) of [15]:

$$(16.4.46)\qquad \langle \phi_H, 1\rangle = \sum_{\substack{b>0\\ b\equiv 0(2)}} (-1)^b (b-1)\, f\,\{J^1_{b-1}\} =$$
$$= \sum_{\substack{b>0\\ b\equiv 0}} \tfrac{1}{2}(-1)^b \int_0^\infty f(y)(J^1_b(y)+J^1_{b-2}(y))dy = \tfrac{1}{2}\int_0^\infty f(y)J_0(y)dy.$$

This settles the equality of the δ-terms in (4.35) of [15] and (16.4.43). Remark that here too I have only derived the sum formula for a smaller class of test functions than Kuznetsov has.

In the case $\varepsilon(\alpha) = -\varepsilon(\beta)$ the relation between ϕ and f in proposition 16.4.6 is

$$(16.4.47)\qquad \phi(s) = 4\pi^{-1}\sin\pi(\tfrac{1}{2}+s+\tfrac{1}{2}\varepsilon(\alpha)\tau)\int_0^\infty f(y)K_{2s}(y)y^{-1}dy,$$

provided this integral converges absolutely; see (14.2.35).
If we work out (16.4.32) for the case $\tau = 0$ we obtain the formula in theorem 2 of [31], p. 133; compare also theorem 7, [15], p. 49.

Literature

[1] Bruggeman, R.W.: Fourier coefficients of cusp forms; Inv. Math. 45, 1-18 (1978).

[2] Elstrodt, J.: Die Resolvente zum Eigenwertproblem der automorphen Formen in der hyperbolischen Ebene I; Math. Ann. 203, 295-330 (1973).

[3] Elstrodt, J.: —— II , Math. Z. 132, 99-134 (1973).

[4] Elstrodt, J.: —— III, Math. Ann. 208, 99-132 (1974).

[5] Erdélyi, A., Magnus, W., Oberhettinger, F., Tricomi, F.G.: Higher Transcendental Functions II; McGraw-Hill, New York-Toronto-London, 1953.

[6] Faddeev, L.: Expansion in eigenfunctions of the Laplace operator on the fundamental domain of a discrete group on the Lobacevskii plane; Trans. Moscow Math. Soc. 17, 357-386 (1967).

[7] Gelbart, S.S.: Weil's Representation and the Spectrum of the Metaplectic Group; Lect. Notes in Math. 530; Springer, Berlin-Heidelberg-New York, 1976.

[8] Gelbart, S., Jacquet, H.: Forms of GL(2) from the analytic point of view; Proc. Symp. Pure Math. 33 (1979), part 1, 213-251; AMS, Providence (R.I.).

[9] Gel'fand, I., Graev, M., Pyateckii-Shapiro, I.: Representation Theory and automorphic Functions; Moscow, 1966; translation, W.B. Saunders, 1969.

[10] Harish Chandra: Discrete series for semisimple Lie groups II; Acta Math. 116, 1-111 (1966).

[11] Harish Chandra: Automorphic Forms on Semisimple Lie Groups; Lect. Notes in Math., 62; Springer, Berlin-Heidelberg-New York, 1968.

[12] Hecke, E.: Lectures on Dirichlet series, modular functions and quadratic forms; Spring Term 1938, the Institute for Advanced Study, Princeton (N.J.).

[13] Jacquet, H., Langlands, R.P.: Automorphic Forms on GL(2); Lect. Notes in Math. 114; Springer, Berlin-Heidelberg-New York, 1970.

[14] Kubota, T.: Elementary theory of Eisenstein series; New York, Halsted Press, 1973.

[15] Kuznetsov, N.V.: The conjecture of Petersson for forms of weight zero and the conjecture of Linnik (Russian); Preprint Khab. KNII.DVNTs. AN.USSR 02-77, Khabarovsk, 1977.

A part of this preprint has appeared in

[15'] Kuznetsov, N.V.: The conjecture of Petersson for forms of weight zero and the conjecture of Linnik. Sums of Kloosterman sums (Russian); Math. Sbornik 111 (153), 334-383 (1980).

[16] Kuznetsov, N.V.: An arithmetical form of the Selberg trace formula and the distribution of the norms of primitive hyperbolic classes of the modular group (Russian); Preprint Khab. KNII. DVNTs. AN. USSR, Khabarovsk , 1978.

[17] Kuznetsov, N.V.: Asymptotic formulas for the eigenvalues of the Laplace operator on the fundamental domain of the modular group (Russian); Preprint Khab. KNII. DVNTs. AN. USSR, Khabarovsk, 1978.

[18] Kuznetsov, N.V.: The distribution of the norms of primitive hyperbolic classes of the modular group and asymptotic formulas for the eigenvalues of the Laplace-Beltrami operator on the fundamental domain of the modular group (Russian); Dokl. AN. USSR 242, No. 1, 40-43 (1978).

[19] Kuznetsov, N.V.: On the conjecture of Lehmer I (Russian); Zap. nauchn. seminarov LOMI 91, 52-80 (1979).

[20] Lang, S.: Analysis II; Addison-Wesley, Reading (Mass.) 1969.

[21] Lang, S.: Sl_2(R): Addison-Wesley, Reading (Mass.) 1975.

[22] Langlangs, R.P.: On the Functional Equations Satisfied by Eisenstein Series; Lect. Notes in Math. 544; Springer, Berlin-Heidelberg-New York, 1976.

[23] Lehner, J.: Discontinuous groups and automorphic functions; Math. Surveys VIII; AMS, Providence, R.I., 1964.

[24] Maass, H.: Lectures on modular functions of one complex variable; Bombay, Tata Inst. of Fund. Research, 1964.

[25] Magnus, W., Oberhettinger, F., Soni, R.P.: Formulas and Theorems for the Special Functions of Mathematical Physics; 3 ed., Springer, Berlin-Heidelberg-New York, 1966.

[26] Neunhöffer, H.: Ueber die analytische Fortsetzung von Poincaréreihen; Sitzungsb. der Heidelberger Ak. der Wiss., Mathematisch-naturwiss. Klasse, 2. Abhandlung, 1973.

[27] Niebur, D.: A class of nonanalytic automorphic functions; Nagoya Math. J. 52, 133-145 (1973).

[28] Petersson, H.: Ueber die Entwicklungskoeffizienten der automorphen Formen; Acta Math. 58, 169-215 (1932).

[29] Petersson, H.: Zur analytischen Theorie der Grenzkreisgruppen I; Math. Ann. 115, 23-67 (1938).

[30] Petersson, H.: Ueber den Bereich absoluter Konvergenz der Poincaréschen Reihen; Acta Math. 80, 23-63 (1948).

[31] Proskurin, N.V.: Sum formulas for general Kloosterman sums (Russian); Zap. nauchn. seminarov LOMI 82, 103-135 (1979).

[32] Proskurin, N.V.: Estimates of eigenvalues of Hecke operators in the space of parabolic forms of weight 0 (Russian); Zap. nauchn. seminarov LOMI 82 136-143 (1979).

[33] Proskurin, N.V.: On the conjecture of Yu.V. Linnik (Russian); Zap. nauchn. seminarov LOMI 91, 94-118 (1979).

[34] Proskurin, N.V.: On general Kloosterman sums (Russian); Preprint LOMI R-3-80, Leningrad 1980.

[35] Pukánszky, L.: The Plancherel Formula for the Universal Covering Group of SL(R,2); Math. Ann. 156, 96-143 (1964).

[36] Roelcke, W.: Das Eigenwertproblem der automorphen Formen in der hyperbolischen Ebene I; Math. Ann. 167, 292-337 (1966).

[37] Roelcke, W.: —— II; Math. Ann. 168, 261-324 (1967).

[38] Sally, P.J.: Analytic continuation of the irreducible unitary representations of the universal covering group of SL(2,R); Memoir 69, AMS, Providence (R.I.), 1967.

[39] Shimura, G.: Introduction to the arithmetic theory of automorphic functions; Iwanami Shoten, Publishers, and Princeton University Press, 1971.

[40] Slater, L.J.: Confluent Hypergeometric Functions; Cambridge, University Press, 1960.

[41] Siegel, C.L.: Some remarks on discontinuous groups; Ann. of Math. 46, 708-718 (1945).

[42] Terras, A.: Fourier analysis on symmetric spaces and applications to number theory; to appear.

[43] Watson, G.N.: A treatise on the theory of Bessel functions; Cambridge, University Press, 1966 (2. edition).

<u>Index of notations</u>

Latin alphabet

Greek alphabet

Index of terminology

Vol. 700: Module Theory, Proceedings, 1977. Edited by C. Faith and S. Wiegand. X, 239 pages. 1979.

Vol. 701: Functional Analysis Methods in Numerical Analysis, Proceedings, 1977. Edited by M. Zuhair Nashed. VII, 333 pages. 1979.

Vol. 702: Yuri N. Bibikov, Local Theory of Nonlinear Analytic Ordinary Differential Equations. IX, 147 pages. 1979.

Vol. 703: Equadiff IV, Proceedings, 1977. Edited by J. Fábera. XIX, 441 pages. 1979.

Vol. 704: Computing Methods in Applied Sciences and Engineering, 1977, I. Proceedings, 1977. Edited by R. Glowinski and J. L. Lions. VI, 391 pages. 1979.

Vol. 705: O. Forster und K. Knorr, Konstruktion verseller Familien kompakter komplexer Räume. VII, 141 Seiten. 1979.

Vol. 706: Probability Measures on Groups, Proceedings, 1978. Edited by H. Heyer. XIII, 348 pages. 1979.

Vol. 707: R. Zielke, Discontinuous Čebyšev Systems. VI, 111 pages. 1979.

Vol. 708: J. P. Jouanolou, Equations de Pfaff algébriques. V, 255 pages. 1979.

Vol. 709: Probability in Banach Spaces II. Proceedings, 1978. Edited by A. Beck. V, 205 pages. 1979.

Vol. 710: Séminaire Bourbaki vol. 1977/78, Exposés 507–524. IV, 328 pages. 1979.

Vol. 711: Asymptotic Analysis. Edited by F. Verhulst. V, 240 pages. 1979.

Vol. 712: Equations Différentielles et Systèmes de Pfaff dans le Champ Complexe. Edité par R. Gérard et J.-P. Ramis. V, 364 pages. 1979.

Vol. 713: Séminaire de Théorie du Potentiel, Paris No. 4. Edité par F. Hirsch et G. Mokobodzki. VII, 281 pages. 1979.

Vol. 714: J. Jacod, Calcul Stochastique et Problèmes de Martingales X, 539 pages. 1979.

Vol. 715: Inder Bir S. Passi, Group Rings and Their Augmentation Ideals. VI, 137 pages. 1979.

Vol. 716: M. A. Scheunert, The Theory of Lie Superalgebras. X, 271 pages. 1979.

Vol. 717: Grosser, Bidualräume und Vervollständigungen von Banachmoduln. III, 209 pages. 1979.

Vol. 718: J. Ferrante and C. W. Rackoff, The Computational Complexity of Logical Theories. X, 243 pages. 1979.

Vol. 719: Categorial Topology, Proceedings, 1978. Edited by H. Herrlich and G. Preuß. XII, 420 pages. 1979.

Vol. 720: E. Dubinsky, The Structure of Nuclear Fréchet Spaces. V, 187 pages. 1979.

Vol. 721: Séminaire de Probabilités XIII. Proceedings, Strasbourg, 1977/78. Edité par C. Dellacherie, P. A. Meyer et M. Weil. VII, 647 pages. 1979.

Vol. 722: Topology of Low-Dimensional Manifolds. Proceedings, 1977. Edited by R. Fenn. VI, 154 pages. 1979.

Vol. 723: W. Brandal, Commutative Rings whose Finitely Generated Modules Decompose. II, 116 pages. 1979.

Vol. 724: D. Griffeath, Additive and Cancellative Interacting Particle Systems. V, 108 pages. 1979.

Vol. 725: Algèbres d'Opérateurs. Proceedings, 1978. Edité par P. de la Harpe. VII, 309 pages. 1979.

Vol. 726: Y.-C. Wong, Schwartz Spaces, Nuclear Spaces and Tensor Products. VI, 418 pages. 1979.

Vol. 727: Y. Saito, Spectral Representations for Schrödinger Operators With Long-Range Potentials. V, 149 pages. 1979.

Vol. 728: Non-Commutative Harmonic Analysis. Proceedings, 1978. Edited by J. Carmona and M. Vergne. V, 244 pages. 1979.

Vol. 729: Ergodic Theory. Proceedings, 1978. Edited by M. Denker and K. Jacobs. XII, 209 pages. 1979.

Vol. 730: Functional Differential Equations and Approximation of Fixed Points. Proceedings, 1978. Edited by H.-O. Peitgen and H.-O. Walther. XV, 503 pages. 1979.

Vol. 731: Y. Nakagami and M. Takesaki, Duality for Crossed Products of von Neumann Algebras. IX, 139 pages. 1979.

Vol. 732: Algebraic Geometry. Proceedings, 1978. Edited by K. Lønsted. IV, 658 pages. 1979.

Vol. 733: F. Bloom, Modern Differential Geometric Techniques in the Theory of Continuous Distributions of Dislocations. XII, 206 pages. 1979.

Vol. 734: Ring Theory, Waterloo, 1978. Proceedings, 1978. Edited by D. Handelman and J. Lawrence. XI, 352 pages. 1979.

Vol. 735: B. Aupetit, Propriétés Spectrales des Algèbres de Banach. XII, 192 pages. 1979.

Vol. 736: E. Behrends, M-Structure and the Banach-Stone Theorem. X, 217 pages. 1979.

Vol. 737: Volterra Equations. Proceedings 1978. Edited by S.-O. Londen and O. J. Staffans. VIII, 314 pages. 1979.

Vol. 738: P. E. Conner, Differentiable Periodic Maps. 2nd edition, IV, 181 pages. 1979.

Vol. 739: Analyse Harmonique sur les Groupes de Lie II. Proceedings, 1976–78. Edited by P. Eymard et al. VI, 646 pages. 1979.

Vol. 740: Séminaire d'Algèbre Paul Dubreil. Proceedings, 1977–78. Edited by M.-P. Malliavin. V, 456 pages. 1979.

Vol. 741: Algebraic Topology, Waterloo 1978. Proceedings. Edited by P. Hoffman and V. Snaith. XI, 655 pages. 1979.

Vol. 742: K. Clancey, Seminormal Operators. VII, 125 pages. 1979.

Vol. 743: Romanian-Finnish Seminar on Complex Analysis. Proceedings, 1976. Edited by C. Andreian Cazacu et al. XVI, 713 pages. 1979.

Vol. 744: I. Reiner and K. W. Roggenkamp, Integral Representations. VIII, 275 pages. 1979.

Vol. 745: D. K. Haley, Equational Compactness in Rings. III, 167 pages. 1979.

Vol. 746: P. Hoffman, τ-Rings and Wreath Product Representations. V, 148 pages. 1979.

Vol. 747: Complex Analysis, Joensuu 1978. Proceedings, 1978. Edited by I. Laine, O. Lehto and T. Sorvali. XV, 450 pages. 1979.

Vol. 748: Combinatorial Mathematics VI. Proceedings, 1978. Edited by A. F. Horadam and W. D. Wallis. IX, 206 pages. 1979.

Vol. 749: V. Girault and P.-A. Raviart, Finite Element Approximation of the Navier-Stokes Equations. VII, 200 pages. 1979.

Vol. 750: J. C. Jantzen, Moduln mit einem höchsten Gewicht. III, 195 Seiten. 1979.

Vol. 751: Number Theory, Carbondale 1979. Proceedings. Edited by M. B. Nathanson. V, 342 pages. 1979.

Vol. 752: M. Barr, *-Autonomous Categories. VI, 140 pages. 1979.

Vol. 753: Applications of Sheaves. Proceedings, 1977. Edited by M. Fourman, C. Mulvey and D. Scott. XIV, 779 pages. 1979.

Vol. 754: O. A. Laudal, Formal Moduli of Algebraic Structures. III, 161 pages. 1979.

Vol. 755: Global Analysis. Proceedings, 1978. Edited by M. Grmela and J. E. Marsden. VII, 377 pages. 1979.

Vol. 756: H. O. Cordes, Elliptic Pseudo-Differential Operators – An Abstract Theory. IX, 331 pages. 1979.

Vol. 757: Smoothing Techniques for Curve Estimation. Proceedings, 1979. Edited by Th. Gasser and M. Rosenblatt. V, 245 pages. 1979.

Vol. 758: C. Năstăsescu and F. Van Oystaeyen; Graded and Filtered Rings and Modules. X, 148 pages. 1979.

Vol. 759: R. L. Epstein, Degrees of Unsolvability: Structure and Theory. XIV, 216 pages. 1979.

Vol. 760: H.-O. Georgii, Canonical Gibbs Measures. VIII, 190 pages. 1979.

Vol. 761: K. Johannson, Homotopy Equivalences of 3-Manifolds with Boundaries. 2, 303 pages. 1979.

Vol. 762: D. H. Sattinger, Group Theoretic Methods in Bifurcation Theory. V, 241 pages. 1979.

Vol. 763: Algebraic Topology, Aarhus 1978. Proceedings, 1978. Edited by J. L. Dupont and H. Madsen. VI, 695 pages. 1979.

Vol. 764: B. Srinivasan, Representations of Finite Chevalley Groups. XI, 177 pages. 1979.

Vol. 765: Padé Approximation and its Applications. Proceedings, 1979. Edited by L. Wuytack. VI, 392 pages. 1979.

Vol. 766: T. tom Dieck, Transformation Groups and Representation Theory. VIII, 309 pages. 1979.

Vol. 767: M. Namba, Families of Meromorphic Functions on Compact Riemann Surfaces. XII, 284 pages. 1979.

Vol. 768: R. S. Doran and J. Wichmann, Approximate Identities and Factorization in Banach Modules. X, 305 pages. 1979.

Vol. 769: J. Flum, M. Ziegler, Topological Model Theory. X, 151 pages. 1980.

Vol. 770: Séminaire Bourbaki vol. 1978/79 Exposés 525–542. IV, 341 pages. 1980.

Vol. 771: Approximation Methods for Navier-Stokes Problems. Proceedings, 1979. Edited by R. Rautmann. XVI, 581 pages. 1980.

Vol. 772: J. P. Levine, Algebraic Structure of Knot Modules. XI, 104 pages. 1980.

Vol. 773: Numerical Analysis. Proceedings, 1979. Edited by G. A. Watson. X, 184 pages. 1980.

Vol. 774: R. Azencott, Y. Guivarc'h, R. F. Gundy, Ecole d'Eté de Probabilités de Saint-Flour VIII-1978. Edited by P. L. Hennequin. XIII, 334 pages. 1980.

Vol. 775: Geometric Methods in Mathematical Physics. Proceedings, 1979. Edited by G. Kaiser and J. E. Marsden. VII, 257 pages. 1980.

Vol. 776: B. Gross, Arithmetic on Elliptic Curves with Complex Multiplication. V, 95 pages. 1980.

Vol. 777: Séminaire sur les Singularités des Surfaces. Proceedings, 1976-1977. Edited by M. Demazure, H. Pinkham and B. Teissier. IX, 339 pages. 1980.

Vol. 778: SK1 von Schiefkörpern. Proceedings, 1976. Edited by P. Draxl and M. Kneser. II, 124 pages. 1980.

Vol. 779: Euclidean Harmonic Analysis. Proceedings, 1979. Edited by J. J. Benedetto. III, 177 pages. 1980.

Vol. 780: L. Schwartz, Semi-Martingales sur des Variétés, et Martingales Conformes sur des Variétés Analytiques Complexes. XV, 132 pages. 1980.

Vol. 781: Harmonic Analysis Iraklion 1978. Proceedings 1978. Edited by N. Petridis, S. K. Pichorides and N. Varopoulos. V, 213 pages. 1980.

Vol. 782: Bifurcation and Nonlinear Eigenvalue Problems. Proceedings, 1978. Edited by C. Bardos, J. M. Lasry and M. Schatzman. VIII, 296 pages. 1980.

Vol. 783: A. Dinghas, Wertverteilung meromorpher Funktionen in ein- und mehrfach zusammenhängenden Gebieten. Edited by R. Nevanlinna and C. Andreian Cazacu. XIII, 145 pages. 1980.

Vol. 784: Séminaire de Probabilités XIV. Proceedings, 1978/79. Edited by J. Azéma and M. Yor. VIII, 546 pages. 1980.

Vol. 785: W. M. Schmidt, Diophantine Approximation. X, 299 pages. 1980.

Vol. 786: I. J. Maddox, Infinite Matrices of Operators. V, 122 pages. 1980.

Vol. 787: Potential Theory, Copenhagen 1979. Proceedings, 1979. Edited by C. Berg, G. Forst and B. Fuglede. VIII, 319 pages. 1980.

Vol. 788: Topology Symposium, Siegen 1979. Proceedings, 1979. Edited by U. Koschorke and W. D. Neumann. VIII, 495 pages. 1980.

Vol. 789: J. E. Humphreys, Arithmetic Groups. VII, 158 pages. 1980.

Vol. 790: W. Dicks, Groups, Trees and Projective Modules. IX, 127 pages. 1980.

Vol. 791: K. W. Bauer and S. Ruscheweyh, Differential Operators for Partial Differential Equations and Function Theoretic Applications. V, 258 pages. 1980.

Vol. 792: Geometry and Differential Geometry. Proceedings, 1979. Edited by R. Artzy and I. Vaisman. VI, 443 pages. 1980.

Vol. 793: J. Renault, A Groupoid Approach to C*-Algebras. III, 160 pages. 1980.

Vol. 794: Measure Theory, Oberwolfach 1979. Proceedings 1979. Edited by D. Kölzow. XV, 573 pages. 1980.

Vol. 795: Séminaire d'Algèbre Paul Dubreil et Marie-Paule Malliavin. Proceedings 1979. Edited by M. P. Malliavin. V, 433 pages. 1980.

Vol. 796: C. Constantinescu, Duality in Measure Theory. IV, 197 pages. 1980.

Vol. 797: S. Mäki, The Determination of Units in Real Cyclic Sextic Fields. III, 198 pages. 1980.

Vol. 798: Analytic Functions, Kozubnik 1979. Proceedings. Edited by J. Ławrynowicz. X, 476 pages. 1980.

Vol. 799: Functional Differential Equations and Bifurcation. Proceedings 1979. Edited by A. F. Izé. XXII, 409 pages. 1980.

Vol. 800: M.-F. Vignéras, Arithmétique des Algèbres de Quaternions. VII, 169 pages. 1980.

Vol. 801: K. Floret, Weakly Compact Sets. VII, 123 pages. 1980.

Vol. 802: J. Bair, R. Fourneau, Etude Géometrique des Espaces Vectoriels II. VII, 283 pages. 1980.

Vol. 803: F.-Y. Maeda, Dirichlet Integrals on Harmonic Spaces. X, 180 pages. 1980.

Vol. 804: M. Matsuda, First Order Algebraic Differential Equations. VII, 111 pages. 1980.

Vol. 805: O. Kowalski, Generalized Symmetric Spaces. XII, 187 pages. 1980.

Vol. 806: Burnside Groups. Proceedings, 1977. Edited by J. L. Mennicke. V, 274 pages. 1980.

Vol. 807: Fonctions de Plusieurs Variables Complexes IV. Proceedings, 1979. Edited by F. Norguet. IX, 198 pages. 1980.

Vol. 808: G. Maury et J. Raynaud, Ordres Maximaux au Sens de K. Asano. VIII, 192 pages. 1980.

Vol. 809: I. Gumowski and Ch. Mira, Recurences and Discrete Dynamic Systems. VI, 272 pages. 1980.

Vol. 810: Geometrical Approaches to Differential Equations. Proceedings 1979. Edited by R. Martini. VII, 339 pages. 1980.

Vol. 811: D. Normann, Recursion on the Countable Functionals. VIII, 191 pages. 1980.

Vol. 812: Y. Namikawa, Toroidal Compactification of Siegel Spaces. VIII, 162 pages. 1980.

Vol. 813: A. Campillo, Algebroid Curves in Positive Characteristic. V, 168 pages. 1980.

Vol. 814: Séminaire de Théorie du Potentiel, Paris, No. 5. Proceedings. Edited by F. Hirsch et G. Mokobodzki. IV, 239 pages. 1980.

Vol. 815: P. J. Slodowy, Simple Singularities and Simple Algebraic Groups. XI, 175 pages. 1980.

Vol. 816: L. Stoica, Local Operators and Markov Processes. VIII, 104 pages. 1980.